General Industry and Technology

General Industry and Technology

Dr. John R. Lindbeck
Professor of Engineering Technology
Western Michigan University

Dr. Irvin T. Lathrop
Professor Industrial Arts
California State University
Long Beach, California

Consulting Editor
Dr. John L. Feirer
Distinguished Faculty Scholar
Western Michigan University

GLENCOE PUBLISHING COMPANY
BENNETT & McKNIGHT DIVISION

Published by Glencoe Publishing Company, a division of Macmillan, Inc.

Send all inquiries to:
Glencoe Publishing Company
15319 Chatsworth Street
Mission Hills, California 91345

Printed in the United States of America

Library of Congress Catalogue Card Number (84:71737)

ISBN 0-02-667380-0

    5 6 7 8   92 91 90 89 88

Cover Art:
  McDonnell Douglas Automation Company, a division of McDonnell
    Douglas Corporation
  General Motors Corporation
  NASA
  Polaroid Corporation
  Eastman Kodak Company
  Heath Company, Benton Harbor, MI 49022
  The Boeing Company

Plan of This Book

A general course should be an integral part of all modern industrial education programs. Such a course has the following major objectives:

1. Students learn about the fields of computer technology, graphic communications, metalworking, plastics, woodworking, building construction, power, and transportation at the introductory level. Furthermore they are led to explore these areas in order to learn how products are designed and manufactured.

2. Students learn how to design and make projects related to these areas. They are encouraged to experiment and do further study to increase their understanding of the principles which underlie materials processing.

3. Students learn basic hand, machine, and drawing skills which are helpful to all people living in an industrial society.

4. Students are encouraged to think seriously about their futures by familiarizing themselves with the various opportunities in industry. Such an exploratory program aids students both vocationally and avocationally.

5. Students learn how industries function and that industries are essentially the same with respect to planning and manufacturing. They also are helped to understand how such areas as power, electricity, and computer technology relate to the broad field of industry.

General Industry and Technology is designed for such a general industrial education course at the beginning level. It introduces students to modern industry and technology, its products, and the world of work. This book is divided into seven Parts: High Technology; Basic Shop Practices; Graphic Communications; Production (Manufacturing and Construction); Power and Energy; Transportation; and Modern Industry. Some of the Parts are further divided into Sections. Graphic Communications, for example, has two Sections: Drawing and Graphic Arts.

General Industry and Technology incorporates numerous features that facilitate teaching and learning:

• Because the book is long and ranges over a wide variety of topics, it is divided into Parts, Sections, and Chapters. Individual chapters are short, presenting the information in easily manageable amounts.

• Within the chapters, material is subdivided according to topic and relative importance. Three levels of subheads are used. **NUMBER 1 SUBHEADS** are printed in capital letters and centered. **Number 2 Subheads** are printed in capital and lowercase letters and centered. NUMBER 3 SUBHEADS are in capitals and small capitals, flush with the left margin.

• Information of special interest has been placed in sidebars set off by rules. Some sidebars give information related to the main topic of the chapter. For example, Chapter 6, "Laser and Fiber Optics Technology," includes the sidebar "What Is Light?" Other sidebars provide useful information about new technologies (as in "Superplastics," Chapter 58) or about issues of concern to modern society ("The Future of Work" in Chapter 107).

• Wherever a particular tool or procedure is discussed, safety rules are included. These are set off by the heading "For Your Safety . . ." and by color bars to make them easy to find.

• Sentence structure and vocabulary have been controlled to keep this book within the reading abilities of most high school students. If a word is likely to be unfamiliar to the student, it is defined in the text where first used. In addition, there is a glossary of high-tech terms at the back of the book.

• Each chapter is followed by review questions which reinforce key information found in the chapter.

• The book is generously illustrated with drawings, tables, and photographs. Many of the illustrations use color to emphasize features or facilitate comprehension. Layout has been organized to keep illustrations as close as possible to their text references.

• Every attempt has been made to point out similarities and close relationships among the industrial areas covered, so that the student can realize how much the major American industries have in common. Subconcepts are listed under each major concept to show further the many similarities among the methods of transforming materials into products. The subconcepts of mechanical linkage, adhesion, and cohesion which fall under the major concept of metal-fastening theory are examples of this.

The authors realize that the information in this book could have been organized in various ways; they feel, however, that the organization they have chosen is valid, functional, and understandable.

The authors wish to thank all of the many individuals and companies who have made valuable contributions to this book. (Specific acknowledgments are listed below.) It would have been difficult, if not impossible, to realize this publication without the generous assistance of those named. The authors also welcome any suggestions or comments from individuals who may use this book.

Acknowledgments

A B Dick Co.
Ace-Sycamore, Incorporated
Addressograph-Multigraph
Aluminum Company of America
America House
American Electrical Heater Corporation
American Forest Products Industries
American Honda Motor Company, Incorporated
American Machine and Foundry, Inc.
American of Martinsville
American Seating Company
American Textile Manufacturers Institute, Inc.
American Trucking Association
American Welding Society
Amtrak
AM International
Anaconda Company
Association of American Railroads
AT&T
Atari Incorporated
Baldor Electric
Bell Aerospace Division of Textron
Bell & Howell
Bell Systems News Features
The Beloit Corporation
Bendix Corporation
Besly-Welles Corporation
Bethlehem Steel Corporation
The Beverly Shear Mfg. Corp.
Billings and Crescent Tools
Bi-Modal Corporation
Binks Manufacturing
Boeing Airplane Company
Branson
British Airways
Brown Manufacturing Co.

Bruning Division, Addressograph Multigraph Corporation
Burlington Northern
California Ink Co.
Canadian National Corporation
Canpro
Carpenter Technology
Caterpillar Tractor Co.
Central Vermont Public Service Corp.
Cincinnati Milacron
Civil Aeronautics Board
Clausing Corporation
Collins Radio Company
Columbian Vise and Manufacturing Company
Corning Glass Works
The Cousteau Society, Inc.
Creative Packaging Co.
Crescent Tool Company
Data Composition, Inc.
Deere and Company
Department of the Air Force
Design Center, Copenhagen
Di-Acro Houdaille
Digital Equipment Corp.
Disston Co.
Dover Publications
Dow Corning
Dunbar Furniture
DuPont Magazine
Eagle Manufacturing Company
Eastman Kodak Company
Edwards Laboratories
E. I. duPont de Nemours & Co., Inc.
Electric Storage Battery Company
Evans and Sutherland
Federal Aviation Administration
The Firestone Tire & Rubber Company
Ford Motor Company

Forest Products Laboratory
Frederick Post Company
Fry Plastics International
GAF
Gemaco
General Electric Corporation
General Motors Corporation
Genevro Machine Company
Georg Jensen Silversmiths, Ltd., Copenhagen, Denmark
Gerber Systems Technology, Inc.
GMC Truck & Coach Public Relations
Graphic Systems Group, Rockwell International
Great Lakes Screw Corporation
Greyhound Bus Lines, Inc.
Gulf Oil Corporation
Hall Enterprises
Harvey Hubbell, Incorporated
Heath Company, Benton Harbor, MI
H. M. Harper Company
Honeywell Photographic Products
Hossfeld Manufacturing Company
Houston Instrument Division of Bausch & Lomb
Hughes Electronics
IBM
Illinois Bell Telephone
Inland Manufacturing Company
Iowa Electric Light and Power Co.
Jackson Products Company
James F. Lincoln Arc Welding Foundation
J. A. Richards Company
Jens Risom Design, Inc.
Jervis B. Webb Co.
Johnson Gas Appliance Company
Jones & Lamson

Kaiser Aluminum
Keiper USA
Kennecott Copper Corporation
Kenworth Truck Co.
Kester Solder Company
Keuffel and Esser Company
Kreuger, Inc.
Lead Industries Association
Lea Manufacturing Company
Leatherwood Manufacturing Company, Incorporated
The Lincoln Electric Company
Lindberg Company
Lockheed Missiles and Space Company, Inc.
Los Angeles Times
The L. S. Starrett Company
Lufkin Rule Company
Lyon Metal Products, Incorporated
McAuto
McCullough Corporation
McDonnell Douglas Automation Company
McDonnell Douglas Corporation
McEnglevan Heat Treating and Manufacturing Company
Machine Design Magazine
Malleable Founders Society
Manufacturers Brush Company
Matson Navigational Company
Michigan Laser Cut
Mid-West Conveyor Co., Inc.
Millers Falls Company
Murray Ohio Manufacturing Company
NASA
National Particleboard Association
National Safety Council
National Welding Equipment Company
New Jersey Zinc Company
Niagara Tools
Nicholson File Company
The Nolan Company
Nordson Corporation
North American Products
Northland Ski Company
Northwest Orient Airlines
Norton Company
NuArc Co., Inc.
Pacific Far East Lines

Pacific Gas and Electric Company
Pan American
Peck, Stow, and Wilcox Company
Pinso Sports Ltd., Quebec
Porsche-Audi
Porter-Cable Corporation
Port of Long Beach
PPG Industries, Inc.
Precast/Schokbeton, Inc.
Precision Metal Products
Radio Shack, a Division of Tandy Corp.
Remington Rand Univac
Republic Steel Corporation
Reynolds Aluminum
Robert P. Gersin Associates, Inc.
Rockwell-Delta Company
Rockwell International Power Tool Division
Rohm and Haas
Rotor Way, Inc.
Sally Dickson Associates
Sandia Laboratories
San Diego Gas and Electric Company
Santa Fe Railway
The Science Museum, London
Scott Foresman and Company
Sea-Land Service
Sears, Roebuck and Company
Seattle Center
S.H.M. Marine International, Inc.
Simonds File Company
Simonds Saw and Steel Company
Simpson Timber Company
Smithsonian Institution
Smith Welding Equipment Company
Snap-on Tools Corporation
The Society of the Plastics Industry, Inc.
Soiltest, Inc.
Southern California Edison
Southern Pacific Company
SpaN Magazine, a publication of Standard Oil of Indiana
Standard Tool Company
Stanley Tools
Structural Dynamics Research Corporation, Milford, OH
Teledyne Post, Des Plaines, IL

Texas Instruments
Thomas C. Thompson Company
3M Company
Todd Shipyard Corporation
Union Carbide
Union Mechling Corporation
United Airlines
United States Navy
U.S. Department of Labor
U.S. Department of Transportation
U.S. Divers Company
U.S. Office of Education
U.S. Steel Corporation
Vandercook Co.
The Vecta Group, Inc.
Welch Scientific Company
Western Electric Corporation
Western Michigan University
Westinghouse Electric Corporation
Weston Instruments
Weyerhaeuser Company
Wheelabrator Corporation
Whirlpool Corporation
Willard Storage Battery Co.
J. H. Williams and Company
Wilson Mechanical Instrument Division
Workrite Products Co.
Wrather Corp.
Xerox
Zinc Institute, Inc.

Thanks are particularly due to the following persons for their valued assistance:
Kalman J. Chany
Mrs. Esther Cowdery
Terry Dykstra
Fred W. Gillman
Osmer Gorton
Charles Honeywell
James Hotary
Bernice Q. Johnson
Daniel Kurmas
Marshall La Cour
James Lathrop
Jerold Harmon Saper
Paul M. Schrock
James L. Shaffer
Floyd Stannard

Table of Contents

PART 1

High Technology

Science is the study of the laws of nature. Technology is the application of science for some practical purpose. Industry, in turn, uses technology to produce the products we use every day.

Since earliest times, humans have been builders. Nearly 5000 years ago the Egyptians built the Great Pyramid of Khufu (or Cheops) near Cairo. One of the largest structures ever built, it covers 13 acres. The pyramid contains 2 300 000 blocks of granite and limestone, each weighing up to 2.5 tons. The Egyptians built all of it without the wheel or iron tools. This is one of the best examples of early technology, and it still stands.

The methods used by people to make things are constantly changing. For example, in ancient times, if someone needed a chair, he or she had to make it. Later on, the people who were especially good at making chairs and other furniture became cabinetmakers. They did nothing else but make wooden articles which they sold or traded to other people. These craftspeople had to be very skilled. They had to know about the different kinds of wood and how to season it. They had to design furniture and therefore know joinery methods. They had to know how to use tools to work

1-1. In early times, a carpenter used hand tools to build homes, cabinets, and furniture.

Dover Publications, Inc.

the wood and what kind of finishes to use and how to apply them. In other words, they had to know about all aspects of their craft. Fig. 1-1.

A cabinetmaker with a successful business could not do all the work alone. Other people were hired and trained to help, and a "cottage industry" began. *Cottage industries* were small factories, with a few workers, located in or near a craftsperson's home. Here labor was divided up according to skill. For example, an apprentice might cut logs into boards and smooth them into lumber. The craftsperson was then free to use his or her special

skills to design and make furniture or cabinets.

With the industrial revolution in the late 1700s, the factory system of production began. Now the cabinetmaker started a furniture factory where large numbers of people worked at highly specialized jobs. One person made only chair legs, another only the seat. Machines were developed to aid the people in their work and to make them more productive. As more and better machines, newer methods of joining and working woods, and better, more efficient mass-production techniques were developed, the modern factory emerged.

HIGH TECHNOLOGY

Today we are experiencing a very exciting change in the way we produce goods—"high technology." *High technology* can be simply defined as the use of computers and other modern technical machines to produce goods and control the production system. In the photo essay on pages 32A-H, you will see some outstanding examples of high technology in American industry. Other examples, along with technical information and descriptions, are found throughout this book.

You will see from these high-tech examples that American industry is changing dramatically in two ways—in the *way* things are made and in the *kinds* of things that are made. Perhaps even more important is the fact that the industries of the future will need people with very special technical skills. Fig. 1-2.

ABOUT THIS BOOK

The purpose of this book is to introduce you to modern industry and technology, its products, and the world of work. Through this book you will be given a beginning course in general industry. You will gain an appreciation of the importance of crafting and technical skills and of industrial products. You will also have a chance to explore your interests in making and doing things. In addition, you will learn about career opportunities in industry and how your interests may lead you to a challenging job.

Every learning activity has objectives or aims. The objectives of a general industry course are as follows:

1. To develop an understanding of industry and technology. You will learn of the marvels of modern technology as well as some of industry's problems. It is important for you to gain an understanding of the place of industry and technology in our society.

2. To develop technical skill and knowledge. As a part of your work in this course, you will learn skills in metalwork, woodwork, electronics, drafting, graphic arts, and other areas. You will explore the materials and measuring devices used. You will learn safe work habits and tool skills and how to make project drawings. You will learn how the computer is used in product design and manufacture. This exploration can help you become aware of your talents and interests. It will also help you become an informed consumer. By learning how products are made, and making some yourself, you will be better able to judge the quality of commercially made goods.

3. To develop creative problem-solving skills using the materials and processes of industry. The joys of crafting are important to you as a human being. You should explore, design, create, and invent. You will be faced with many problems in your work which require intelligent solutions. Through shop experimentation and planning you can know the pleasure of successful solutions. These skills can lead to satisfying hobbies as well as successful careers.

4. To learn about the effects of industry on our environment. Technology aids people in many ways, but it also causes some problems. Perhaps more serious than any other is pollution.

Pollution is a problem in all technical societies. One reason for this is that such societies create a large amount of waste. This may seem surprising at first, since efficient production would seem to eliminate waste. However, such efficiency is not the whole story. There is a point at which it becomes cheaper and more profitable to waste a material than to reuse it.

For example, labor costs have become so high that it is cheaper to throw away "tin" cans than it is to collect them, melt them down, and reuse the metal. Likewise, it is easier for a beverage firm to use throwaway bottles instead of the type that must be returned to the store, hauled to the bottling factory, washed, inspected, and refilled.

Such practices pose a big problem for society: what to do with the huge accumulation of cans, bottles, and other junk. It is interesting to note that in nonindustrial, developing nations, waste disposal problems of this sort are practically nonexistent. Every material is too valuable to throw away. By contrast, every year Americans discard millions of autos, tires, bottles, and cans and many tons of paper. We also dirty the air we breathe by gushing vast amounts of smoke and fumes into it. Because of our carelessness, we dump many tons of waste into our waterways, polluting the water we need for drinking, washing, transportation, and recreation. In this course you will learn about some

1-2. *Jobs in industry are becoming highly specialized, requiring technical training. This worker is performing a calibrating operation on a heart pacemaker.*

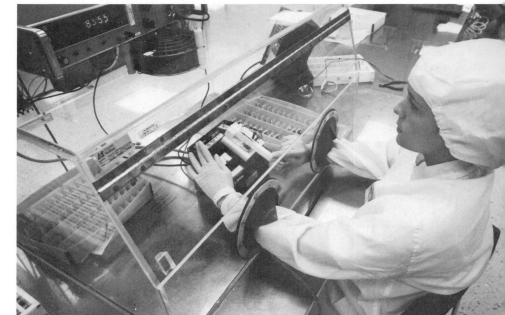

of the solutions to the problems of industrial wastes.

5. To develop your leadership potential. A good leader motivates others to work efficiently and to do their best. A good leader also helps others work together effectively. In a general shop class, there are many opportunities for you to develop leadership abilities.

Everyday activities in the shop can help develop leadership skills. For example, you may be asked to serve as the safety officer or be placed in charge of the tool room. Both of these jobs involve responsibility, and responsibility helps develop leadership.

Perhaps your class will set up a student company to mass-produce a product. You can practice leadership by serving as one of the officers in the company, by helping to plan and set up the production line, or by taking charge of some other activity, such as advertising.

You may also have the opportunity to join a student club. By taking part in club activities, you can learn skills in planning, or-

---

Consumer Choices and the Free Enterprise System

In a free enterprise system, private businesses compete with each other for profit. The government does not interfere beyond what is necessary to protect public interest and keep the national economy in balance.

In such a system supplies of goods and services, and their prices, will respond to consumer demand for them. Consider the microcomputers. At first, these were built from kits by hobbyists. Supplies were low and prices high. In the mid-1970s the first mass-produced microcomputers were introduced. They transformed the micro from a hobbyist's toy into a tool for businesses, schools, and the home. As sales (demand) increased, production became more economical and prices went down. Other companies began manufacturing microcomputers. The competition drove prices down still further. The lower prices meant more people could afford a computer. Thus demand increased.

From this example, you can see that price, supply, and demand are related to one another. You can also see that the products people buy help decide what products will be made.

---

ganization, and getting along with others that will help you become an effective leader.

6. To explore career opportunities in industry. Industry needs people trained in a variety of skills. Skilled craftspeople, technicians, engineers, scientists, managers, and clerical workers— these are all typical of the kinds of jobs available in business and industry. High technology demands highly skilled workers.

7. A general industry course requires that you work safely and in cooperation with both the teacher and other students. Good work habits and attitudes are important if you are to get as much as possible from your shopwork experiences.

QUESTIONS AND ACTIVITIES

1. List seven objectives for a general industry course.

2. What is a cottage industry? In what ways is it like today's factories? In what ways is it different?

3. What is *high technology*?

4. American industry is changing dramatically in what two ways?

A *computer* is an automatic machine that makes calculations and processes information at very high speeds. Today computers have many and varied uses. They bring us video games and war games. They can tell when to pick oranges or which team to pick in a football game. They are used to keep track of all the items a company has in its warehouse and all the satellites that are orbiting the earth. They help writers prepare novels and homemakers prepare menus. Computers are even being used to design other computers. Fig. 2-1.

HISTORY OF COMPUTERS

From earliest times people have had the need to compute (make calculations). At first they probably used their fingers and toes. Then devices were invented to help compute. One of the earliest was the abacus. The abacus consists of rows of parallel wires, rods, or grooves on or in which slide small beads or blocks. Calculations are performed by moving the beads or blocks. Fig. 2-2. The abacus was invented over 5000 years ago, probably in Babylonia. It is still used in parts of Asia and the Middle East.

In 1642 the French scientist and philosopher Blaise Pascal built an adding machine to help his father with his tax collecting duties. This machine used a mechanical gear system to add and subtract numbers with as many as eight digits. Fig. 2-3.

Pascal's machine could only count. Later, other machines were invented which could also multiply, divide, and figure square roots.

Charles Babbage, an English inventor, designed the first true computer. His "analytical engine" combined arithmetic processes with decisions based on its own computations. In other words, the answer that the machine computed could be used by it to solve subsequent steps of a complex problem. Babbage worked on the analytical engine between 1834 and 1854. He was never able to finish building it because of a lack of money and

because the technology of the 1800s was not refined enough to make metal parts to the required precision.

In the late 1800s, Herman Hollerith was working for the United States Bureau of the Census. He developed a way to quickly tabulate the huge amount of information from the 1890 census. Information about each person was recorded on a card by a system of punched holes. A machine read the cards and totalled up the statistics. Fig. 2-4. Hollerith later formed his own company, which eventually became IBM.

Digital Equipment Corp.

2-1. Computers play a big role in modern life. Their ability to process information at high speeds has many applications. Can you think of some in addition to those already mentioned?

2-2. *An abacus. This was one of the earliest calculating instruments.*

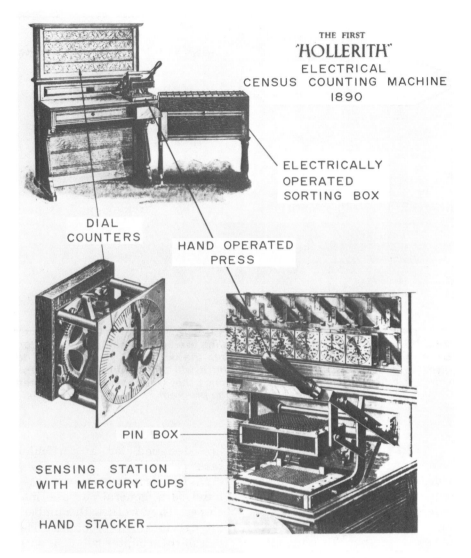

THE FIRST
"HOLLERITH"
ELECTRICAL
CENSUS COUNTING MACHINE
1890

ELECTRICALLY
OPERATED
SORTING BOX

DIAL
COUNTERS

HAND OPERATED
PRESS

PIN BOX

SENSING STATION
WITH MERCURY CUPS

HAND STACKER

2-4. *This is the machine Herman Hollerith invented to tabulate information from the 1890 census.*

In 1939 a young man named Howard Aiken came to IBM with an idea for a fully automatic calculator. During the next five years, he and a team of engineers developed what came to be known as the Mark I. The Mark I was used during World War II to calculate mathematical tables used for navigation and targeting.

The Mark I was slow compared to today's computers. It completed one instruction in one one-hundredth of a second. Modern computers can complete instructions in billionths of seconds. This means that if the two computers did the same task, the modern computer would finish ten million times faster.

The Mark I had no electronic parts. Its internal structure was

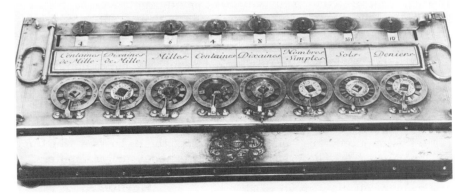

2-3. *Pascal's adding machine. The wheel at the far right was for units; the one next to it, for tens; the third from the right, hundreds; and so on.*

a maze of gears and machinery. The first electronic computer was designed and built at the University of Pennsylvania in 1946. This computer was named ENIAC (Electronic Numerical Integrator and Calculator). ENIAC had 18 000 vacuum tubes, which failed on the average of one every seven minutes. After the ENIAC came the Univac. The Univac was the first commercially available computer. The first customer was the United States Bureau of the Census.

```
5    HOME
6    I = 0:S = 0:Q = 0
10   PRINT "HOW MANY NUMBERS DO YOU WISH TO AVERAGE";
20   INPUT N
30   IF N > = 1 AND N < = 100 THEN 80
40   PRINT
50   PRINT "YOU CAN ONLY AVERAGE BETWEEN 1 AND 100 NUMBERS"
60   PRINT
70   GOTO 5
80   PRINT
90   PRINT "INPUT YOUR NUMBERS, 1 PER QUESTION MARK"
100  PRINT
110  FOR I = 1 TO N
120  PRINT I;
130  INPUT Q
140  S = S + Q
150  NEXT I
160  PRINT
170  PRINT
180  PRINT "AVERAGE=";S / N
190  PRINT
200  PRINT "WOULD YOU LIKE TO AVERAGE ANOTHER SET";
210  INPUT A$
220  PRINT
230  IF A$ = "YES" THEN 5
240  END
```

2-5. *This is a BASIC program for calculating averages on an Apple computer. Once this program has been entered into the computer, you only need to give the computer a set of numbers and it will tell you their average. The words inside quotation marks are the ones that will appear on the screen when you use the program.*

TYPES OF COMPUTERS

Modern computers can be divided into three types: analog, digital, and hybrid. *Analog computers* solve problems by expressing one physical quantity in terms of another. To understand this principle, think of an analog watch. In this type of watch, the passage of time is indicated by the movement of the hands around the dial. Thus one quantity (time) is measured in terms of another (the position of the hands).

Analog computers are used to solve problems in which several quantities vary continuously over a period of time. For example, analog computers are used in airplane simulators because they can describe the relationships of speed, altitude, and lift and thus imitate the behavior of a real aircraft. Most analog computers are special-purpose machines. They are designed for a particular task.

Digital computers, on the other hand, are general-purpose machines. They work with numbers (digits) to solve problems. The rest of this chapter will deal with digital computers.

Hybrid computers combine analog and digital operations. One example of a hybrid system is numerical control (NC). You will learn more about numerical control in Chapter 4.

DIGITAL COMPUTERS

Unlike analog computers, digital computers respond only to exact signals, not to varying degrees of signals. The signals for a digital computer are electric pulses called *bits*. These electric pulses can be in only one of two states: on or off; they either exist or do not exist.

People use the number 1 to represent an on-bit and 0 to represent an off-bit. Patterns of bits are used to form characters called *bytes*. One byte usually contains eight bits. Computers recognize a value for each byte. For example, in one computer language, the letter A is represented by 11000001. A byte does not always stand for a letter, however. It may be a number, a punctuation mark, or a command.

These patterns of ones and zeros are a kind of language. They give the computer data (information) and instructions. This language is called *machine language*. It is easy for computers to work with but difficult for humans to use. Therefore, other computer languages have been developed that are more like human language. One of the most widely used is BASIC (Beginner's All-purpose Symbolic Instruction Code). BASIC was developed by John Kemeny and others at Dartmouth College in the 1960s. It uses English words, punctuation marks, and algebraic notations. People write programs (instructions for the computer) in BASIC, and the computer then translates these programs into machine language. Fig. 2-5.

BASIC is the most popular language for microcomputers. The *microcomputer* is a computer that is small enough to be portable. Some will even fit into a briefcase. Even though they are small, microcomputers can be as powerful as some computers that are much larger. In recent years, microcomputers have become very popular for use in small businesses, homes, and schools. Most of the discussion which follows is about microcomputers.

Hardware

The equipment that makes up a computer system is called the

INPUT

KEYBOARD
CASSETTE TAPE
  RECORDER
DISK DRIVE
PUNCHED CARD
  READER
PUNCHED TAPE
  READER
OPTICAL SCANNER
CRT

OUTPUT

CASSETTE TAPE
  RECORDER
DISK DRIVE
PRINTER
PLOTTER
CARD PUNCHER
TAPE PUNCHER
CRT

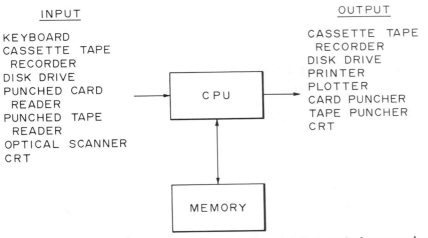

2-6. *The flow of data through a computer system. The central processing unit and memory are inside the computer. The input/output devices are peripheral (external) equipment. Some I/O devices are used for data storage as well as for communicating with the computer.*

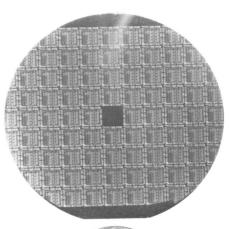

2-8. *Chips are made on a "wafer" of silicon. After manufacture, the chips are separated. Each chip contains one integrated circuit, or IC, which consists of many thousands of transistors. You can see how small one chip is compared to a quarter.*

hardware. All computers have four types of hardware: a central processing unit (CPU), memory, storage, and input/output (I/O). Figs. 2-6 & 2-7.

CENTRAL PROCESSING UNIT

The CPU is the heart and brain of the computer. Here all the control functions of the machine take place and all the data are processed. In a microcomputer the CPU is contained on a single chip. A computer *chip* is a small piece of silicon, about ¼″ square, on which there are thousands of tiny transistors. Figs. 2-8 through 2-10. A CPU that is

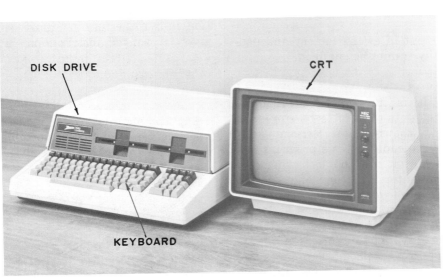

Heath Company, Benton Harbor, MI 49022

2-7. *A typical microcomputer system. Data are entered in the computer using either the keyboard or the disk drive, which in this system has been placed above and behind the keyboard. The output device is a CRT (cathode ray tube), similar to a television set. The CPU and memory are housed below the disk drive.*

2-9. *A single chip. The wires that stick out around the chip connect it to the pins on its case (see Fig. 2-10). Again, notice the size of a chip compared to the size of a quarter.*

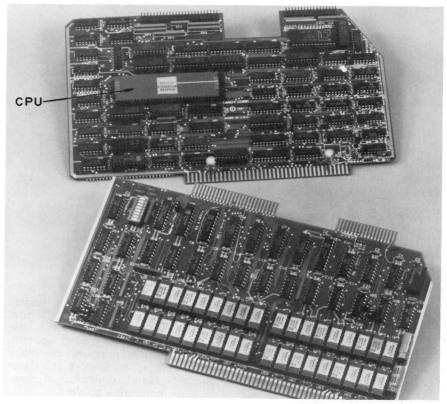

CPU

Radio Shack, a division of Tandy Corp.

2-10. *Circuit boards. Each rectangular case on the boards contains a single chip. Pins projecting from the bottom of the case connect it to the wires on the board. The arrow points to the case which holds the CPU.*

sizes. Fig. 2-11. To use tapes, a standard cassette tape recorder is needed. To use disks, a disk drive is required. The tape recorder or disk drive is connected to the computer. When a tape or disk is inserted and the proper instructions are given to the computer, a copy of the program on the tape or disk is sent to the computer's RAM. The opposite can also be done; you can send a copy of a program from RAM to the tape or disk.

INPUT/OUTPUT

The input/output devices make it possible for people to communicate with the computer. *Input* devices are used to enter data or instructions into the computer. A keyboard is an example of an input device. *Output* devices move data out of the computer. A printer is an output device. Fig. 2-12. Some devices, such as disk drives, tape recorders, and CRTs, handle both input and output.

Software

By itself, computer hardware can do nothing. For any operation it is to perform, it must be given a set of instructions—a

contained on a single chip is called a *microprocessor.*

MEMORY

Memory is also located on chips, but these chips are different from microprocessors. The memory chips hold information for quick access by the CPU. There are two types of memory: RAM and ROM. RAM (random-access memory) stores information temporarily, while the computer is working. Data can be loaded ("written") to or retrieved ("read") from RAM. Once the computer is shut off, though, any data in RAM are erased. ROM (read-only memory) is permanent, but, as the name implies, the CPU can only read it. ROM

is put in at the factory and usually cannot be changed.

STORAGE

It is not practical for large amounts of data to be stored inside the computer. For one thing, if someone writes a program into RAM, it will be lost when the computer is turned off. Another form of memory is needed for long-term mass storage.

Two types of long-term storage devices for microcomputers are tapes and disks. The tapes are the same as the audio tapes on which music is recorded. Disks can be compared to records; both are flat and circular. There are both hard and flexible (floppy) disks, and they come in various

2-11a. *A floppy disk, or diskette, in its protective case. Never remove the disk from this case.*

2-11b. *When not in use, floppy disks are stored in envelopes like these.*

American Textile Manufacturers Institute, Inc.

What Do Looms Have in Common with Computers?

program. Another word for program is *software.*

There are two classes of programs: application programs and operating system programs. *Application* programs solve problems, play games, and so forth. Application programs always reside in the computer's random-access memory (RAM), not read-only memory (ROM). They are transferred to RAM from a tape or disk. For example, when you want the computer to act as a word processor, you use the disk or tape with the word processing

In the early 1800s Joseph-Marie Jacquard of France invented an automated loom for weaving cloth. Cards with punched holes were used to control the operation of the loom. Different cards produced different patterns in the cloth. Herman Hollerith applied the principle of punched cards to his census counting machine, a forerunner of the computer. The first computers also obtained their instructions from punched cards, but most of today's computers use magnetic tape or disks.

Digital Equipment Corp.

2-12a. *A printer.*

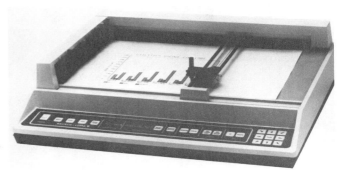

Houston Instrument Division of Bausch & Lomb

2-12b. *For charts or drawings, a plotter is used. Many plotters can print in several colors.*

application program on it and transfer that program to RAM.

Operating system programs are used to manage the computer and its input, output, and storage devices. For example, they are used to transfer application programs from a tape or disk into the computer's memory.

COMPUTER LANGUAGES

Programs can be written in various languages. (In this sense, a language is a related set of commands that can be translated into machine-language instructions the computer can use.) Each programming language has its own vocabulary and grammar. There are low-level and high-level languages. Low-level languages are close to machine language. High-level languages resemble human language. Some high-level languages are:

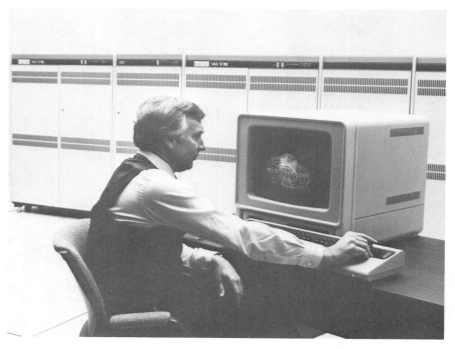

Digital Equipment Corp.

2-14. Computers are used by engineers to help them design products. Computer-aided design (CAD) will be discussed in the next chapter.

ATARI

2-13. Many families are buying microcomputers for home use. How might you use a microcomputer in your home? (ATARI® and 800 are trademarks of Atari Corp.)

- FORTRAN (Formula Translator). Developed for writing science and engineering programs. Now also used in business and education.
- COBOL (Common Business Oriented Language). Designed for business uses.
- PL/1 (Program Language 1). Developed for solving both scientific and business problems.
- BASIC (Beginner's All-purpose Symbolic Instruction Code). Widely used in business, education, and home computers.

USES OF COMPUTERS

Computers have changed greatly since the days of the Mark I, which was eight feet high and fifty feet long. Today's microcomputers fit on a desktop and are faster and more powerful than the Mark I. In the future, computers will be still smaller and more powerful.

Computers are part of your life. Your class schedule was probably written up by a computer. There are small, nonprogrammable computers in video games, digital watches—even in greeting cards. Perhaps you have a microcomputer in your school or home. Fig. 2-13.

Computers are part of industry, too. Fig. 2-14. In the next few chapters, you will learn about the ways computers are used in design and manufacturing.

QUESTIONS AND ACTIVITIES

1. Define computer. Discuss some various uses of computers.

2. Who designed the first true computer? What did the inventor call his machine? During what period did he work on it?

3. From what source did the first computers obtain their instructions? What contains the instructions for today's computers? What is another name for these "instructions" for the computer?

4. Modern computers can be divided into three basic types. Name them and briefly describe each type.

5. What is BASIC? Why was it developed?

6. What is computer hardware? Name the four types of computer hardware discussed in this chapter and describe the job of each.

7. What are RAM and ROM and what is the difference between the two?

8. What is an input device for? An output device? Give an example of each.

9. What is another name for software? What does software do?

10. What is "high-level" language? Give two examples and tell what areas they are used in.

CHAPTER 3

Computer-Aided Design

Many industries use computers in design and drafting. The use of computers in this area of work is called *computer-aided design,* or CAD.† Modern CAD systems are *interactive.* This means that a person (the "user") directs (interacts with) a computer to create drawings. The person supplies the ideas; the computer does the drawing. Fig. 3-1.

The CAD user sends commands to the computer through a typewriter-like keyboard. The computer understands and fol-

†This is also called computer-aided design and drafting (CADD).

3-1. *This CAD operator is creating and studying a geometric model by sending commands to the computer.*

McDonnell Douglas Automation Company, a division of McDonnell Douglas Corporation

lows these instructions because it has been programmed to do so. (See Chapter 2, "Computer Technology.") It shows the results of its work on the computer screen, also called a *CRT* (cathode ray tube). The graphics display (drawing) on the CRT is made up of geometric elements (parts) such as points, lines, circles, and planes—the same kinds of parts used to make an ordinary mechanical drawing. By using the keyboard or special controls, such as a light pen or joystick, the user can command the computer to change a drawing in some way. For example, the drawing can be made larger or smaller, moved to another part of the screen, or changed from a three-view to a pictorial drawing. (A three-view drawing usually shows the front, top, and right side of an object. See Chapter 14. A pictorial drawing looks much like a picture or photograph of an object. See Chapter 16.) The user can also add inch or metric dimensions to the drawing. (See Chapters 17 and 18.)

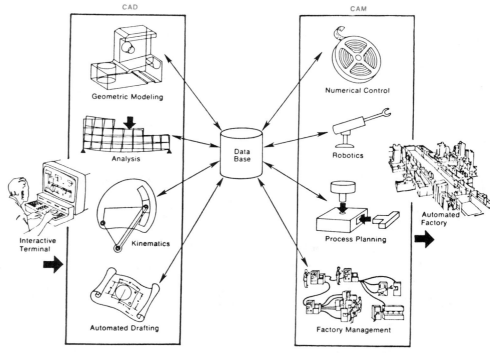

Machine Design Magazine

3-3. *The main uses of CAD are shown here. As you can see, the data base is used for both CAD and CAM functions. A data base is a collection of information organized for rapid computer search and retrieval. Ideally, the CAD/CAM functions can be combined to run an automated factory. (You'll learn about CAM in the next chapter.)*

HARDWARE FOR CAD

Before we see some examples of the work of a CAD system, let's look at the hardware. You already know that the typical CAD system has a **CRT** to display graphics and a **keyboard** and other controls to direct the computer. Like any computer system, it must also have a central processing unit, or **CPU.** (See Chapter 2.) For storing data, most systems use **disks,** although some use **tapes.** Some systems also have:

• A **digitizer,** or graphics tablet, which changes line drawings and other graphics into numerical (digital) data. These data are sent to the computer, which changes them back to drawings. In this way, a drawing on paper can be entered in the computer.

• A **special functions keyboard** to call up stored data.

• **Two CRTs:** one to display alphanumeric (written) data and one for graphics.

• A **printer** or photo plotter to produce hard copies (paper printouts) of graphics from the CRT.

McDonnell Douglas Automation Company, a division of McDonnell Douglas Corporation

3-2. *These are the parts of a typical CAD system: (A) CRT. (B) Alphanumeric CRT. (C) Alphanumeric keyboard. (D) Functions keyboard. (E) Joystick. (F) Disk drive.*

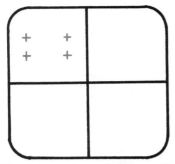

1. The screen is split into sections for top, front, side, and isometric views. Four points are set in the top view to locate the four corners of the pyramid.

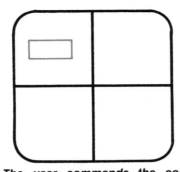

2. The user commands the computer to connect the points with straight lines to outline the top view, or face, of the pyramid.

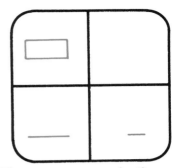

3. The user commands the computer to project the image into the front and side views.

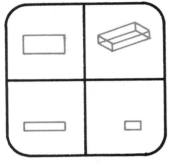

4. The user commands the computer to project the face into the second dimension (height), making a rectangular block. The computer is also commanded to create an isometric view of the block. (An isometric view is a pictorial view in which one corner of the object appears closest to you.)

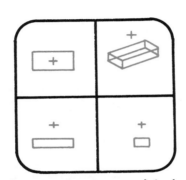

5. The user locates a point above the block, and commands the computer to project the point into all views.

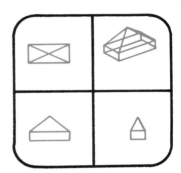

6. The user commands the computer to connect the point to the top corners of the block with straight lines. This creates a pyramid, and the model is complete.

Machine Design Magazine

3-4. This is how the computer is used to create a wire frame geometric model of a pyramid. It is called a wire frame because the model looks as though it is made from wire pieces joined together to make a frame.

● A **pen plotter** to make ink drawings of the CRT display.

The kinds of systems vary a great deal, depending on where and how they are to be used. Fig. 3-2.

SOFTWARE FOR CAD

The CAD software consists of computer programs stored on disks or tapes. Normally, a drafter or designer does not need to write the programs. He or she uses ready-made programs developed by others. The drafter or designer only has to learn how to use the programs.

WHAT CAD SYSTEMS CAN DO

There are four main applications of CAD: geometric modeling, engineering analysis, kinematics, and automated drafting. Fig. 3-3. Each of these is explained and illustrated in the following pages.

Geometric Modeling

A CAD user can model (create the shape of) a geometric object on the CRT. The object can be shown as either a wire frame model (Fig. 3-4) or a solid model (Fig. 3-5). The computer can change this model into digital (numerical) data and store it for later use.

The geometric modeling function is very important because it serves as a basis for engineering analysis and automated drafting. It can also be used to create numerical control tapes for computer-aided manufacturing, or

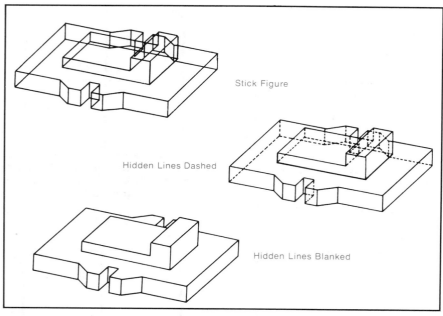

Machine Design Magazine

3-5. *This is how the computer creates a solid model from a wire frame model ("stick figure").*

shock absorber, or the movement in a riveted steel beam joint. Engineers and technicians can study a part on the CRT and change its design before the part is manufactured. This practice saves time and money, and it results in better products.

Kinematics

Some CAD systems can show the motion of such moving parts as automobile door hinges or hood linkages, the bucket on an end loader, or a farm plow blade. The study of such motion is called *kinematics*. Before computers, the action of such mechanisms often was tested with pin-and-cardboard models. These trial-and-error methods were often inaccurate and could take many hours of work. The com-

CAM. (Computer-aided manufacturing is discussed in Chapter 4.)

Engineering Analysis

The geometric model information is stored in the computer. With keyboard commands, the user can have the computer figure the surface area, weight, volume, or other characteristics of the part represented by the model. It also can do a "finite element analysis" of the part. In this analysis method, the part is broken down graphically into a "mesh" of sections, or elements. The computer then uses these elements as a base to figure stresses and deflections (bending) in the part. Fig. 3-6. Such an analysis would take many hours of work by ordinary "hand" methods. The computer can do it in minutes, and more accurately.

Engineering analysis has many applications. For example, it is used to figure the stress on a truck frame, the deflection of a

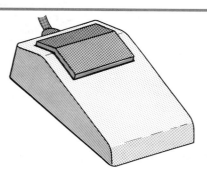

CAD and Mouse Game

Graphics information sent to the computer is based on the X-Y coordinate system—horizontal and vertical lines like the ones you've used on graphs. Any point on a drawing can be described by its X-Y coordinates and transferred to the computer. This process is called digitizing.

Digitizing can be done on a keyboard, but a faster way is to use a graphics tablet (digitizer). This is a flat surface in which is embedded a grid of wires. A special tool is used to move over the surface of the tablet. One such tool is called a mouse.

Here's how a single line of a

drawing would be transferred to the computer:

The paper drawing is placed on the graphics tablet. The mouse is moved over one point of the drawing, and a button on the mouse is pressed. This completes an electrical circuit between the grid and the mouse. The electrical pulse sends the point's coordinates to the computer. The result appears on the screen as a bright mark. The mouse is moved to another point. Another mark appears on the screen. The CAD user directs the computer to connect the marks, and the straight line is completed.

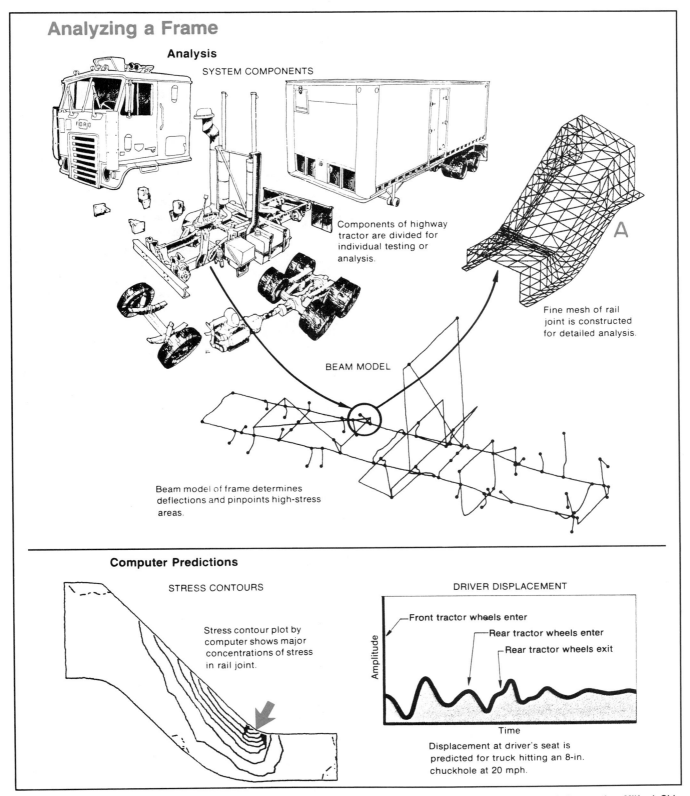

Analyzing a Frame

Analysis

SYSTEM COMPONENTS

Components of highway tractor are divided for individual testing or analysis.

Fine mesh of rail joint is constructed for detailed analysis.

A

BEAM MODEL

Beam model of frame determines deflections and pinpoints high-stress areas.

Computer Predictions

STRESS CONTOURS

Stress contour plot by computer shows major concentrations of stress in rail joint.

DRIVER DISPLACEMENT

Front tractor wheels enter

Rear tractor wheels enter

Rear tractor wheels exit

Amplitude

Time

Displacement at driver's seat is predicted for truck hitting an 8-in. chuckhole at 20 mph.

Structural Dynamics Research Corporation, Milford, Ohio

3-6. *This is how finite element analysis is done. Note the mesh pattern on the rail section of the truck frame (A). The computer prediction shows the stresses on this part. If the part is to break, it will break at the point of the arrow.*

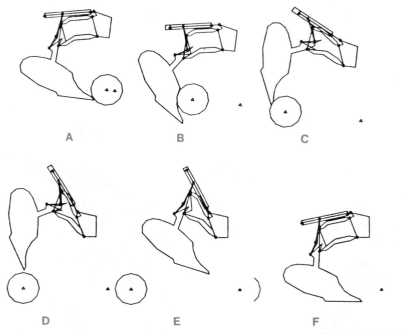

Deere and Company

3-7. *These step-by-step, computer-generated pictures show how a spring-loaded plow would unhook from a rock buried in the ground. This is an example of kinematics. The computer simulation was done by engineers at John Deere in Moline, Illinois.*

made in every other view. Both detail and assembly drawings can be made, complete with dimensions. (See Chapter 19.) A hard copy can be made of the drawing on a plotter. Fig. 3-9.

USE IN INDUSTRY

Computer-aided design systems are important in industry because they can be used to create designs quickly and accurately. Look at the sheet metal layout problem in Fig. 3-10. If you were to draw the stretchout (pattern) for this rectangle-to-round transition piece by hand at a drawing board, it could take you hours. Also, if you were not very careful in figuring the true lengths of the bend lines, the sheet metal piece would not fit properly when you bent it. Using a CAD system with the right program, all you would need to do is

puter can do it in minutes, and, again, more accurately. A typical kinematic analysis is shown in Fig. 3-7.

Automated Drafting

With computerized drafting, engineering drawings can be made automatically. Fig. 3-8. A CAD system is used to generate (draw) up to six views on the CRT. For example, it might draw the front, top, and right side of an object. A design change made in one view is automatically

McDonnell Douglas Automation Company, a division of McDonnell Douglas Corporation

3-8. *This mechanical part was designed and drawn with a computer. It is an example of automated drafting.*

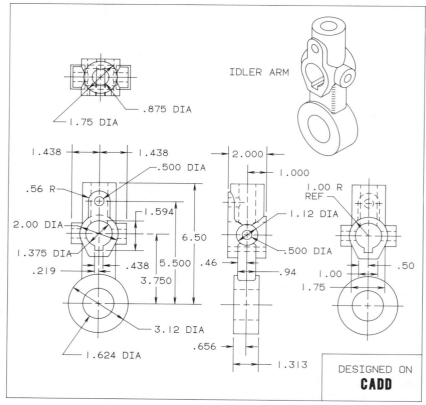

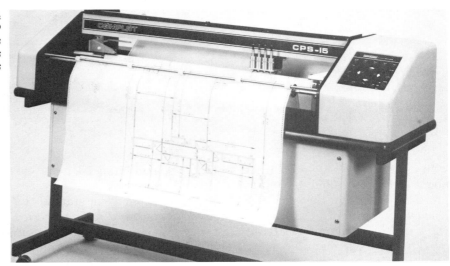

3-9. *An architectural drawing is being produced on this plotter. This machine draws the lines and does the lettering very rapidly.*

punch in the sizes of the rectangle and circle and the height of the piece. The computer would produce the drawing you see in Fig. 3-10. The whole process would take only a few minutes. The convenience and efficiency of the CAD system make it a valuable design and drafting tool.

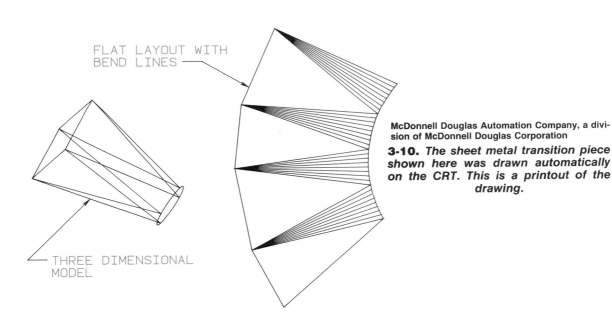

FLAT LAYOUT WITH BEND LINES

3-10. *The sheet metal transition piece shown here was drawn automatically on the CRT. This is a printout of the drawing.*

THREE DIMENSIONAL MODEL

QUESTIONS AND ACTIVITIES

1. What is the CRT and what do the initials stand for?

2. Define graphics display.

3. Name five types of hardware for CAD other than the keyboard and the CPU. Tell the job of each.

4. What is done in geometric modeling? Why is the geometric modeling function important?

5. How is finite element analysis done? What might it be used for?

6. What is kinematics? Name some parts that might be designed with the aid of kinematic analysis.

7. How is the CAD system used in automated drafting?

8. Why are CAD systems important in industry?

CHAPTER 4

Computer-Aided Manufacturing

In earlier chapters you learned about computers and how they are used in design and drafting. However, there also are many ways in which computers are applied in modern manufacturing. The use of computers to assist in the manufacture of products is called *computer-aided manufacturing,* or CAM. Three important applications of CAM are numerical control, process planning, and factory management. They will be explained in this chapter. Robotics and lasers are other examples of CAM. They will be described in Chapters 5 and 6. Fig. 4-1.

NUMERICAL CONTROL

A common way to make a machined part is for a person to make it with a machine tool. A newer way is to use machines that make the part automatically. The machines are controlled by coded information on tape. Fig. 4-2. Controlling a machine tool with a coded information tape to produce a part is called *numerical control,* or NC. The simplest NC systems use instructions punched on paper tape. Making the paper tape is a hand process, where the NC instructions are prepared from information on engineering drawings.

Newer systems use *computer numerical control,* or CNC. The machine tool is wired to a computer which stores the NC instructions on tape or disks. This allows instructions to be stored,

handled, and changed more easily and effectively.

The most modern, state-of-the-art system is *direct numerical control,* or DNC. In this system, also known as distributed numerical control, the machine control is distributed among a network of computers. Through these computers, the machines are also linked to a central mainframe (large-capacity) computer that supplies instructions. The system also permits feedback from machine to computer. Thus the machine can inform the computer

about production and machine tool status.

Computer-aided design can be effectively combined with DNC in the manufacturing process. The designer can use CAD to draw the metal part and test it for accuracy and fit. Then he or she can go on to create a tool-cutting pattern and send the information to a milling machine which automatically cuts the part out of metal. Fig. 4-3. This same computer information can be used to test the quality of the finished piece.

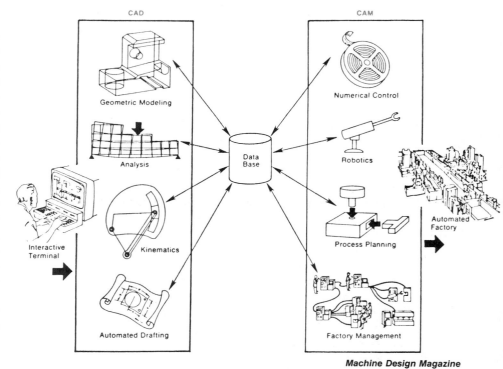

Machine Design Magazine

4-1. The main parts of a CAM system are shown here. Note how they relate to the CAD systems.

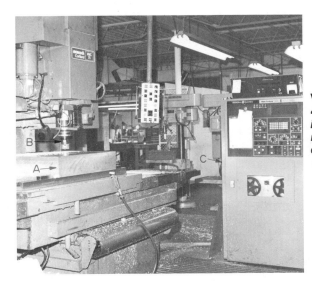

Western Michigan University

4-2. This workpiece (A) is being machined automatically. The milling cutter (B) is controlled by an NC tape machine (C).

remind the machine operator to change a dull cutting tool. Any machine operation can be programmed: drilling, milling, turning, grinding, bending, or others.

The information is punched into paper tape or coded onto magnetic tape or disks. The instructions are "read" by the NC unit, which then sends commands to the machine to make the part. Programming can be done manually, or with the help of a computer. In DNC, the information is fed directly into the computer data bank.

PROCESS PLANNING

Very simply, *process planning* is figuring out the best methods to use to make a product, and then organizing this information into a series of steps to follow. A process plan is the middle step between designing and manufacturing a product. The process plan must include such information as the function of the product; the materials needed; the

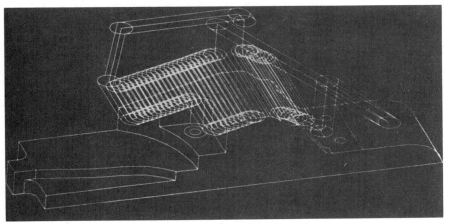

Gerber Systems Technology, Inc.

4-3a. This image was made by a computer. It shows the cutting path for making a metal part.

A term commonly used to describe this automatic manufacturing process is *computer-aided design/computer-aided manufacturing*, or CAD/CAM. These systems will be used more and more in the future. The result will be totally automated factories which manufacture hundreds of different products. All the functions of the factory will be controlled by computers. Fig. 4-4. People will continue to be very important in keeping these factories running.

However, they will have to be trained in computers and other high-technology skills.

NC Programming

The success of NC depends upon the part programs. A *part program* is a detailed set of instructions to be followed by the machines used to make a part. The program specifies the cutting tools to be used, the cutting order, and the speeds and feeds of the tools. The program can also

Gerber Systems Technology, Inc.

4-3b. Here the part is being cut by the numerically controlled machine.

29

SKETCH OF COMPUTER INTEGRATED FACTORY

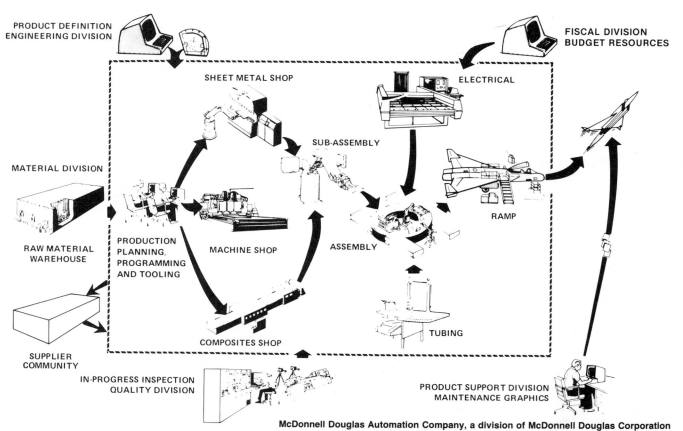

PRODUCT DEFINITION
ENGINEERING DIVISION

FISCAL DIVISION
BUDGET RESOURCES

SHEET METAL SHOP

ELECTRICAL

SUB-ASSEMBLY

MATERIAL DIVISION

RAMP

RAW MATERIAL
WAREHOUSE

PRODUCTION
PLANNING,
PROGRAMMING
AND TOOLING

MACHINE SHOP

ASSEMBLY

SUPPLIER
COMMUNITY

COMPOSITES SHOP

TUBING

IN-PROGRESS INSPECTION
QUALITY DIVISION

PRODUCT SUPPORT DIVISION
MAINTENANCE GRAPHICS

McDonnell Douglas Automation Company, a division of McDonnell Douglas Corporation

4-4. This diagram shows a computer-controlled factory which manufactures airplanes.

number of products to be made; the required operations, tools, and machines to make them; when the products must be ready; and what the costs will be.

A process plan is a simple matter for a simple product, such as a three-legged wooden footstool. But a complex product such as a car or a microwave oven involves hundreds of different parts, materials, and processes. These parts may be made in different factories in different states, or even in foreign countries. Working out suitable process plans in these cases requires a computer.

In most manufacturing operations, there are products or parts that have similar shapes and require similar processes. A computer is used to classify and code such items so that they can be grouped together for efficient manufacturing. To see how this would work, let's go back to our example of the footstool. If your factory produced ten different models of the stool, it would be wise to "group" the tops which were round. Then you could make them and drill their leg holes with as few machine setups as possible. Modern process planning involves the same kind of thinking. You can see how important this is when trying to automate a factory.

FACTORY MANAGEMENT

Factory management systems tie together all of the operations of a factory. Inventory control, scheduling, production control, record keeping, purchasing, sales, and material handling are some of the operations which can be computerized into a total factory management system.

For example, groups of similar parts can be made in individual manufacturing "cells," or factory units. The computer will link all these cells for efficient operation.

Moving parts from place to place on an assembly line and storing the finished goods are important aspects of manufactur-

Jervis B. Webb Co.

4-5. *On this assembly line, huge truck frames are moved along automatically, and the wheels and tires are set in place. This is material handling, which is a key part of factory management.*

ing. Whether one is assembling radios or automobiles or bicycles, a steady flow of parts is essential. Materials must get to the right place at the right time. Also, raw materials, parts, and finished products must be stored until they are needed. This is part of "materials requirements handling," and today much of it is done by computer.

In the truck assembly line shown in Fig. 4-5, you can see many heavy parts being moved automatically. Parts are also moved with parts carriers. Fig. 4-6. Many of these carriers are computer-controlled, requiring no operator. Such vehicles are

General Motors Corporation

4-6. *This motorized robot with forklift delivers inventory to the assembly line.*

guided by electric wires buried beneath the floor.

Computers also are used to control the automated storage and retrieval (AS/R) system in Fig. 4-7. The operator of an AS/R system punches in instructions, telling the computer which stored product is wanted. Fig. 4-8. The computer in turn orders the mobile hoists to put the parts

on the conveyors. The computer then orders the conveyors to send the parts to specified shipping areas. You can appreciate the problems a company can have which has to store thousands of parts or products. Without a computerized system, one could spend hours looking for something. The AS/R does it quickly, with little effort, and it makes few mistakes.

This chapter has shown you some of the ways high technology is changing the methods used to produce goods. The automated factory is a reality. There are few now, but we will see many more in the future. This means that you will also have to "gear-up" for high technology if you want to be a skilled participant in the future factory.

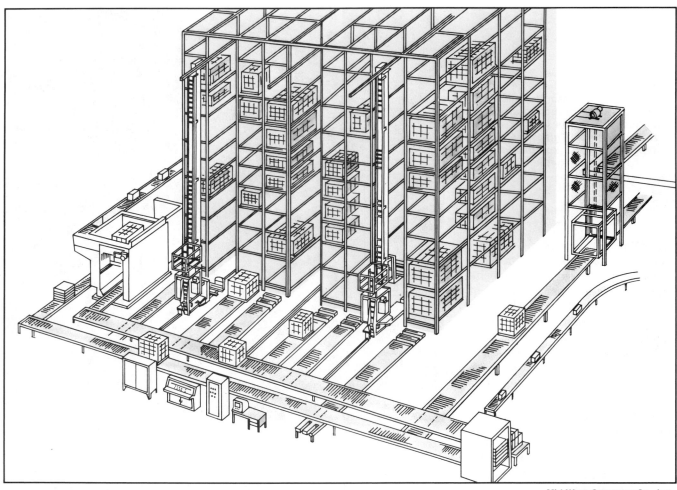

Mid-West Conveyor Co., Inc.

4-7. *This automated storage and retrieval system is controlled by a computer. It is an efficient way of moving materials and products.*

Jervis B. Webb Co.

4-8. *This operator is controlling an AS/R system.*

HIGH TECHNOLOGY
New Directions for Industry

TRW

From drafting tools to telephones, high technology is transforming American industry. Computers, robots, and lasers are changing the way products are designed, built, and tested. The pictures in this photo essay show some of the applications of high technology to graphic communications, manufacturing, power and energy, transportation, and communication. You'll learn more about these industries as you read this book.

Graphic Communications

Overleaf: A technician uses a light pen to modify a drawing on the computer screen.

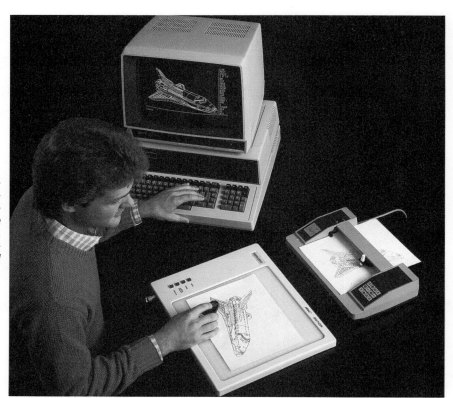

In industry, computers are replacing drafting boards. Here the operator is using a digitizer and a keyboard to input a drawing into the computer. Once on the computer, drawn objects can be moved, altered, erased, rotated, or printed out on a plotter, as shown at right.

Heath Company, Benton Harbor, MI 49022

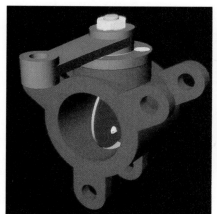

The valve shown in these two photos is not real. It is a computer image constructed by a solid modeling system. These systems, which involve both hardware (equipment) and software (programs), build and display mathematically accurate models of three-dimensional objects. The models can be analyzed and modified to arrive at the best design. They are also used to generate information needed for computer-aided manufacturing.

McDonnell Douglas Automation Company, a
division of McDonnell Douglas Corporation

McDonnell Douglas Automation Company, a
division of McDonnell Douglas Corporation

Read about computer-aided design and manufacturing in Chapters 3 and 4.

For large-volume data processing, computer output microfilmers (COMs) provide an alternative to paper printouts. Instead of printing computer data on paper, the Kodak COMs shown here use a helium-neon laser to record data onto microfilm or microfiche at rates up to 20 000 characters per second. The film is processed automatically in about 5 seconds. These COMs can deliver processed microfilm or microfiche at up to 10 000 pages per hour.

Courtesy of Eastman Kodak Company

Document storage and retrieval are made more efficient with the intelligent microfilmer and the microimage terminal. As it copies documents onto microfilm, the intelligent microfilmer (background, left) encodes the images with data that will make retrieval possible. To retrieve a document, the microimage terminal (foreground, right) scans the microfilm, reading the codes to locate the desired document within seconds.

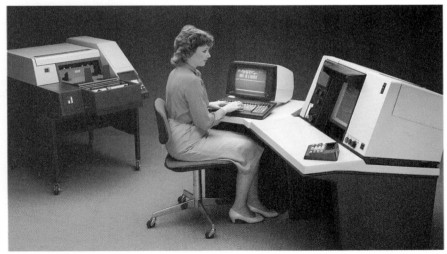

Courtesy of Eastman Kodak Company

Read about other developments in electronic printing and duplicating in Chapter 30.

Manufacturing

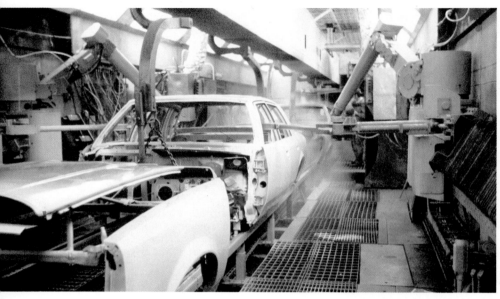

General Motors Corporation

General Motors Corporation is one of the industries utilizing robots to do jobs which would be boring or hazardous to human workers, such as welding and spray painting. Tireless, precise, and consistent, these machines can speed production, improve product quality, and decrease costs.

Pitney Bowes 1983 Annual Report/Photographer: Gabe Palmer

Robots have been developed which are capable of precision assembly work. Pictured here is a computerized robot at Pitney Bowes' electronic assembly operations facility. The robot places a printed wiring board in the tray and then assembles keyboards for the company's Model 6500 series electronic postage meter by placing the keycaps into the electronic switch housing.

General Motors Corporation

Read more about robots in Chapter 5

High technology can revitalize existing technologies. In the late 1950s propeller-driven aircraft were being replaced by faster, smoother-flying jet airplanes. Today the high cost of fuel has led to a renewed interest in the fuel-efficient turboprop engine. New propeller designs being tested consist of eight or ten short, thin blades constructed of high-strength composite materials. The advanced turboprops with their unusual blade shapes are designed to be quieter but still produce the needed thrust. The new designs are tested in wind tunnels to determine performance at various rotation rates and wind speeds. The photo shows a laser being used to measure the speed of air as it flows past the propeller blades.

NASA

Read about composite materials in Chapter 58.

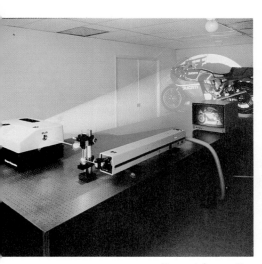

Newport Corporation

Newport Corporation

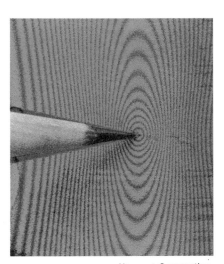

Newport Corporation

To produce high-quality products, designers and manufacturers must know how parts react to stress or vibration. A new technique called holographic interferometry is being used to detect deformation caused by vibration or stress (mechanical, electrical, thermal, etc.) In this technique, interacting waves of light from laser beams form interference "fringe" patterns which can be viewed by eye or on a video monitor, or stored on video tape for later playback. Each fringe represents a 12.4 microinch change in elevation. (One microinch = one millionth of an inch.) Thus even microscopic changes in shape are easily seen.

The photo above shows the Newport Corporation's HC-1000 holographic camera being used to analyze a motorcycle. The object is illuminated with an argon laser (foreground), while the hologram may be viewed on the television monitor. In the photos at right, a pop top can is being deformed. Note the bull's-eye pattern where the pencil touches the can.

Read about holography in Chapter 28.

Power and Energy

Read about solar cells and fuel cells in Chapter 91.

Extending like a wing from the Space Shuttle Discovery, the Solar Array Experiment (SAE) converts sunlight to electricity. The array, developed by Lockheed Missiles & Space Company, contains 84 panels made from a lightweight material called Kapton. Solar cells are welded directly to this material. An array with all 84 panels outfitted with active cells will produce 12.5 kilowatts, an amount of power equal to the electrical usage of several average households. In the future, similar systems will provide electricity for extended Shuttle missions and orbiting space stations.

Here, the SAE has been unfolded to 73 feet, about 70% of its full length. For the trip back to Earth, the array will be refolded, accordion-style, into its containment box, a package only 3½ inches thick.

NASA

Electrical power for the Space Shuttle orbiter is supplied by a fuel cell system that produces power by the electrochemical conversion of hydrogen and oxygen. Shown here is one of the three fuel cells that make up the generating system. Each unit measures 14 inches high, 15 inches wide, and 40 inches long, and it weighs 200 pounds. Each fuel cell is capable of generating 12 kilowatts at peak and 7 kilowatts average power.

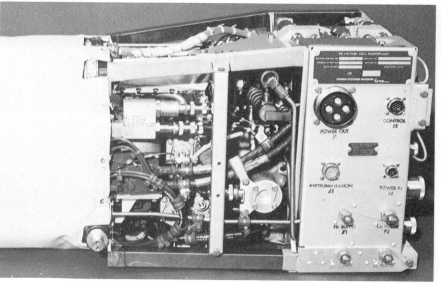

NASA

Transportation

The new 757 and 767 aircraft feature many technical advances: fuel-efficient engines; lightweight, tough plastic composite frame and skin sections; computerized guidance and control systems; improved radar and communications; and comfortable interiors. Advanced technology makes a fast, efficient, and safe air transportation system possible.

Read about air transportation in Chapter 102.

Computer simulations enable students to practice piloting an airplane without ever leaving the ground. This photo shows pilots being trained for night flying in a computer-controlled flight simulator. The view outside the window is a computer-created picture of the airport runway, as they approach it at night. The view changes constantly as they get closer.

Communication

Fiber optics is the science of transmitting light through thin, flexible strands of glass or plastic. These strands, called optical fibers, can be used to send light around corners, through underground cables—anywhere the fiber can go. In many parts of the country, copper telephone wires are being replaced by the smaller, more efficient optical fibers.

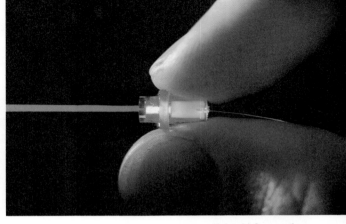

Polaroid Corporation

Read about fiber optics in Chapter 6.

Read about communications satellites in Chapter 96.

Read about facsimile transceivers in Chapter 96.

High tech on the high frontier: the SBS-4 communications satellite rises from the Space Shuttle Discovery to begin its life in space. Communications satellites orbit the Earth at an altitude of about 22 300 miles, receiving and transmitting signals from Earth stations or from other satellites.

NASA

High technology is part of today's business world. One example is the facsimile transceiver, a device which transmits and receives copies of documents over telephone lines. Here, a sales representative reviews facsimile transmission procedures on a Pitney Bowes Model 8800 facsimile machine with a corporate client's manager of telecommunication services.

Pitney Bowes 1983 Annual Report/
Photographer: Gabe Palmer

NASA

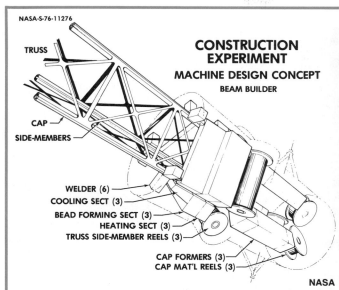

NASA-S-76-11276

TRUSS

CONSTRUCTION EXPERIMENT
MACHINE DESIGN CONCEPT
BEAM BUILDER

CAP

SIDE-MEMBERS

WELDER (6)
COOLING SECT (3)
BEAD FORMING SECT (3)
HEATING SECT (3)
TRUSS SIDE-MEMBER REELS (3)

CAP FORMERS (3)
CAP MAT'L REELS (3)

NASA

This is one method of building beams automatically in space. Sheet metal, stored on reels, is formed into hollow cap bars, and the side members are welded to them to form a truss or beam.

NASA

An artist's drawing of the automated beam builder being used in space to build a solar power station.

Building in Space

The exciting voyages of the Space Shuttle are familiar to all of us. This is the world's first reusable aerospace vehicle. It leaves Earth as a rocket, is a spacecraft while in orbit, and flies back to Earth as an airplane. The Shuttle will haul equipment needed for scientific experiments and industrial processes. It also can permit, for the first time, the assembly of very large structures in space, such as antennas, solar power collectors, and space platforms.

One machine that will be used to build in space is the *automated beam builder*. It will be stored in the Shuttle's cargo bay. Spools of very light material, such as graphite-epoxy or metal composites, will be loaded into the machine on Earth and carried into orbit. Once at the space construction site, the beam builder will heat, shape, and weld the material into metre-wide triangular beams. These beams can be cut to any length and then latched together to build large

structures. By loading the cargo bay with extra spools, enough material could be carried up in one trip to build thousands of metres of beam.

The beam builder is computer controlled, an example of CAM in space. This same technology could be used for building similar structures on Earth.

QUESTIONS AND ACTIVITIES

1. Define numerical control.
2. Briefly describe how CAD can be useful in CAM.
3. Define part program and tell what a part program might include.
4. What is process planning? What information should it include?

5. What information should be computerized in a total factory management system?
6. What is an AS/R system and what does it do?
7. What is an example of CAM in space? What might it be used to build?

One of the newest and most exciting CAM systems is the *robot*. According to the Robot Institute of America, "A robot is a reprogrammable multifunctional manipulator designed to move materials, parts, tools, or specialized devices, through variable programmed motions for the performance of a variety of tasks." Put simply, this means that a robot is a machine that can be directed to move materials, parts, or tools and do work. Fig. 5-1. It is important to know that the main purpose of robots is to do jobs which require the handling of physical objects. As the computer extends the mental abilities of humans,

so does the robot extend their physical abilities.

Robotics is the technology of designing, building, and using robots. This new technology speeds production, improves product quality, and lowers costs. Robots can work 24 hours a day, every day of the year. These machines can do the boring and tiring jobs now done by humans. They can also free humans from very dangerous jobs. Some examples are: inspecting nuclear reactor welds; handling hot workpieces during heat-treating; placing parts in an acid bath; and working in areas where toxic fumes are present.

KINDS OF ROBOTS

Real robots do not necessarily look like C3PO or R2D2 from *Star Wars*. And they cannot do all the things those robots could. Most of today's robots are just complicated machines that do fairly simple, routine tasks. Most industrial robots do one, or a combination, of four kinds of jobs:

● They do actual work, such as welding or painting. Fig. 5-2.

● They transport or move parts or tools.

● They hold parts in position so that they can be worked on by other robots or machines.

● They assemble parts.

Some robots are very simple "pick and place" devices. They pick up a part, whether it's a tiny glass bulb or a huge axle, and move it from one location to another. These are called *non-servo*

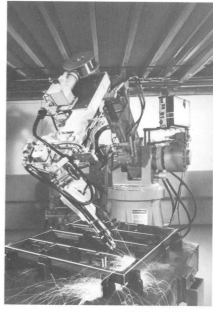

Cincinnati Milacron

5-2. This robot is welding a computer frame. It welds 44 two-inch seams, from many angles, in under 12 minutes. It would take a human welder 45 minutes to weld this frame.

robots, and they are relatively simple to program and to operate. (A *servo mechanism* is an automatic control system that uses feedback to direct the action of a robot.) The non-servo robots move until their limits of travel are reached (go and stop positions). They perform easy, repeatable tasks at high speeds. They are also less expensive than servo-controlled robots.

Servo-controlled robots can move anywhere within their work areas. They are therefore ideal for jobs such as spray paint-

General Motors Corporation

5-1. This robot installs light bulbs in back of an automobile instrument panel.

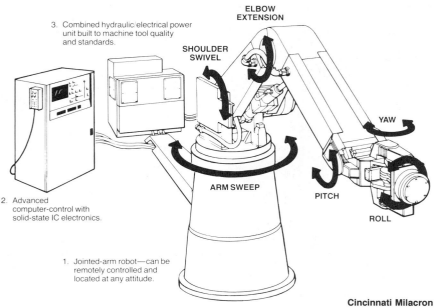

3. Combined hydraulic/electrical power unit built to machine tool quality and standards.

ELBOW EXTENSION

SHOULDER SWIVEL

YAW

ARM SWEEP

PITCH

ROLL

2. Advanced computer-control with solid-state IC electronics.

1. Jointed-arm robot—can be remotely controlled and located at any attitude.

Cincinnati Milacron

5-3. The parts of a typical robot. On some robots these parts are in a single unit.

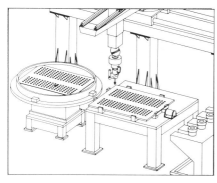

Westinghouse Electric Corp.

5-4. This robot slides along an overhead track. Its "fingers" hold tools to drill, ream (widen), and deburr (smooth) holes in metal workpieces.

ing or arc welding. Such robots are more complex, more expensive, and more difficult to program. They perform with smooth motions and accurate control of speed, and they can exactly duplicate the movements of a human being.

PARTS OF A ROBOT

The typical industrial robot has three main parts: a jointed arm unit, a computer control unit, and a power unit. Fig. 5-3.

Jointed Arm Unit

In Fig. 5-3, note that the jointed arm unit has a shoulder, elbow, arm, and wrist which can move, much like human joints can, to do work. The robot's "fingers" (called "end effectors") can be designed to do any kind of job. For example, some robot fingers are suction cups used to pick up and install automobile glass or to move a wooden door onto a spray-painting rack. Others are grippers which pick up parts and place them in machines to be

turned or drilled. In other robots, the fingers are chucks to hold tools for drilling or welding. Fig. 5-4.

The size and design of the ro-

bot arm unit control its work area. This means simply that a robot can stretch, turn, raise, lower, or twist its jointed parts only so far. It has its limits. The area within these limits is called the "working envelope." Fig. 5-5. This envelope is important. For example, if a robot were to be

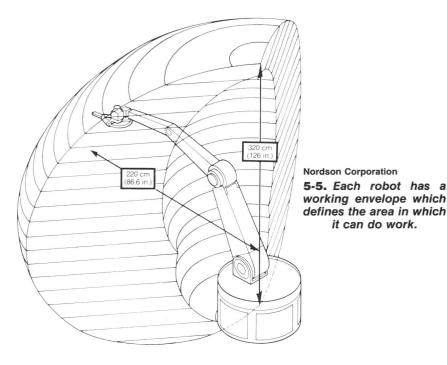

320 cm (126 in.)

220 cm (86.6 in.)

Nordson Corporation

5-5. Each robot has a working envelope which defines the area in which it can do work.

Working Envelope: Isometric View

35

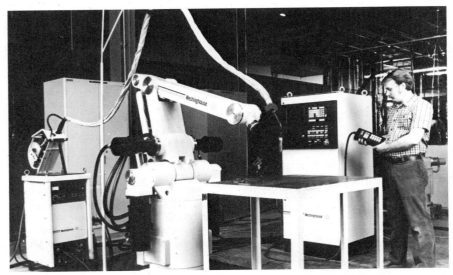

Westinghouse Electric Corp.

5-6. *Using the teaching pendant to establish the positions of a robot weld head.*

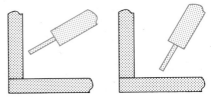

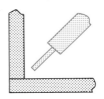

Westinghouse Electric Corp.

5-7. *The search function of a computer program allows the robot welder to determine the welding starting point.*

used to spray paint an auto body, the body would have to lie within the limits of the working envelope.

Computer Control Unit

All but the simplest of robots are controlled by computers. A computer program is prepared for each job to be done. The typical robot has a small computer to guide its movements. Signals are sent to the power unit by the computer control unit. These signals open and close valves or switches to drive the robot movements, such as rotation; raise-and-lower; in-and-out; and wrist yaw, pitch, or roll. These movements in turn direct the robot to assemble parts; spray paint; weld; solder; machine; heat treat; fabricate sheet metal; do casting, stamping, or forging; inspect; handle material; rivet; or drill.

To better understand how a robot is made to perform its tasks, let's follow the programming pro-

5-8. *The HERO 1 robot is an inexpensive, "smart" robot. It is shown here with and without its cover.*

Heath Company, Benton Harbor, MI 49022

cess of a welding robot. After a basic welding program is entered into the computer control unit, the robot welder must still be shown where to weld. This is done by a skilled operator using a *remote teaching pendant*. This pendant guides the robot welder to its welding program points. This new information is then entered into the computer control unit, so that the robot can repeat the same welds on identical workpieces. Fig. 5-6.

A *search function* is also en-

tered into the program. Fig. 5-7. This allows the robot to find the weld seam area and fix the starting point of the weld. In Fig. 5-7, the welding head searches for (and finds) the vertical section, then moves to find the horizontal section. Next the control system calculates the intersection of these two sections. The welding head moves to this new position, ready to weld.

The robot is now programmed to weld as many parts as it finds in its working envelope. It may

not be C3PO, but it is a remarkable machine.

Some robots even control themselves—by sound, sight, or heat. These are "sensory" robots. Robots may also use "incipient sensing," which means they will stop a movement just before they make an error. This prevents damage to the robot and avoids ruined parts. An example of such a robot is shown in Fig. 5-8. Its functions and operation are explained in Fig. 5-9. This intelligent robot is inexpensive and

5-9. *The parts and functions of HERO 1 are shown here.*

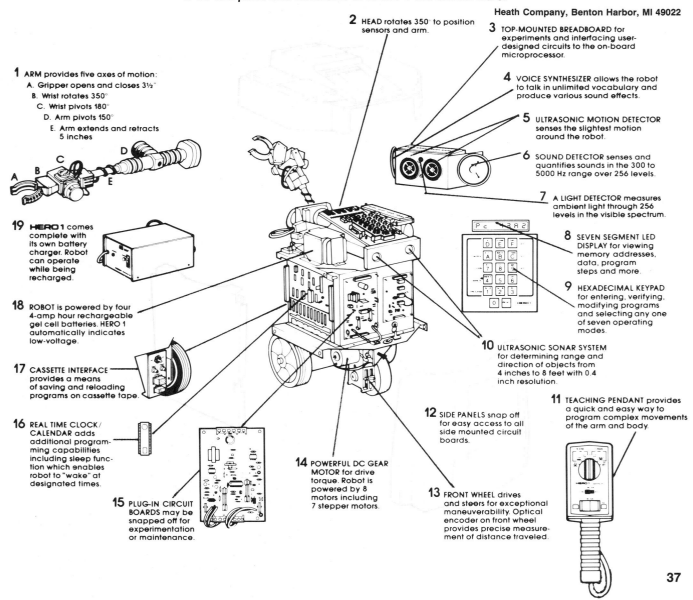

Heath Company, Benton Harbor, MI 49022

1 ARM provides five axes of motion:
A. Gripper opens and closes 3½"
B. Wrist rotates 350°
C. Wrist pivots 180°
D. Arm pivots 150°
E. Arm extends and retracts 5 inches

2 HEAD rotates 350° to position sensors and arm.

3 TOP-MOUNTED BREADBOARD for experiments and interfacing user-designed circuits to the on-board microprocessor.

4 VOICE SYNTHESIZER allows the robot to talk in unlimited vocabulary and produce various sound effects.

5 ULTRASONIC MOTION DETECTOR senses the slightest motion around the robot.

6 SOUND DETECTOR senses and quantifies sounds in the 300 to 5000 Hz range over 256 levels.

7 A LIGHT DETECTOR measures ambient light through 256 levels in the visible spectrum.

8 SEVEN SEGMENT LED DISPLAY for viewing memory addresses, data, program steps and more.

9 HEXADECIMAL KEYPAD for entering, verifying, modifying programs and selecting any one of seven operating modes.

10 ULTRASONIC SONAR SYSTEM for determining range and direction of objects from 4 inches to 8 feet with 0.4 inch resolution.

11 TEACHING PENDANT provides a quick and easy way to program complex movements of the arm and body.

12 SIDE PANELS snap off for easy access to all side mounted circuit boards.

13 FRONT WHEEL drives and steers for exceptional maneuverability. Optical encoder on front wheel provides precise measurement of distance traveled.

14 POWERFUL DC GEAR MOTOR for drive torque. Robot is powered by 8 motors including 7 stepper motors.

15 PLUG-IN CIRCUIT BOARDS may be snapped off for experimentation or maintenance.

16 REAL TIME CLOCK/CALENDAR adds additional programming capabilities including sleep function which enables robot to "wake" at designated times.

17 CASSETTE INTERFACE provides a means of saving and reloading programs on cassette tape.

18 ROBOT is powered by four 4-amp hour rechargeable gel cell batteries. HERO 1 automatically indicates low-voltage.

19 HERO 1 comes complete with its own battery charger. Robot can operate while being recharged.

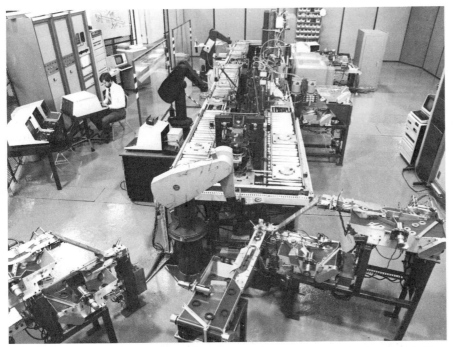

Westinghouse Electric Corp.

5-10. *Robots are being studied in this experimental laboratory to create a fully automated factory.*

ROBOTS AND THE FUTURE

Much work is going on in industry to combine robots with other parts of an automated factory. Fig. 5-10. People have only begun to explore the many uses for robots in the factory and the home. Robots of the future might clean your house, guard it, wash your car, or mow your lawn.

The exploration of space also has led to proposed uses for robots. An idea for a mining robot on the moon is shown in Fig. 5-11. This robot clears a path with a grader and then scoops ore from the ground with a bucket. This ore is transported to a factory for processing. Fig. 5-12. Such a moon-mining program would be entirely automatic. There would be no humans there to operate it. An exciting thought, isn't it?

simple to program and use. It may well be one of the first to be used in a home to help with the household chores.

Power Unit

The third major part of a robot is its power unit. This unit provides the hydraulic, electric, or pneumatic power which enables the robot to move. Sometimes the power unit is a separate part, and sometimes it is built into the robot.

REMOTE CAMERA ARM

NAVIGATIONAL RECEIVER (TRANSPONDER)

NAVIGATION

TWO LENSES

CAMERA

ELECTROSTATIC SOLAR CELL WIPER

ATTACHMENTS DRIVES & CONTROLS TOWLINE

SOLAR CELLS

ELECTROSTATIC CAMERA LENS WIPER

ARM DRIVES

COMPUTER "SMARTS"

SCOOP DRIVES

FRONT END BUCKET

CONNECTORS

FUEL CELLS

PRECISION GRADER

GRADING BLADE

GRADE SENSORS

MOTOR

SPUN BASALT TIRES

MOTOR

DOZER BLADE ATTACHMENT

MASS ~ 2200 kg

OPTIONAL ATTACHMENTS

NASA

5-11. *This "robot of the future" is designed to mine minerals on the moon. Note that it travels on a programmed path, which it keeps level with its "grader." It uses solar power and can be programmed from Earth.*

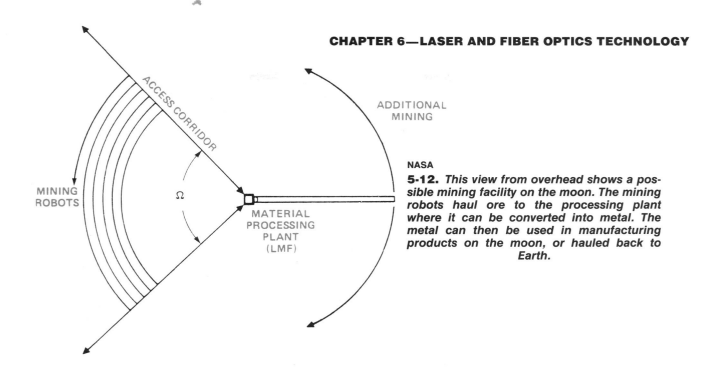

NASA

5-12. *This view from overhead shows a possible mining facility on the moon. The mining robots haul ore to the processing plant where it can be converted into metal. The metal can then be used in manufacturing products on the moon, or hauled back to Earth.*

QUESTIONS AND ACTIVITIES

1. Simply speaking, what is a robot?

2. Define robotics. What are some of its advantages?

3. Name the four kinds of jobs industrial robots are generally used for.

4. Name the three main parts of an industrial robot and describe the job or function of each.

5. What is a working envelope?

6. Discuss what you think might be some possible uses of robots in the future.

7. Have a class debate about the pros and cons of the increasing use of industrial robots and other computerized aids to manufacturing.

CHAPTER 6

Laser and Fiber Optics Technology

The use of light in new and important ways is one of the most exciting developments in technology. You have seen the light-emitting diodes (LEDs) on the faces of digital clocks and pocket calculators. These "light up" to display the time of day or the sum of an addition problem. Lasers and fiber optics also are part of this technology, often referred to as "optoelectronics." They are the subjects of this chapter.

LASERS

Lasers are becoming very important in industry. A *laser* is a device that amplifies, or strengthens, light. The word stands for Light Amplification by Stimulated Emission of Radiation.

One common laser, and the first to be developed, is the ruby laser. It consists of a ruby rod surrounded by a coiled flash lamp. The ends of the ruby rod are flat, and they are parallel with each other. One end of the rod is completely coated with reflective material. The other end is only partially coated. Fig. 6-1.

Every time the flash lamp is set off, it excites atoms in the ruby rod. (An excited atom is one with a high energy level.) The atoms seek to return to a lower energy level; so they give off their excess energy in the form of light. Much of this light escapes through the sides of the ruby rod. But some of it hits the mirrored ends of the rod and bounces back. As this light travels back through the rod, it strikes other excited atoms, causing them to give off light of the same *wavelength* and *phase* (see inset). The original light continues on its path, joined by the new light. This process is repeated, back and forth between the mirrors, and the light becomes more and more intense. When it becomes strong enough, the light "breaks through" the partially mirrored end of the rod, emerging as a laser beam. Thus a laser is a device in which **l**ight is **a**mplified (strengthened) because atoms are **s**timulated (excited) to **e**mit (give off) **r**adiation in the form of light.

This is a very simple description of how a laser works. In practice it is much more technical and complex. But this explanation should help you under-

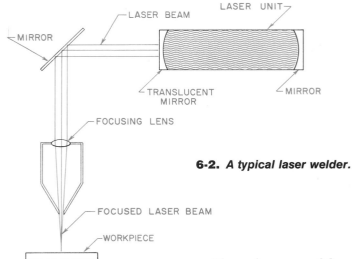

6-2. *A typical laser welder.*

stand how light can be treated in such a way that it can become strong enough to weld metal or burn through a diamond.

Lasers can do these remarkable things because laser light differs from ordinary light in several important ways.

● Laser light is *coherent*. This means it is ordered and directed, not scattered like the light from a light bulb. A laser beam diffuses (scatters) far less with distance than ordinary light. For example, ordinary light beamed from Earth to the moon would spread out to an area 40 000 kilometres (about 24 800 miles) in diameter. A laser beam would produce a spot only 3 kilometres (about 2 miles) in diameter.

The coherence of laser beams makes them useful as alignment and measuring tools. One such tool is the laser beam aligner. It is used in damage assessment and repair of automobile underbodies. A laser gun is mounted on a pivot axis. Scanning targets are placed at various points on the vehicle. The laser beam measures the distance between these targets. In this way, it can be determined whether the parts of the underbody are in correct position in relation to each other.

● Lasers produce very narrow beams of light which can be easily controlled. This property makes lasers useful for precision cutting. For example, ruby lasers are used to drill holes in sapphires for watch bearings. In biology research, lasers are used to separate chromosomes.

● High-energy power levels can be produced by lasers, and this energy can be focused with lenses or mirrors. This is something like focusing the rays of the sun with a magnifying glass to start a camp fire, but with much more effective results. The laser beam quickly vaporizes the material it touches, without affecting the surrounding material. For example, it is possible to make a laser eraser. The laser heats and burns away the ink

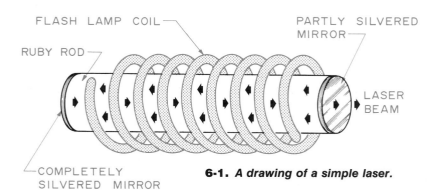

6-1. *A drawing of a simple laser.*

Michigan Laser Cut

6-3. *This is a fine example of laser cutting of wood.*

but leaves the paper beneath it intact.

There are many types of lasers, and they are used to do many things. Lasers vary greatly in size, power, and type of light they emit. Thus different lasers are used for different purposes. Some typical examples of how lasers are used follow.

Laser Welding

Figure 6-2 shows how a typical metal welding laser works. The laser chamber is filled with carbon dioxide gas. Electrical charges are fired into the chamber to excite the atoms and start the process of creating a laser beam. The laser beam that emerges from the chamber is reflected by mirrors to a focusing lens. This lens narrows the beam to a fine, very hot point and directs it to the work surface for welding.

Laser welding focuses the heat exactly where it's needed, and it does not warp the metal as some other types of welding do. It is also very fast, and it's easy to control by blocking the beam as it leaves the mirror. This system

also can be used for cutting and heat-treating metals. (You will learn more about laser cutting of metals in Chapter 42.)

Lasers work very well with computer-controlled equipment. The narrow beam can be moved a space as tiny as a millionth of an inch. Also, the pulse of the beam can be controlled very precisely to a billionth of a second.

Laser Cutting of Wood

Wood can be engraved or cut to produce an extremely delicate wood carving. First, a sketch of the design to be engraved is made. Using this sketch, a brass stencil is made. The brass stencil is mounted on a piece of wood which has been sanded and lacquered. A laser beam is used to burn away the exposed wood to the desired depth. What actually happens is that the wood is vaporized, rather than burned, by the 30 400°F laser beam. This leaves an extremely clean cut, with no smoke or ashes. The beam is very narrow—0.010 inch—so very fine cuts are possible. The brass stencil reflects the laser beam, protecting the wood beneath it. Therefore burning takes place only where the wood is not protected by the stencil. The result is a precisely cut piece of wood. Fig. 6-3.

Some lasers produce beams so intense and powerful they can cut metal. However, an unusual feature of laser cutting is that steel can be cut much faster than wood. The reason is that the laser beam is generally set to make a thinner cut (0.008 to 0.009 inch) in steel, while wood cuts are slightly wider (0.010 to 0.028 inch). Steel plate ¼ inch thick can be cut at the rate of 80 inches per minute.

Laser Micrometer

Laser micrometers are used in industry to measure the diameters of workpieces—such as wires, rods, or bars—made of almost any material. In a typical laser micrometer, the workpiece is passed through a laser beam. Fig. 6-4. A special data analyzer "reads" the diameter and displays it in digital (numerical)

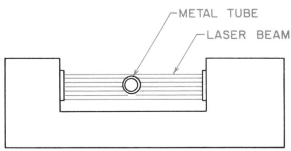

METAL TUBE
LASER BEAM

6-4. *This is how a typical laser micrometer works.*

Pioneer Video, Inc.

6-5. *A laser disk looks much like a phonograph record.*

form. A printer can also print the data for further analysis.

The advantage of the laser (or optical) micrometer is that it measures without having to touch the workpiece. This means that measuring can take place continuously, while the workpiece (such as a metal tube) is being manufactured. Any errors in the diameter can be detected immediately. This information can be sent to the production machine, where the error is corrected. This saves time and reduces the amount of "scrap" produced.

Other uses of the micrometer include measuring the sizes of fiber optic strands, bearings, films, and drill bits.

Lasers in Printing

Lasers are widely used in the printing industry for color separation, cylinder engraving, exposing film, typesetting, and cutting paper. These uses will be discussed in the graphic arts chapters (22-30). A newer printing application is the laser scanner and recorder.

In this system, lasers are used to scan a paste-up. (A *paste-up* is an arrangement of words and illustrations on a page.) The reflected laser light is converted into digital information that represents the images on the paste-up. The information can then be sent by cable, telephone lines,

microwaves, or satellite to another location. This other location could be in the same city, or thousands of miles away. At the receiving end, the information goes through a recorder which uses the digital information to control a low-power, air-cooled laser. This laser changes the information back to the original images, either onto film or onto a printing plate.

Newspaper and magazine publishers find this system useful. A daily newspaper, for example, can be set in type and arranged into pages, then transmitted to several printing presses in different regions of the country. At each printing press, local news is added. Then the paper is printed and distributed.

Lasers in Medicine

Lasers are used as surgical tools. As the laser beam contacts human tissue, the cells are su-

perheated and changed instantly to vapor. This technique causes little damage to the surrounding tissue, and there is very little, if any, bleeding. The tiny laser beam is especially useful in eye surgery.

Laser Audio and Video Systems

Laser technology has been applied to the home entertainment industry. Movies and other programs are now available on laser video disks. The disks look like, and can be handled and stored like, an ordinary phonograph record. Fig. 6-5. Program information is stored in billions of microscopic "pits" etched into the disk. Fig. 6-6. The pits are arranged in circular tracks, as many as 54 000 of them, and each carries a separate video frame. A thin layer of plastic protects the disk from fingerprints and dust. Similar laser disks are used in audio

MAGNIFIED VIEW OF A VIDEODISC

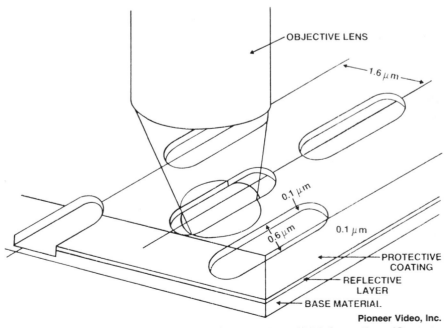

OBJECTIVE LENS

1.6 μm

0.1 μm

0.6 μm

0.1 μm

PROTECTIVE COATING

REFLECTIVE LAYER

BASE MATERIAL

Pioneer Video, Inc.

6-6. *The tiny pits etched into the disk carry recorded information. (One μm is equal to 0.000039 inch.)*

reproduction, as you would use a phonograph record or tape recording.

A laser video disk player looks much like a phonograph turntable. Fig. 6-7. It is different in that there is no stylus (needle) to wear away the disk. Instead, a low-power laser beam is focused by a lens onto the disk. Fig. 6-6. The pits in the disk cause variations in the amount of light that is reflected as the disk passes under the laser beam. These variations are sensed by the machine and converted into pictures and sound for reproduction in an ordinary television set.

These laser audio and video systems can replace records and audio and video tapes. They are very durable and will continue to reproduce excellent images and sounds for many years.

Lasers in Communication

Like radio waves, laser beams can be used for communication. The amount of information that an electromagnetic wave can carry increases as its frequency increases. Since the frequency of a laser beam is much higher than that of a radio wave, much more information can be carried. In theory, one laser beam could carry as much information as all the world's radio channels. But there's a drawback. Like all light, laser light is blocked by clouds, rain, and snow. This is no problem in outer space, but for effective communication on Earth, the laser beam must be protected. This is done by enclosing the beam in a "pipe"—an optical fiber. You will read about optical fibers in the next section.

FIBER OPTICS

Fiber optics is the science of transmitting light through very thin, flexible strands of glass or plastic. These strands are called

6-7. *A laser disk player.*

optical fibers. Fig. 6-8. The light travels from one end of the fiber to the other, much the same way as electricity travels down a wire. Optical fibers are transparent (can be seen through). You may wonder, then, why the light travels along the fiber instead of escaping through the sides.

Every substance through which light can pass deflects it from its straight path. That is why a drinking straw placed in a glass of water looks bent. The amount by which a substance de-

flects light is called its *refractive index*. In theory, it is possible to have a substance deflect light so much that it sends the light back in the direction it came from. Optical fibers come close to this ideal. Therefore very little light escapes through the sides of the fiber.

This means the light can be made to go wherever the fiber goes—around corners, inside machines, even inside the human body. In the space program, a bundle of optical fibers is used as a camera lens. The fibers extend from the camera to a place on the outside of the spacecraft. The camera itself stays safely inside, recording the images the optical fibers send it.

Fiber Optics in Communication

You read earlier that laser beams can be transmitted along optical fibers. Telephone companies are making use of this technology to replace heavy copper-wire cables in large cities.

This system uses a laser the

6-8. *Note how light is transmitted through this tiny glass fiber.*

6-9. *A UPC symbol.*

Other Uses of Fiber Optics

Probably one of the most familiar uses of fiber optics and lasers is to read UPC symbols on grocery products. The UPC (Universal Product Code) symbol is the series of bars found on most packaged items. Fig. 6-9. The code identifies the item. As the clerk passes the packages over a scanner, a laser reads the codes. Optical fibers transmit the data

size of a grain of sand. The beam from this laser is thinner than a strand of human hair. This tiny laser is cemented to the end of a long strand of optical fiber, which takes the place of the copper wire.

Here is how it works in an ordinary telephone conversation. The human voice is changed into an electric signal. Instead of traveling through copper wire to the receiving phone, the electric signal activates the laser. The laser beam, turning on and off 44 million times a second, sends the conversation through the optical fiber as a light signal. Then, at the receiving end, the light is changed back into electric signals, and then to voice signals.

One of the advantages of this system is its compactness. A single fiber can carry 672 telephone conversations at the same time. A bundle of fibers made into a cable only half an inch in diameter can carry the same load as six copper-wire cables, each several inches thick. Another advantage is that the laser signal can travel at least four miles without having to be boosted, and there's no electrical interference. A copper-wire signal can only travel about one mile.

Data Composition, Inc.

6-10. *These are examples of bar codes used in industrial information control systems.*

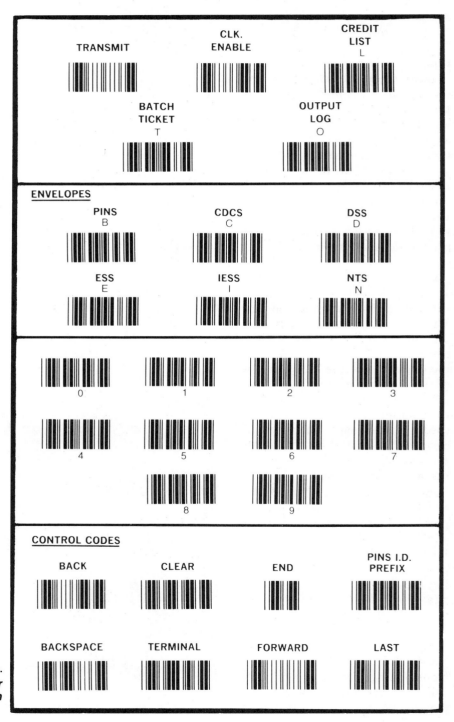

What Is Light?

That may sound like a very simple question, but the answer is complex. Light is a form of energy. Unlike some other forms of energy, light is always moving. If light is absorbed by some object, and thus stops moving, it ceases to be light. Light moves very fast: 186 000 miles per second.

The light that reaches us carries information about the places it's been because those places have partly absorbed, reflected, or deflected it. Our brains interpret the information and thus we see blue sky or yellow school buses or the face of a friend.

For many years, scientists debated whether light exists as waves or particles. The answer seems to be that it is both, depending on which characteristics of light one wants to examine. When light is emitted or absorbed, it behaves like particles. But when light interacts with other light beams, it behaves like waves.

A light wave can be thought of as looking like the sketch shown here. Ordinary light, such as sunlight, is a mixture of wavelengths. Have

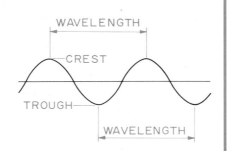

you ever used a prism? The prism separates light into its various wavelengths. We see these different wavelengths as colors.

In ordinary light the waves are not in phase ("in step") with each other. In laser light, they are. Laser light is an intense beam of light of a pure single color, or wavelength, and it is in phase.

Visible light is only part of a whole range of wavelengths called the *electromagnetic spectrum*. The spectrum also includes gamma rays, X-rays, ultraviolet and infrared light, microwaves, and radio waves.

to a central computer into which all the prices have been programmed. The computer sends the information back to the cash register, adds the total bill, and also keeps inventory records. Next time you are in a grocery store, look for the UPC symbols on the packages.

Industry also uses bar coding as part of an information control system. A simple, machine-readable, self-adhesive bar code label is applied to workpieces, cartons, or documents. This permits plant managers to track the progress of each job, record the time spent at each work station, and know the location of each part at any given time. This control system is an important part of automatic, computerized factories. Typical examples of these bar codes are shown in Fig. 6-10.

QUESTIONS AND ACTIVITIES

1. What is a laser? What does the word *laser* stand for?

2. Name three ways laser light differs from ordinary light and tell how each of these properties makes laser light useful.

3. There are many types of lasers, and they are used to do many things. Name five typical examples of how lasers can be used.

4. What is fiber optics?

5. Name two uses of fiber optics and tell what advantages they provide.

PART 2

Basic Shop Practices

CHAPTER 7

Product Design

The next time you use a tool to do a job around the house, take the time to study it. Does the tool work as it should? Does the wrench hold the nut securely, or the chisel cut as it should? Have the right materials been used to make the tool strong enough? Is the tool comfortable to hold and well balanced? If you answer yes, the object is probably well designed.

Tools, like all other products of industry—sports equipment, automobiles, chairs, knives and forks, and dozens of other items that you use each day—were planned and made to do certain jobs and to make your life better. The people who plan these products are called *designers*. They work in the research and development departments of indus-

tries. Their job is to design new and better products and to improve old ones. Fig. 7-1. If these products work as they should, if they are made of the right materials and are interesting to look at, they are well designed. If not, the products will be poor. You will be doing some design work in this general industry course. It is important that you learn something about design so that you will be proud of the things you make.

Homes and other buildings are also designed. The people who design these structures are called *architects*. The buildings they design must meet the needs of the people who are to use them. The materials to be used, the size and arrangement of the rooms, the efficient use of energy—all these must be considered as the architect plans the building.

WHAT IS DESIGN?

Very simply, designing is planning. To do it well you must think carefully about the product you wish to make. It should be original work—your work and not someone else's. In other words, you must be creative. *Designing* is creative planning to meet some special human need.

In order to be good, the product that you are designing should meet three requirements:
- It should work properly (functional requirement).
- It must be made of the correct materials (material requirement).

Canpro

7-2. The correct materials were used to make this skate blade—tough plastic and steel. Also, the blade's appearance matches its function.

- It should be pleasing to look at (visual requirement).

Look at the hockey skate blade and support in Fig. 7-2. This unit is attached to a hockey shoe and must move the player through sharp turns and long glides. It has a functional shape that allows the player to move with ease. The blade is made of carbon steel, and the holder is made of tough, high-impact plastic. These materials allow the skate to withstand hard use. The unit also has a nice appearance, with a clean and graceful shape. This skate blade and support illustrate an object which is functionally, materially, and visually correct—a well-designed product of industry. Let's take a closer look at these three requirements as they apply to other kinds of products.

Function

A product is functional if it works as it is supposed to work.

Tool handles must be functional in order to be held easily and safely. The tools are used

Zinc Institute, Inc.

7-1. In order to plan products, designers must know how to draw. You will learn more about drawing in Part Three of this book.

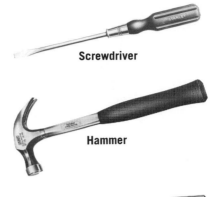

Screwdriver

Hammer

Hacksaw

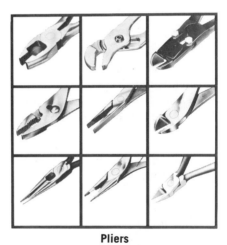

Pliers

7-3. Tools are designed for function—the right tool for the right job.

and held differently; so their handles must be different. Fig. 7-3.

The same is true of other things you hold. Knives and forks, fishing poles, baseball bats, golf clubs, and steering wheels should all be easy to grip.

Since plier jaws also have special uses, or functions, they must be designed with these uses in mind. Some are used for holding delicate parts, some for cutting heavy or thin wire, still others for heavy gripping and turning.

They must differ in design so that they can perform different functions.

Chairs must be designed or planned for different uses. They must be comfortable and be right for each use. Think how difficult it would be to work at a drafting table while sitting at a tablet armchair! Function in design means making sure a product works as it is supposed to work. Fig. 7-4.

These are just a few examples of the part function plays in product design. You can probably think of many others. While you are designing, you should be thinking about the purpose or use for your product. You must design it to be functional.

Material

A product must be made from the proper kind and amount of material. If you design something to be used outdoors, it must be made to withstand water, wind, and sun. Materials must be chosen according to the product's use.

Many kinds of materials are used to make products. You must know something about the materials before you start designing with them. Find out about such things as cost, durability, and strength. Every material has certain characteristics all its own. You should take full advantage of this fact as you design.

Learn what to do and what not to do with a material. Fig. 7-5. For example, plastic should not automatically be used as a substitute for wood. It is expensive and will not be as good as wood for some projects. Instead, plastic should be used in ways that take advantage of its own characteristics. For instance, plastic laminates (the materials used to cover kitchen counters) are strong and heat-resistant. Other

plastics can be bent, lathe-turned, or blow-formed.

Some plastic is tough and transparent, while other kinds scratch easily and are opaque (can't be seen through). These are the kinds of things you should know about materials.

Also keep in mind that you should use only enough material to do the job and no more. Don't waste material. Your project will not only cost less but will also

Tablet armchair

Drafting stool

7-4. Chairs are also designed for function—the right chair for the right job. A tablet armchair and a drafting stool have different functions and must therefore be of different heights.

Materials

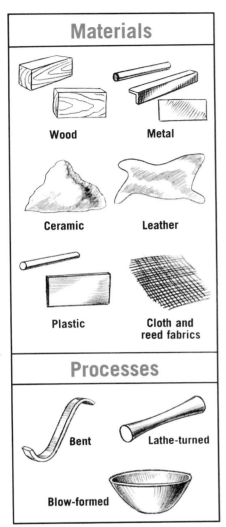

Wood   Metal

Ceramic   Leather

Plastic   Cloth and reed fabrics

Processes

Bent   Lathe-turned

Blow-formed

7-5. Materials should be studied so that they are used wisely in designing products. Which of the materials can be worked by the three process examples?

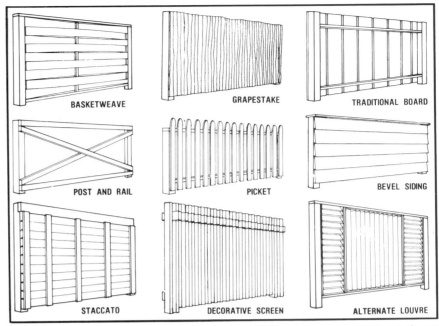

BASKETWEAVE   GRAPESTAKE   TRADITIONAL BOARD

POST AND RAIL   PICKET   BEVEL SIDING

STACCATO   DECORATIVE SCREEN   ALTERNATE LOUVRE

7-6. A product's appearance is important. Which of these fence styles do you think is most attractive?

look better if material is used in proper amounts.

Appearance

A good product is pleasing to the eye. Everyone prefers things which are beautiful to those which are ugly. Designers must keep this in mind as they design products. Fig. 7-6.

ELEMENTS AND PRINCIPLES OF DESIGN

A designer must know certain design elements and principles. The "building blocks" of design are called the *design elements.* These are the lines, the forms, the shapes (spatial forms, or solids), and the surface treatment (color, texture) that make up any three-dimensional product. The lines can appear as graceful curves or straight and strong. These grow into forms and solid shapes which display color and texture. Fig. 7-7.

The following principles of design should help you create products that will have a nice appearance.

1. The different parts of a product should look good together. This is called *unity.* The salt shaker in Fig. 7-7 has unity because the bowl and the cap flow together in a clean, unbroken curve.

2. The shaker also has *variety* since the cap is made of stainless steel and the bowl of teakwood. The different colors and textures give the product a more interesting appearance.

3. Products should be *well proportioned,* as is the table in Fig. 7-8. Notice that the legs are of a proper size compared with the top. The table does not look awkward, as it would if the top were very heavy and supported by thin, spindly legs.

4. If all parts of the table are well proportioned, this product will have *balance.*

Unity, variety, proportion, and balance are called *principles of design* because they deal with ways of organizing the shapes and materials of products.

7-7. All the design elements are present in this salt shaker—line, shape, form, and texture.

49

Krueger, Inc.

7-8. *This table is attractive because the designer paid attention to design principles.*

DESIGN-ANALYSIS METHOD

As stated earlier, designing is a kind of planning. You should plan a project as you would plan any other kind of detailed work. The project that you are designing should be thought of as a problem to be solved as you would solve a problem in mathematics or science. It is helpful to break the work into a series of steps to make it easier to solve. A good way of doing this is to use the design-analysis method. In order to explain this method more clearly, let's take a certain problem. Imagine that you will be designing a rack to hold a few books on your study desk. You must first state your problem so that you know exactly what you are trying to solve.

Step 1. State the problem: to design a bookrack to be used on a desk.

You must next ask some questions about function and materials for the bookrack. How many books must it hold? How large are the books? Are there any special materials you should use? (Perhaps you would want to match a note pad holder or a pen set.)

Step 2. Analysis and research. List ideas pertaining to function and materials.

a. Should hold six books.

b. My books measure about 1″ thick, 9″ high, and 6″ wide.

c. Make of material to match my desk set.

d. Should be easy to remove and replace books.

e. Books shouldn't topple over when one or two are removed.

f. Should not scratch desk top.

Step 3. Developing solutions. Sketch four or five ideas on paper. What might this rack look like? Improve your sketches and rework some of them. Think of the visual requirements of the unit. Be inventive and original. Figure 7-9 shows some of the possibilities.

Step 4. Experimentation. Make a paper or wooden model of the rack. Try out some of the ideas to see if they will work.

Step 5. Final solution. This should be a final working sketch or drawing. From this sketch you will proceed to make the rack. Feel free to make changes in the drawing—and in the rack—as you are making the project, if you feel you can improve it. When you are finished, go back and correct the drawing. The illustration in Fig. 7-10 is the final solution for the bookrack problem.

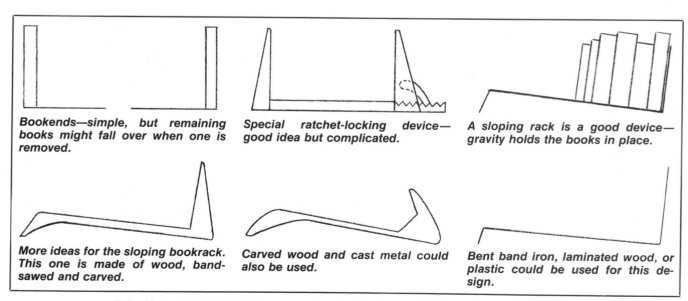

Bookends—simple, but remaining books might fall over when one is removed.

Special ratchet-locking device—good idea but complicated.

A sloping rack is a good device—gravity holds the books in place.

More ideas for the sloping bookrack. This one is made of wood, bandsawed and carved.

Carved wood and cast metal could also be used.

Bent band iron, laminated wood, or plastic could be used for this design.

7-9. *Here are several design ideas for the bookrack. Which do you think is best?*

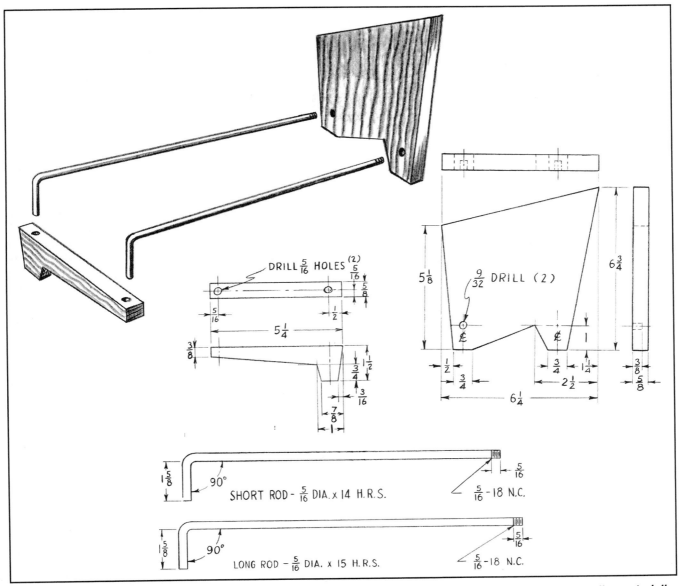

7-10. *A working sketch of the final solution to the bookrack problem. It is a good solution—functionally, materially, and visually correct.*

QUESTIONS AND ACTIVITIES

1. What are the three design requirements?
2. List the four design elements.
3. List the four principles of design.
4. Make a bulletin board display illustrating one or more of the principles of design.

5. Design some simple project using the design-analysis method.

CHAPTER 8

Measurement

Measurement is important in daily living. All the products you use have measurements: lumber for houses, metal in bicycles, paper in books—all must be measured before the product is sold. Milk is measured in quarts, fishing line in yards, and butter in pounds.

THE MEANING OF MEASUREMENT

Measurements are made to answer questions such as: How heavy is this steel bar? How long is it? How much carbon does it contain? To learn these answers, we must compare the bar to fixed, standard quantities. To measure something is to compare it with a standard unit.

Standard Units—Customary

In order to make measurements, we must have suitable units of measurement. The weight of the steel bar will be in pounds or ounces, the length in feet (') or inches ("). These are units of the *customary* (or English) system used in a few countries. All other countries, including England, use, or are planning to use, the *metric* system. The metric system is described in this chapter.

In ancient times the standard units were based upon the human body. For example, a cubit was the distance from the tip of the middle finger to the end of the elbow, about 18". A digit was the width of the middle finger, about ¾". But because human measurements were different from person to person, more accurate systems had to be invented.

Because of the need for accuracy, standard units of measurement were set. Measuring devices had to conform to these standards. This is still true today. At the National Bureau of Standards in Washington, D.C., the standards for the inch, foot, and yard measures are available. Some other standard units are the ounce, the pound, and the gallon. The crude measuring devices of ancient times have been replaced by more accurate instruments such as the steel rule, the micrometer, and the weighing scale.

Standard Units—Metric

As stated earlier, most of the world uses the metric system of measurement. This system began in France in 1790 and was based upon the *metre*.† Originally the metre was one ten-millionth of the distance from the North Pole to the equator. Today a more accurate measure of the metre is used.

The original metric system was not perfect. Some of the measures were confusing and were not convenient for modern technology. In addition, many metric countries created their own different metric units. In 1960, a simplified metric system was developed. The name International System of Units, abbreviated "SI" (for the French, "Système International d'Unités"), was adopted for this modernized metric system.

In the SI metric system, larger and smaller measures of each unit are made by multiplying or dividing the basic units by 10 and its multiples. This feature is convenient because such calculations as dividing by 16 (to convert ounces to pounds) or by 12 (to convert inches to feet) are no longer necessary. Similar calculations in the metric system can be made by simply shifting the decimal point. Thus the metric system is a "base-10" or "decimal" system.

The SI system is built upon a foundation of seven base units: metre, kilogram, second, kelvin, ampere, candela, and mole. Fig. 8-1. There are also two supplementary units, the radian and steradian, used to measure plane and solid angles. However, geometric plane angles will ordinarily continue to be measured in degrees and their decimal fractions. There are also many derived units such as the newton (to measure force) and the pascal (to measure pressure). These derived units are based upon combinations of base and supplementary units. But for everyday use,

---

†In this book the spelling *metre* is used. It is recognized that many authors prefer the *er* spelling. There are a number of good arguments supporting both spellings. However spelled, the important thing to remember is that the symbol, *m*, remains the same.

8-1.
Base Units of the SI Metric System

Unit	Symbol	Definition
metre	m	The measure of length. The metre is equal to a specific number of wavelengths of the light given off by the atom krypton-86. Commonly used related measures are the kilometre (km) = 1000 metres; the centimetre (cm) = 0.01 (one-hundredth) metre; and the millimetre (mm) = 0.001 (one-thousandth) metre.
kilogram	kg	The measure of mass. The kilogram is equal to the mass of the standard kilogram cylinder located at the International Bureau of Weights and Measures in France. A copy of this kilogram is located at the National Bureau of Standards in Washington, D.C. The kilogram is often used to measure what we commonly call weight. However, weight is actually based upon mass and the pull of gravity. Common related measures are the gram (g) = 0.001 (one-thousandth) kilogram and the milligram (mg) = 0.001 (one-thousandth) gram.
second	s	The measure of time. The second is equal to a specific number of movements of the cesium atom in a device known as an atomic clock. A common related measure is the millisecond (ms) = 0.001 (one-thousandth) second. The minute, hour, day, and year are also used, although they are not SI units because they are not based upon ten.
kelvin	K	The measure of temperature. The kelvin is equal to a specific fraction of the temperature at which water exists as a solid, liquid, and vapor. This is called the triple point of water. The kelvin is used mainly for scientific measurements. For practical, everyday purposes, the degree Celsius (°C) is used. Water boils at 100°C, and it freezes at 0°C. The Celsius scale is equal to, but replaces, the old Centigrade temperature scale.
ampere	A	The measure of electrical current. The ampere is equal to the amount of current in two parallel wires one metre apart, that results in a specific force between the two wires. The milliampere (mA) = 0.001 (one-thousandth) ampere, is a common related measurement.
candela	cd	The measure of luminous intensity. The candela is equal to the amount of light given off by platinum at its freezing point, under pressure. At this freezing point, platinum is glowing hot. The candela is used to measure an amount of light.
mole	mol	The measure of amount of substance. The mole is equal to the number of particles contained in a specific amount of carbon. This unit is used mainly in special scientific measurements.

only four measurements are important. Fig. 8-2.
- The *metre* for length.
- The *litre* and the *cubic metre* for volume.
- The *kilogram* for mass (weight).

- Degree *Celsius* (formerly centigrade) for temperature.

Other metric measures are described and used throughout this book.

In the metric system, there are special terms or prefixes used to

make calculations easier. Fig. 8-3. These are the same whether used with the gram or the metre or any other unit. For example, the prefix *kilo* means "1000." Thus a kilogram is 1000 grams; a kilometre is 1000 metres. For everyday measurements the prefixes *kilo*, *centi*, and *milli* are used most. Other information on the metric system is found in the Appendix.

MEASURING INSTRUMENTS

In order to measure simply and accurately, instruments are required. There are many instruments for measuring. Here are some examples.

Linear Measurement

The rule (often called a "scale") is used to make linear measurements such as length, width, or height. Most customary rules are divided fractionally. In woodworking, the bench rule is accurate to 1/16", the metalworker's rule to 1/64". Note the divisions on the one-inch rule shown in Fig. 8-4. The longest line between the

Conversion Chart for Common Units

Length	1 metre (m) = 39.37 inches 25.4 millimetres (mm) = 1 inch 304 mm = 1 foot 1 kilometre (km) = 0.6 mile
Volume	1 litre (L) = 1.05 quarts [1 litre is the same as one cubic decimetre (dm^3)] 1 gallon = 3.79 litres 1 quart = 0.9 litre 1 cubic metre (m^3) = 35.3 cubic feet or 1.3 cubic yards
Weight	1 ounce = 28.35 grams (g) 1 kilogram (kg) = 2.20 pounds 1 pound = 0.45 kg 1 000 kg = 1 metric ton (t) 1 metric ton = 2204 pounds
Temperature	degree Celsius (°C) = °F − 32, divided by 1.8 degree Fahrenheit (°F) = °C × 1.8, plus 32

8-2. *These units are used most often in everyday life.*

SI Unit Prefixes

Multiple or Submultiple	Prefix	Symbol	Pronunciation*	Means
1 000 000 000 = 10^9 1 000 000 = 10^6 1 000 = 10^3 100 = 10^2 10 = 10^1	giga mega kilo hecto deka	G** M** k** h da	jig'a (a as in about) as in *mega*phone as in *kilo*watt heck'toe deck'a (a as in about)	One billion times One million times One thousand times One hundred times Ten times
Base unit 1 = 10^0				
0.1 = 10^{-1} 0.01 = 10^{-2} 0.001 = 10^{-3} 0.000 001 = 10^{-6} 0.000 000 001 = 10^{-9}	deci centi milli micro nano	d c** m** μ** n**	as in *deci*mal as in *centi*pede as in military as in microphone nan'oh (a as in ant)	One-tenth of One-hundredth of One-thousandth of One-millionth of One-billionth of

*The first syllable of every prefix is accented to make sure that the prefix will keep its identity. For example, the preferred pronunciation of kilometre places the accent on the first syllable, not the second.

**Most commonly used and preferred prefixes. Centimetre is used mainly for measuring the body, clothing, sporting goods, and some household articles.

8-3. *Some multiples and prefixes. These may be used with all SI metric units.*

0 and the 1 is the ½″ mark. The next longest lines are the ¼″ marks. The shortest lines are the ¹⁄₁₆″ marks. Note that ⁸⁄₁₆″ equals ⁴⁄₈ or ²⁄₄ or ½.

To measure a line exactly you can count the number of sixteenths. In Fig. 8-5 line d is

eleven sixteenths past the 2″ mark; it is 2¹¹⁄₁₆″ long. Read the other measurements in Fig. 8-5.

Metric rules are also read by counting the divisions. These rules usually come in metre, half-metre, 300-millimetre and 150-millimetre lengths. Longer

metric tape measures are also available. Remember that a metre is divided into 100 centimetres and 1000 millimetres. The rule is usually marked in one of two ways: centimetre (cm) divisions (1, 2, 3, 4, 5, etc.) or millimetre (mm) divisions (10, 20, 30, 40, 50, etc.). Remember that 10 mm equal 1 cm, and that each of the divisions between centimetres equals one millimetre. Note that one inch equals 25.4 millimetres (or about 25 millimetres). Two inches equal 50.8 millimetres (or about 50 millimetres).

To read the metric rule in Fig. 8-6 look at distance AB. Notice that it is 2 mm spaces or divisions past the 10 mark. This means that line AB is 12 mm long. Always give the measurement as "12 mm," not "2 cm, 2 mm." Measure the other lines to make sure you know how to read the metric scale. Remember that in the metric system, shop measurements are usually given in millimetres.

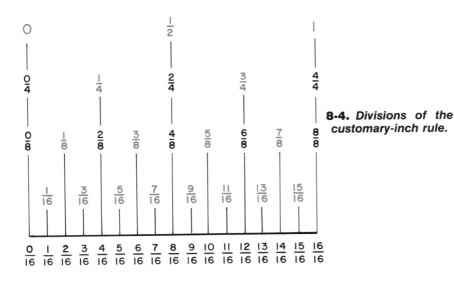

8-4. *Divisions of the customary-inch rule.*

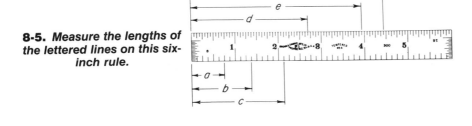

8-5. *Measure the lengths of the lettered lines on this six-inch rule.*

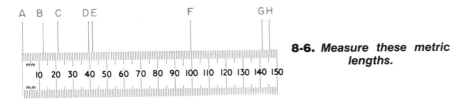

8-6. *Measure these metric lengths.*

8-7. *This balance is used in school science and industrial arts laboratories. It weighs accurately in grams.*

Weight Measurement

When we weigh something, we balance it against standard units of weight. Actually, what we are doing is comparing the mass of the object with the mass of the standard unit. You can use an ordinary weighing scale to weigh yourself. Other kinds of scales are used for technical and scientific purposes. Fig. 8-7.

Customary scales weigh in ounces and pounds; metric scales weigh in grams and kilograms.

Temperature Measurement

A thermometer is used to measure temperature. The U.S. customary temperature unit is the degree Fahrenheit (°F). The metric unit is the degree Celsius (°C). A comparison of these temperature units is shown in Fig. 8-8.

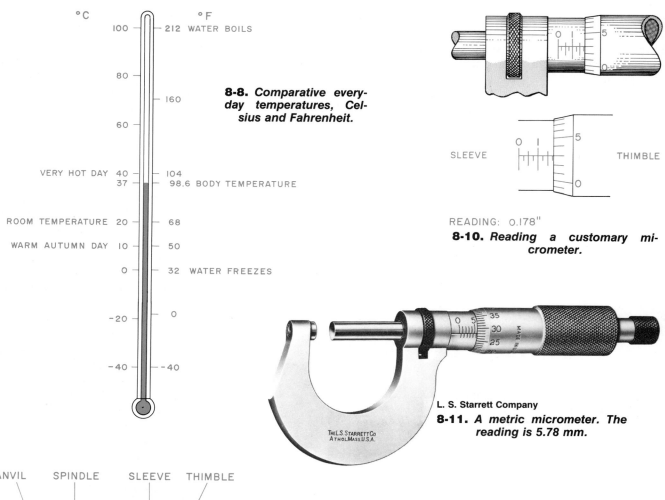

°C   °F

100 — 212 WATER BOILS

80

— 160

60

8-8. *Comparative every-day temperatures, Celsius and Fahrenheit.*

VERY HOT DAY 40 — 104

37 — 98.6 BODY TEMPERATURE

ROOM TEMPERATURE 20 — 68

WARM AUTUMN DAY 10 — 50

0 — 32 WATER FREEZES

—20 — 0

—40 — —40

READING: 0.178"

8-10. *Reading a customary micrometer.*

SLEEVE   THIMBLE

L. S. Starrett Company

8-11. *A metric micrometer. The reading is 5.78 mm.*

ANVIL   SPINDLE   SLEEVE   THIMBLE

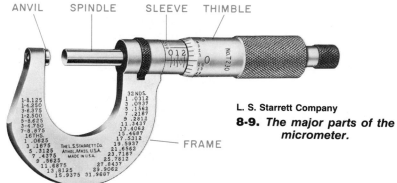

FRAME

L. S. Starrett Company

8-9. *The major parts of the micrometer.*

Other metric temperatures are shown elsewhere in this book.

ACCURACY

It is hard to get exact measurements of objects. For example, you could measure the diameter of a steel rod with a woodwork-

er's rule and get one figure. You could measure it with calipers and a steel rule and get a more accurate figure. Finally you could use a micrometer and get a more accurate measure still. In industry there are electronic instruments for precise measurements.

With precision instruments, measurements can be made with remarkable accuracy. The mi-

crometer is an example of such an instrument. It can make measurements finer than 1/20 the thickness of this page.

Study the micrometer in Fig. 8-9. Each vertical line on the sleeve (or hub) is equal to 0.025" (25 thousandths of an inch). Every fourth vertical line is numbered, indicating 0.100" (100 thousandths, or 1/10 of an inch). The scale on the thimble is divided into 25 equal parts, each representing 0.001" (one thousandth of an inch). To use the micrometer, place the object to be measured between the anvil and the spindle and gently draw the measuring surfaces together by rotating the thimble. Do not force the thimble.

The following example will

Approximate Customary—Metric Conversions

	When you know:	You can find:	If you multiply by:
LENGTH	inches	millimetres	25
	feet	millimetres	300
	yards	metres	0.9
	miles	kilometres	1.6
	millimetres	inches	0.04
	metres	yards	1.1
	kilometres	miles	0.6
AREA	square inches	square centimetres	6.5
	square feet	square metres	0.09
	square yards	square metres	0.8
	square miles	square kilometres	2.6
	acres	square hectometres (hectares)	0.4
	square centimetres	square inches	0.16
	square metres	square yards	1.2
	square kilometres	square miles	0.4
	hectares	acres	2.5
MASS	ounces	grams	2.8
	pounds	kilograms	0.45
	tons	metric tons	0.9
	grams	ounces	0.04
	kilograms	pounds	2.2
	metric tons	tons	1.1
LIQUID VOLUME	ounces	millilitres	30
	pints	litres	0.47
	quarts	litres	0.95
	gallons	litres	3.8
	millilitres	ounces	0.03
	litres	pints	2.1
	litres	quarts	1.06
	litres	gallons	0.26
TEMPERATURE	degrees Fahrenheit	degrees Celsius	0.6 (after subtracting 32)
	degrees Celsius	degrees Fahrenheit	1.8 (then add 32)
POWER	horsepower	kilowatts (kw)	0.75
	kilowatts	horsepower	1.34
PRESSURE	pounds per square inch (psi)	kilopascals (kPa)	6.9
	kPa	psi	0.15
VELOCITY (SPEED)	miles per hour (mph)	kilometres per hour (km/h)	1.6
	km/h	mph	0.6

8-12. *Use this chart to convert from customary to metric units or from metric to customary units. Conversions will be approximate.*

show you how to read a micrometer. (Refer to Fig. 8-11.)

1. The number 1 line on the sleeve is visible, and it equals 0.100″.

2. Three additional lines on the sleeve are visible, each representing 0.025″; 3 × 0.025″ equals 0.075″.

3. Line 3 on the thimble is lined up with the horizontal line on the sleeve. Each line represents 0.001″; 3 × 0.001″ equals 0.003″.

4. The total reading is 0.178″ (0.100 + 0.075 + 0.003).

To read a metric micrometer (Fig. 8-11) do the following:

1. The number 5 line on the sleeve is visible and equals 5 mm.

2. One short line past the number 5 line is visible. This equals 0.5 mm.

3. Line 28 on the thimble is lined up with the horizontal line on the sleeve. Each line represents 0.010 mm; and 28 × 0.010 = 0.28 mm.

4. The total reading is 5.78 mm (5 + 0.5 + 0.28).

The uses of other measuring instruments, both customary and metric, are covered elsewhere in this book.

The chart in Fig. 8-12 shows how to convert from customary to metric units and from metric to customary units.

QUESTIONS AND ACTIVITIES

1. Why do we need fixed, standard units of measurement?

2. Select four small pieces of wood. Measure these in inches *and* in millimetres.

3. What is the smallest inch division on a woodworker's rule? On a metalworker's rule?

4. What are the seven base units of the metric system?

5. Approximately how many millimetres are in an inch?

CHAPTER 9

Planning Procedures

Planning Sheet

Name .... Grade ....

.... Bookrack ....

(Name of Project)   (Date started)   (Date completed)

Bill of Materials:

No.	Size			Name of part	Material	Unit cost	Total cost
	T	W	L				
1	⅝	6¼	6¾	Headpiece	Walnut		
1	⅝	1½	5¼	Footpiece	Walnut		
1	5/16D	—	15	Long Rod	CRS		
1	5/16D	—	14	Short Rod	CRS		
4				Felt Dots			

Tools and Machines: Bench rule, pencil, awl, backsaw, band saw, plane, disc sander, sandpaper, drill press, hacksaw, file, bending jig, steel wool, wipe-on finish, lacquer, lacquer thinner, brushes, die, diestock, felt dots.

Outline of Procedure or Steps:
1. Lay out and cut wooden parts to shape.
2. Sand surfaces and edges of wooden parts.
3. Drill holes in wooden parts.
4. Apply finish to wooden parts.
5. Cut metal rods to length; remove burrs.
6. Make 90-degree bends in metal rods.
7. Cut threads in metal rods.
8. Clean rods with steel wool and thinner.
9. Apply lacquer to metal rods.
10. Assemble headpiece, footpiece, and metal rods.
11. Place felt dots in position.

3. Outline of steps to follow in making the project. To make your outline, first study the individual parts of the project. Next, think of what you must do to make each part. Then think about the assembly and finishing of the project. Listing all of these operations or tasks in proper order will give you the outline. Have your teacher check your list. Steps 1, 2, and 3 are usually placed on a planning sheet, shown in Fig. 9-1.

4. Working drawing. This drawing is necessary for you to learn the sizes and shapes of the project's parts. Both detail and assembly drawings are generally used. Figs. 9-2 & 9-3.

The plan of procedure shown in Fig. 9-1 would be used to make one of the bookracks discussed in Chapter 7. Study the differences between these steps and those in Chapter 106, "Mass Production in the School," where a project has been planned for mass production.

After you have designed a product and made the drawings, you must decide how you are going to make the product. This is necessary to avoid mistakes and waste of materials. Industry does this through very careful tooling and control procedures in production. You must do this in the shop by preparing a plan of procedure for your work. There are four main parts to this plan.

1. Bill of materials. List all of the items that will actually be part of the project. Be sure to use their right names. Look at the

supply catalogs and materials lists in your shop to find these correct terms.

2. List of tools and machines. All of the necessary tools, machines, and other equipment should be listed. This can save you time later if you discover that you don't have a needed tool or machine in the shop.

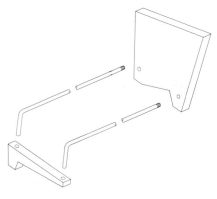

9-2. An assembly drawing showing how the parts of the bookrack will fit together.

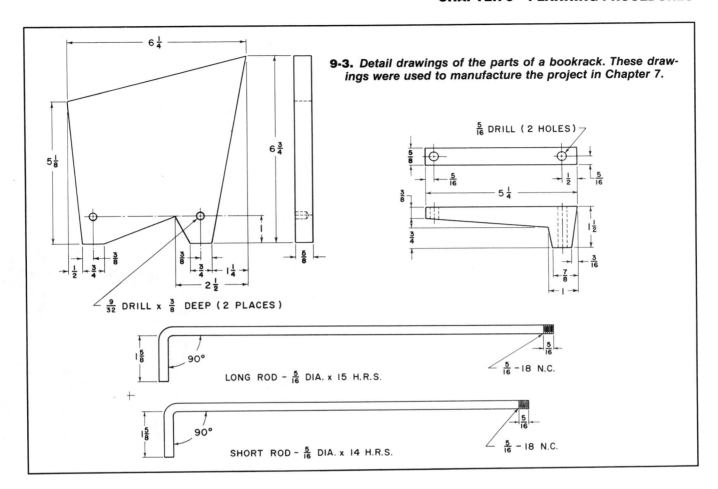

9-3. *Detail drawings of the parts of a bookrack. These drawings were used to manufacture the project in Chapter 7.*

QUESTIONS AND ACTIVITIES

1. List the four parts of a plan of procedure.
2. Why is it important to list the tools and machines that will be needed for a project?
3. How is an "outline of steps" made?
4. What is the purpose of working drawings?

CHAPTER 10

Safety

Safety in the school shop, at home, and in industry is important because it protects people. If people become careless about safety, they can lose eyes, fingers—or lives.

This is why industry is so concerned about safety. Fig. 10-1. Every day there are hundreds of industrial accidents in this country, some of them fatal. Many hours of work are lost, and some workers are crippled for life. Safety programs are a part of every industry, and workers are reminded to develop safe work habits. Fig. 10-2. People in all lines of work must practice safety. Fig. 10-3.

As a part of your work in this course, you too must learn to work safely. A careless act can cause an accident. Observe all safety rules in the school shop, and be sure you have received proper instructions before operating a machine or using a tool. Remember the ABCs of safety— ALWAYS BE CAREFUL. Here is a list of general safety rules for you to follow. Special safety precautions are listed in other sections of this book.

1. Wear the right clothes for the job. Ties and jewelry should be removed and sleeves rolled up. Approved safety glasses must be worn. Fig. 10-4. Wear special protective clothing when working with hot metals or acids.

2. Know your tools and equipment before you use them. Use the proper tool for each job. Always follow instructions for tool use, and when in doubt, ask your teacher. Fig. 10-5.

3. Don't run or act foolishly in the shop. Practical jokes and pranks often lead to accidents; so avoid them.

4. Keep your shop and work area clean. An orderly shop is a safe one. Pick up tools and scraps

NASA

10-1. People in industry must practice safety in order to protect employees. This technician works with high-pressure gases. Notice how well protected he is.

National Safety Council

10-2. Safety posters are a constant reminder to develop safe work habits. Read and remember the safety posters you see in your shop.

NASA

10-3. *Pilots wear protective clothing and practice safety in their flying.*

Fred W. Gillman

10-5. *Your teacher can show you how to use shop equipment safely.*

Jackson Products Co.

10-4. *Wear approved safety glasses or goggles at all times when you are working in the shop. The lenses must be at least 3.0 mm thick and be specially hardened. The frames must also be of special design and strength.*

Eagle Manufacturing Company

10-6. *Safety can for storing liquids.*

of material which might cause falls. Wipe up water or oil spills. Report broken or unsafe tools to your teacher.

5. Store flammable liquids in small safety lab cans. Fig. 10-6. Do not store these dangerous liquids in large, five-gallon cans.

6. Place oily rags and similar materials which can easily catch fire in special containers. Fig. 10-7. These should be emptied daily. All flammable liquid waste should be placed in safety disposal cans. Fig. 10-8. Never pour such liquids down a sink drain.

7. Use common sense when working in the shop. Carry tools safely, and use them so that a slip of a tool won't injure you or your neighbor. If you have an accident, no matter how small, report it right away so that you can get first aid. Most accidents can be prevented if you think about what you are doing.

In 1970, Congress passed the

Eagle Manufacturing Company

10-7. *Container for oily wastes.*

Eagle Manufacturing Company

10-8. *Liquid disposal can.*

OSHA Safety Checklist

Hand and Portable Power Tools

1. All hand and portable power tools are in good operating condition: no defects in wiring; equipped with ground wires.	Yes	No
2. All portable equipment is equipped with necessary guarding devices.	Yes	No
3. All compressed air equipment used for cleaning operations is regulated at 30 psi or less; chip guarding and personal protective equipment are provided.	Yes	No

Machine Guarding and Mechanical Safety

1. Every production machine has been inspected as to the following items and found to be in satisfactory operating condition:

a) Cleanliness of machine and area	Yes	No
b) Securely attached to floor	Yes	No
c) Operations guarded	Yes	No
d) Illumination	Yes	No
e) Effective cutoff devices	Yes	No
f) Noise level	Yes	No
g) Adjustment	Yes	No
h) Material flow	Yes	No

Material Hazards

1. All hazardous gases, liquids, and other materials are properly labeled and stored.	Yes	No
2. Areas where hazardous materials are in use are fire-safe and restricted to authorized employees.	Yes	No
3. Where X-ray is used, the area is properly shielded and dosimeters are used and processed for all authorized employees.	Yes	No
4. Protective clothing is worn by employees when oxidizing agents are being used.	Yes	No
5. All hazard areas are posted with NO SMOKING signs.	Yes	No
6. All areas where caustics or corrosives are used have been provided adequately with eye fountains and deluge showers.	Yes	No

10-10. *Checklists like this one are used to find out whether OSHA safety rules are being followed.*

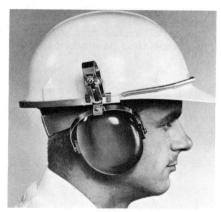

Jackson Products Company

10-9. *Wear ear protection when working near noisy equipment.*

Occupational Safety and Health Act (OSHA). The purpose of this act is to make sure that people will have safe and healthful working conditions. Safety and health standards have been set up to cover all types of safety, from safety glasses and ear protection to the height of railings on factory stairways. Fig. 10-9. These standards must be obeyed by all employers and employees. School shops are also affected by OSHA standards. A typical OSHA safety checklist is shown in Fig. 10-10. How does your school shop measure up?

Remember the safety rules you learned in the school shop. Be safety-smart. Practice these same rules at home, at work, and at play.

QUESTIONS AND ACTIVITIES

1. List six important general safety rules.

2. What is the purpose of the Occupational Safety and Health Act?

3. Why should you remove jewelry before working with machines?

4. Why is it important to keep the shop clean? List some ways to keep the shop clean and orderly.

PART 3
Graphic Communications

There are about 12 000 parts in the average car. Just think how hard it would be if the people who design and build cars had to use words alone to describe the size and shape of each part.

No product, be it a motorcycle or a computer, could be made without accurate drawings. And the more complex the product, the more drawings are needed—about 14 000 of them to build a car. Some of these are body drawings, which picture sheet metal or plastic surfaces. Others are

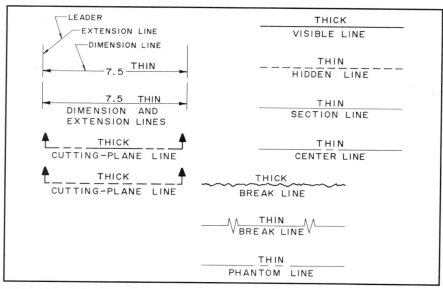

11-2. The alphabet of lines. As you can see, each kind of line has a different use and meaning.

11-1. These signs use no words, just symbols. Can you figure out their meanings? Answers are on page 67.

mechanical drawings, which show the size and shape of parts and assemblies. There are also drawings which show the tools needed to make the parts. The process of making these drawings is called *drafting*.

THE LANGUAGE OF DRAFTING

In a sense, drafting is a *language* because it communicates facts and ideas. As a matter of fact, you might call drafting a kind of picture writing. You can see many examples of picture writing in everyday life. Figure 11-1 shows examples of picture writing you might find in airports, restaurants, and other public places. These symbols can

be understood in any country because they don't use words. Pictures or symbols are just as important in drawings. You will learn about many symbols that are used to make drawings accurate and easy to understand.

Alphabet of Lines

In the language of drawing, different lines have different meanings. It is important that you understand what these various lines stand for. Studying Fig. 11-2 will help you to make and read drawings and sketches correctly.

• A *visible line* (or *object* line) is a thick, solid line used to show the outline of an object.

• A *hidden line* is a thin,

dashed line used to show surfaces that are not normally seen.

● A *dimension line* is a thin line that shows the extent of a dimension. The dimension (size) is written either within or above the line.

● An *extension line* is a thin line that extends out from the object and is used with dimension lines.

● A *leader* is a thin line that points out the part to which a dimension or note refers.

● A *center line* is a thin line that consists of alternating long and short dashes. It is used to show the center of an object.

● A *cutting-plane line* is a thick line. It may be either long dashes or long dashes separated by two short dashes. The cutting-plane line shows an imaginary cut (section) through an object.

● *Section lines* are thin, parallel lines that show cut surfaces in section view. Section lines are drawn at an angle.

● A *break line* is used to show that part of an object has been removed. Thick break lines are used for short breaks; thin lines for long breaks.

● A *phantom line* is a thin line used to draw alternate positions for a part.

Figure 11-3 shows how these lines are used.

Symbols

A drawing in which every screw thread, plumbing fixture, or electric outlet is drawn exactly as it appears would be cluttered and crowded. Such a drawing would also take a long time to make. To simplify drawings so that they are easier to make and to read, drafters use many symbols. There are symbols for building materials, plumbing fixtures, electricity, and home appliances.

Some of them are shown in Fig. 11-4. Study these symbols to understand how time is saved by using them in drawing.

Symbols are also used to show threaded fasteners and tapped holes (holes with internal screw threads). Study Fig. 11-5. See how difficult it is to draw screw threads (Part A). Compare this with the two methods for simplifying thread drawings by using symbols (Parts B & C). Tapped holes can also be shown with symbols. Fig. 11-6.

In Fig. 11-3 you saw one use of break lines. They were used to break an object that was shown in section. Breaks are also used if an object is too long to fit on the paper. The drafter draws only the beginning and end of the object. Figure 11-7 shows ways to break round and rectangular objects.

THE IMPORTANCE OF DRAWING

Drawings are important in school shopwork and in industry because they *show* rather than tell how something is made. Fig. 11-8. It is much easier to understand how to make a part when you are shown a drawing of it than if someone merely *tells* you how it should be made.

In this section you will learn

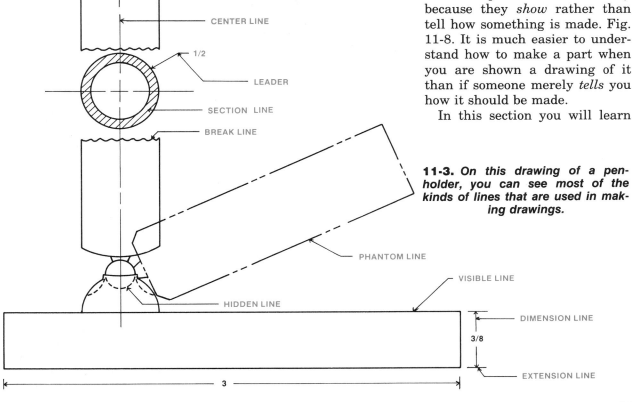

11-3. *On this drawing of a penholder, you can see most of the kinds of lines that are used in making drawings.*

CENTER LINE

1/2

LEADER

SECTION LINE

BREAK LINE

PHANTOM LINE

HIDDEN LINE

VISIBLE LINE

DIMENSION LINE

3/8

EXTENSION LINE

3

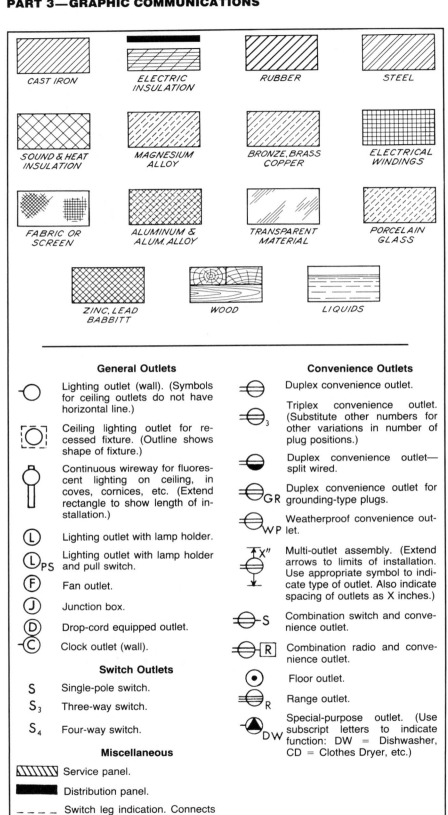

CAST IRON

ELECTRIC INSULATION

RUBBER

STEEL

SOUND & HEAT INSULATION

MAGNESIUM ALLOY

BRONZE, BRASS COPPER

ELECTRICAL WINDINGS

FABRIC OR SCREEN

ALUMINUM & ALUM. ALLOY

TRANSPARENT MATERIAL

PORCELAIN GLASS

ZINC, LEAD BABBITT

WOOD

LIQUIDS

General Outlets

Lighting outlet (wall). (Symbols for ceiling outlets do not have horizontal line.)

Ceiling lighting outlet for recessed fixture. (Outline shows shape of fixture.)

Continuous wireway for fluorescent lighting on ceiling, in coves, cornices, etc. (Extend rectangle to show length of installation.)

L — Lighting outlet with lamp holder.

L$_{PS}$ — Lighting outlet with lamp holder and pull switch.

F — Fan outlet.

J — Junction box.

D — Drop-cord equipped outlet.

C — Clock outlet (wall).

Switch Outlets

S — Single-pole switch.

S$_3$ — Three-way switch.

S$_4$ — Four-way switch.

Miscellaneous

Service panel.

Distribution panel.

Switch leg indication. Connects outlets with control points.

Convenience Outlets

Duplex convenience outlet.

Triplex convenience outlet. (Substitute other numbers for other variations in number of plug positions.)

Duplex convenience outlet— split wired.

GR — Duplex convenience outlet for grounding-type plugs.

WP — Weatherproof convenience outlet.

Multi-outlet assembly. (Extend arrows to limits of installation. Use appropriate symbol to indicate type of outlet. Also indicate spacing of outlets as X inches.)

S — Combination switch and convenience outlet.

R — Combination radio and convenience outlet.

Floor outlet.

R — Range outlet.

DW — Special-purpose outlet. (Use subscript letters to indicate function: DW = Dishwasher, CD = Clothes Dryer, etc.)

11-4. *Selected symbols for materials and electricity.*

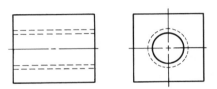

ROOT OF THREAD

A   B

C

11-5. *The three ways of drawing an external thread: (A) Conventional method. (B) Regular thread symbol. (C) Simplified thread symbol.*

11-6. *The symbols for internal threads (two views).*

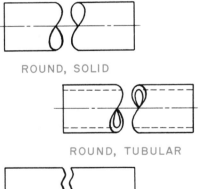

ROUND, SOLID

ROUND, TUBULAR

RECTANGULAR

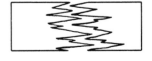

RECTANGULAR, WOOD

CONVENTIONAL BREAKS

11-7. *These are the conventional breaks used to fit long objects on a drawing sheet.*

Answers to Fig. 11-1

1—Women
2—Men
3—Elevator
4—Facilities for handicapped
5—Restaurant
6—Telephone
7—No smoking
8—Vending
9—Fire stairs
10— Litter basket
11—Gifts
12—Information

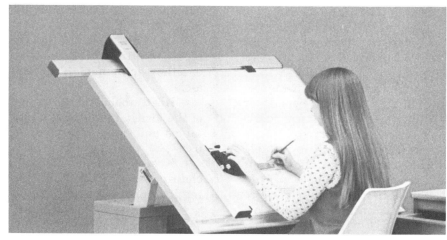

Teledyne Post, Des Plaines, IL

11-9. You will learn how to use drafting machines and other instruments in your study of drawing.

how to make and read many different kinds of drawings. They include pictorials, sections, and multiview and architectural drawings. You will learn how to use instruments and to make prints (copies). Fig. 11-9.

Modern drafting practices include the use of computers. See Chapter 3 for information about computer-aided drafting and design.

Neatness and accuracy are important if you are to make good drawings. The drawing class you take in school may lead you to a career as a drafter. Fig. 11-10. See Chapter 107 for more information on careers in drafting.

Teledyne Post, Des Plaines, IL

11-8. This machine shop student is checking his drawing before drilling a hole in a metal part.

Teledyne Post, Des Plaines, IL

11-10. Your study of drawing may lead you to choose a career in drafting.

FOR YOUR SAFETY . . .

Although drawing is not dangerous, there are a few things you should keep in mind.

● Be careful when using the dividers, compass, pencil, or ruling pen. The sharp points can cause an injury.

● If your drafting table needs adjustment, have your teacher show you how.

● Keep adequate clearance around drafting tables, in accordance with safety and fire regulations.

● Do not let T-square blades, drafting machine arms, or other equipment block the aisles.

● Keep stools and chairs out of the aisles. Place them under tables and desks when not in use.

● Keep all four legs of your chair or stool on the floor. Tilting it can cause you to fall.

● Tape is safer than thumbtacks for fastening paper to drafting boards.

● Keep chemicals (such as ink) away from your nose, eyes, and throat. Use chemicals only where there is adequate ventilation.

● Keep hands and fingers clear of the paper-cutter blade. Place the blade in the down position when not in use.

QUESTIONS AND ACTIVITIES

1. Why is drafting called a "language"?

2. List five kinds of lines in the alphabet of lines.

3. Sketch three different symbols for each of the following: A. Building materials. B. Electricity.

4. Discuss why drawings are important in the school shop and in industry.

CHAPTER 12
Drafting Equipment

As you learned from the last chapter, *drafting* is putting ideas on paper in a graphic, or picture, form. This chapter shows and describes some of the equipment needed to do this.

Drawing boards provide a flat, smooth surface for drawing. Fig. 12-1. They are made of wood, such as pine or bass, with straight working edges. Some boards have special metal inserts in the working edges to keep

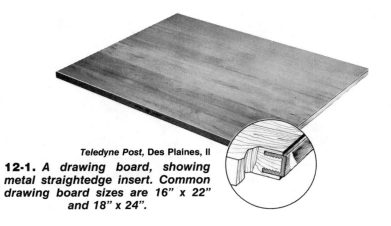

Teledyne Post, Des Plaines, Il

12-1. A drawing board, showing metal straightedge insert. Common drawing board sizes are 16" x 22" and 18" x 24".

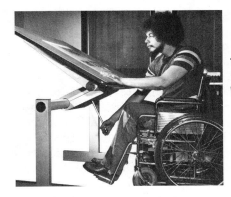

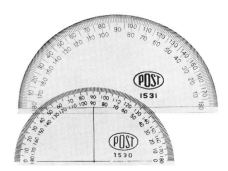

The Huey Company
12-2. Drafting tables are made of wood or metal. The top of this table is adjustable.

Teledyne Post, Des Plaines, IL
12-3. A T square should be handled carefully so that it stays "square." Avoid dropping or twisting it.

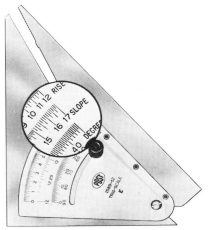

Teledyne Post, Des Plaines, IL
12-5. The protractor and the combination triangle are used to lay out angles which cannot be formed by triangles.

them true (square). Sometimes a drawing table becomes the drawing surface. Fig. 12-2. Drawing boards should be kept free of nicks, dents, and gouges for good drawing quality.

T squares are used for drawing straight horizontal lines. They are also used as guides for holding other instruments, such as triangles. T squares have two parts, the head and the blade. These are firmly fastened at right angles and are made of wood. Better quality T squares have a blade with plastic edges which resist nicks and provide a truer drawing edge. Fig. 12-3.

Triangles are used to draw vertical and inclined (oblique) lines. The two main types of triangles are the 30°–60° and the 45°. The 30°–60° triangle has a 90° angle, a 30° angle, and a 60° angle. The 45° triangle has one 90° angle and two 45° angles. These can be used in combination to form angles of 15° and 75°. Fig. 12-4. To form angles of other sizes, use the protractor and combination triangle. Fig. 12-5.

Irregular curves are used to draw curved shapes which cannot be drawn with other instruments. These tools are made of plastic or wood and are available in many styles and sizes. The instrument is used to connect points which have been laid out in a curve. Lay the instrument

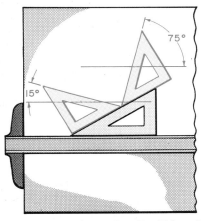

Teledyne Post, Des Plaines, IL
12-4. Using the triangles to lay out 15° and 75° angles. The 45° triangle is at the bottom. The 30°-60° triangle is at the top. The T square is used to support the triangles.

over at least three of the points and draw the line. Continue until the curve is complete. Fig. 12-6.

Drafting scales are important tools for the drafter. They are used for laying out distances and dimensions on drawings. These scales are made in flat or triangular shapes with many different markings, or *graduations*. These graduations are different for each

of the three main types of drafting scales: the architect's, the civil engineer's, and the mechanical engineer's. The wood or plastic, triangular architect's scale is commonly used in the school shop. The architect's scale is used to make working drawings of buildings and machine parts. Fig. 12-7. All scales on this tool are based upon the foot. For ex-

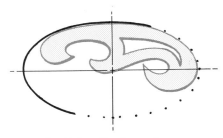

12-6. Using the irregular curve.

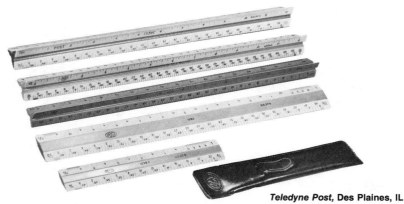

Teledyne Post, Des Plaines, IL

12-7. Drafting scales are available in many sizes and shapes. Architect's scales, commonly used in the school shop, are shown here.

ample, a ¾ scale means that ¾″ = 1′0″, giving a ¹⁄₁₆ size drawing. Note that it does not mean ¾ size. A scale of ¾″ = 1″ means ¾ size. Other scales on this instrument are shown in Fig. 12-8. The ¼ or ⅛ scales are often used for drawing buildings and house plans.

To use an architect's scale, study the drawing in Fig. 12-9. The distance to the left of the zero mark on this ¾ scale is divided into 24 graduated markings (each marking represents ½″) that equal the total distance of one foot. The numbered divisions to the right of the zero represent one foot each. To measure

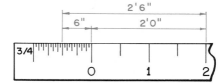

12-9. Reading the architect's scale. Note that you must pick up distances on both sides of the zero mark.

2′6″ on this scale, you must first measure the feet by counting two numbered divisions to the right of the zero. Then measure the inches by counting twelve graduated markings to the left of the zero. Add the distance on the right of the zero mark (2′) to the distance on the left of the zero mark (6″). The total distance equals 2′6″. Other architect's scales are read in a similar way. Always start at the zero mark; measure the number of feet first and then add the additional inches on the other side of the zero mark. Be sure to count the small graduated divisions because their number varies from one scale to another.

Metric scales are used exactly as one would use a customary-inch scale. Fig. 12-10. The metric scale ratios are in some cases identical to the customary. In other cases, they are close. For example, ½″ = 1′ scale means a

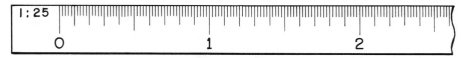

12-10. A metric drafting scale.

12-8.
Architect's Scales

Scale	Meaning	
16	1′ = 1′0″	full size
3	3″ = 1′0″	¼ size
1½	1½″ = 1′0″	⅛ size
1	1″ = 1′0″	¹⁄₁₂ size
¾	¾″ = 1′0″	¹⁄₁₆ size
½	½″ = 1′0″	¹⁄₂₄ size
⅜	⅜″ = 1′0″	¹⁄₃₂ size
¼	¼″ = 1′0″	¹⁄₄₈ size
³⁄₁₆	³⁄₁₆″ = 1′0″	¹⁄₆₄ size
⅛	⅛″ = 1′0″	¹⁄₉₆ size
³⁄₃₂	³⁄₃₂″ = 1′0″	¹⁄₁₂₈ size

12-11.
Customary and Metric Scales

Customary Scale (inch:inch)	Means	Metric Scale (mm:mm)
12:12	12″ = 1′0″ full size	1:1
6:12	6″ = 1′0″ half size	1:2
		1:3*
3:12	3″ = 1′0″ quarter size	1:5
1:12	1″ = 1′0″	1:10
¾:12	¾″ = 1′0″	1:20
½:12	½″ = 1′0″	1:25
¼:12	¼″ = 1′0″	1:50
⅛:12	⅛″ = 1′0″	1:100
¹⁄₁₆:12	¹⁄₁₆″ = 1′0″	1:200
¹⁄₃₂:12	¹⁄₃₂″ = 1′0″	1:500

*Available, but not preferred.

ratio of 1:24 (1″ = 24″). The metric replacement scale is 1:25 (1 mm = 25 mm). The common customary-inch scales are shown with their metric replacements in Fig. 12-11.

Drafting machines combine the scale, protractor, T square, and triangle in one unit. Fig. 12-12. Drafting machines are easy and convenient to use.

Drafting templates are plastic patterns that speed the process of drawing certain shapes. Fig. 12-13. There are many styles available for drawing circles, arcs, squares, bolt heads, nuts, screws, and so forth.

Drawing paper comes in many sizes, weights, and colors. The paper generally used in the school shop is 8½″ × 11″ and is white or manila (buff-colored). It is heavy with a hard finish for good line quality. It also erases easily without smudging. Several grades of tracing paper or vellum

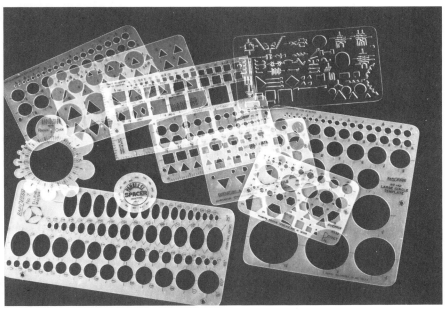

Keuffel & Esser Company

12-13. *These drafting templates are for drawing circles, arcs, and other shapes quickly and easily.*

Teledyne Post, Des Plaines, IL

12-12. *A typical drafting machine setup.*

12-14.
Comparative Sizes of Drawing Paper

Customary (inches)	Metric (millimetres)
A (8½ × 11)	A4 (210 × 297)
B (11 × 17)	A3 (297 × 420)
C (17 × 22)	A2 (420 × 594)
D (22 × 34)	A1 (594 × 841)
E (34 × 44)	A0 (841 × 1189)

are also used. The chart in Fig. 12-14 shows standard inch-dimension paper with metric replacement sizes.

Drawing paper is fastened to the drawing board or table with a special masking tape. Thumbtacks may also be used, but they leave marks on the paper and the drawing board.

Pencils used for drawing are usually the common wood type, but mechanical models are also used. Pencil leads come in different grades of hardness from 6B, the softest, through grades 5B, 4B, 3B, 2B, B, HB, F, H, 2H, 3H, 4H, 5H, 6H, 7H, 8H, to 9H, which is the hardest. A 3H or 4H pencil is used for making light construction lines.† A 2H or H pencil is used for drawing object or finish lines, and the F or HB is used for freehand sketching and shading.

Pencils must have sharp points for good drawing. Use a pencil sharpener, or if you need a strong tip, use a knife. Dress the point with a sandpaper pod or a mechanical pointer.

When erasing, an *erasing shield* is used to protect the lines you wish to keep. This tool is usually made of metal. Fig. 12-15. A good gum eraser should be used. Keep it clean for best results. Electric erasers are also used.

---

†Construction lines are very light lines used to "block in" an object. They serve as a base for darkening in the permanent lines.

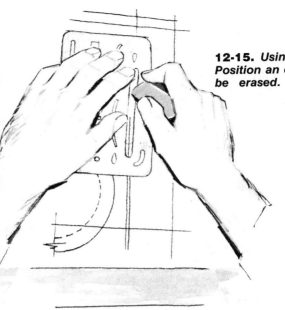

12-15. *Using the erasing shield. Position an opening over the line to be erased. Rub gently with the eraser.*

cil being used on the rest of the drawing. Keep these sharp, and be careful of the needle points on instruments. Inking attachments for drawing instruments are also available. Ruling or inking pens may also be used.

DRAWING AN OCTAGON

One good way to learn how to use drawing instruments is to draw geometric shapes and patterns. A good geometric shape to learn to draw is an octagon. An octagon has eight equal sides and eight equal angles. It is often used as a shape for metal or wood projects such as wastepaper baskets, hot-dish holders, and metal trays. It might also be the shape of a ticket in printing. To draw an octagon, refer to Fig. 12-17 often while following these directions.

1. Draw a square of the same size as the octagon.

2. Draw diagonal lines AB and CD.

3. Adjust the compass so that the metal point is about $1/32''$ longer than the lead point. Then

Drawing instruments have many uses. With them drafters scribe (mark) circles, measure distances, and make inked lines. These are precision instruments. They can be used with great accuracy if they are cared for and used properly. Some common instruments are shown in Fig. 12-16. The large pencil spring bow compass (A) is used for drawing large circles and arcs. The straight pencil compass (B) is also used for this purpose, but can be less accurate unless used with care. The small pencil spring bow compass (D) is used to draw smaller circles and arcs. Dividers (C & E) are used to transfer distances from the drafting scale to the paper. Distances between views (such as between a front and a side view) can also be laid out with this tool. Beam compasses (F) are used, much like trammel points in metalworking, to draw very large circles and arcs.

The pencil points in compasses should be of the same degree of hardness as the point on the pen-

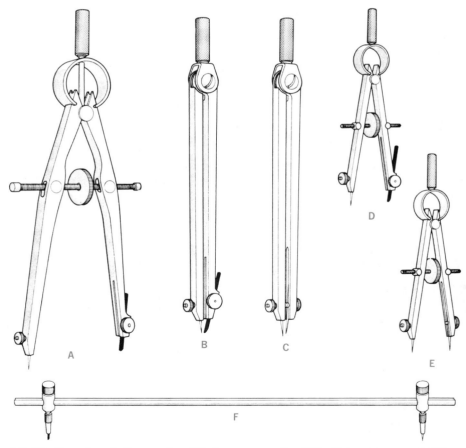

12-16. *Drawing instruments: (A) Large pencil spring bow compass. (B) Straight pencil compass. (C) Straight dividers. (D) Small pencil spring bow compass. (E) Spring bow dividers. (F) Beam compass.*

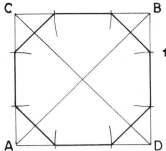

12-17. *Drawing an octagon.*

open the compass until the distance between the metal point and the lead point of the compass equals half the length of one of the diagonal lines.

4. Put the metal point of the compass on point A of the square. Place the pencil end of the compass just to the left of line AC. Holding the top of the compass between your thumb, forefinger, and third finger, turn it clockwise until it passes just to the right of line AC. (This is called "striking an arc.") Keep the com-

pass on point A and strike an arc across line AD. Move the metal point of the compass to point C and strike arcs across lines CA and CB. Move the compass point to point B of the square and strike arcs across lines BC and BD. Move the compass point to point D and strike arcs across lines DA and DB.

5. Draw straight lines to connect the points where the arcs have intersected the lines of the square.

QUESTIONS AND ACTIVITIES

1. To make a drawing that is ¾ the size of the actual object, what scale should be used?

2. List two main types of triangles used in drawings.

3. What grades of pencils are used for making light construction lines? For drawing finish or object lines? For sketching and shading?

4. What instrument is used for drawing large circles and arcs?

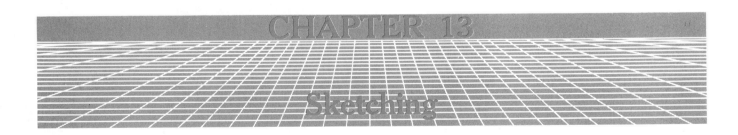

CHAPTER 13

Sketching

In designing, it is necessary to draw pictures which show the ideas you may have for a product. Freehand sketches are better for this than instrument drawings made with T squares and triangles because they are easier and quicker to make. The purpose of this chapter is to explain and show how to sketch quickly and accurately. These techniques are

not difficult for a beginner. You don't have to be an artist to learn to sketch well. All you need is an understanding of the technique, and a lot of practice.

TOOLS FOR SKETCHING

Sketching tools are very simple; pencil and paper are all you need. Any good, tough paper that you can erase on cleanly can be

used. Paper covered with ¼″ squares is convenient for making simple sketches. Fig. 13-1. Isometric grid paper can be used for pictorial sketches. Fig. 13-2. An HB grade drawing pencil is ideal for sketching because of its soft, black lead. Light or dark lines can be made by changing the pressure. Ordinary writing pencils, Nos. 2 and 2½, are

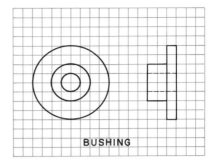

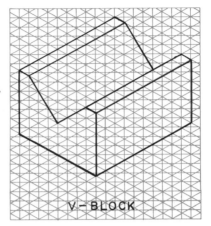

13-1. *Grid papers are helpful in making good sketches.*

13-2. *Isometric grid paper.*

V—BLOCK

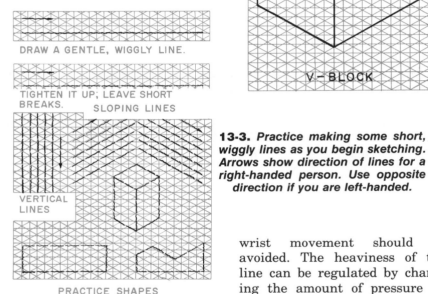

DRAW A GENTLE, WIGGLY LINE.

TIGHTEN IT UP; LEAVE SHORT BREAKS. SLOPING LINES

VERTICAL LINES

PRACTICE SHAPES

13-3. *Practice making some short, wiggly lines as you begin sketching. Arrows show direction of lines for a right-handed person. Use opposite direction if you are left-handed.*

wrist movement should be avoided. The heaviness of the line can be regulated by changing the amount of pressure on the pencil.

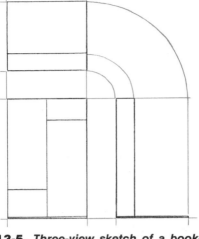

13-5. *Three-view sketch of a book-end.*

Lines

Straight lines should always be drawn between two points. One point is used as a starting place and the other is used as a guide for keeping the line straight. The pencil should be placed firmly on the left-hand point, then pulled slowly to the right-hand point. Avoid any jerky, ragged, or curvy lines.

A good way to learn sketching is to practice making "wiggly" lines. This will help you to sketch straighter, more accurate lines and shapes. Exaggerate the wig-

also good. Don't use ball-point pens. Their lines can't be erased, and minor corrections are difficult to make.

SKETCHING TECHNIQUES

There are no rules about how you should hold the sketching pencil. Hold it whatever way is easiest and most comfortable for you. The main thing to remember is to make free, easy lines and curves, without forcing them. Free arm movement is essential for sketching smooth, neat horizontal lines. Finger and

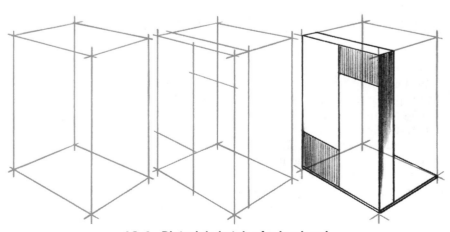

13-4. *Pictorial sketch of a bookend.*

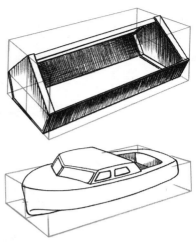

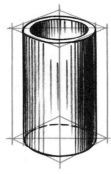

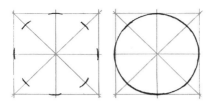

13-8. Circles are made more easily if you use guidelines.

13-6. Pictorial sketches of a tote box and a boat.

13-7. This sketch of a pencil holder is made from two ovals or ellipses and two straight lines. The line shading makes it look more realistic.

gles at first and then make them tighter. Leave short breaks in the lines if you wish. Any technique is all right as long as it helps you to produce a good sketch. Fig. 13-3.

To draw a vertical line, start at the top of the sheet and sketch downward, using a free arm movement. It is poor practice to turn the paper so that vertical lines can be drawn as horizontal ones, or to sketch vertical lines by using the little finger or the paper edge as a guide.

Oblique or angular lines are sketched much the same as horizontal lines. Circles and arcs can be sketched with a quick arm or wrist movement as one becomes more skilled. However, the beginner should use guidelines to make accurate circles and shapes. This is described in the following paragraphs.

Sketching Objects

A *pictorial sketch* is like a picture, and it has width, depth, and height. To begin pictorial sketching, first form a box of light construction lines, or guidelines, using the same scale as will be used to sketch the object. Next sketch in the top, front, and side views in their proper positions. Then study the object and darken the construction lines. Fig. 13-4. You will learn more about pictorial drawing in Chapter 16.

An *orthographic* sketch is a drawing which shows an object from several views, usually the front, top, and right side. It is made by lightly sketching or blocking in the views in their correct positions and then darkening the lines. Fig. 13-5. See Chapters 14 and 16 for information on orthographic and pictorial theory.

With practice you can acquire the skill to sketch more difficult objects such as the tote box and the boat in Fig. 13-6. For a more professional-looking sketch, try shading the surfaces as shown on the box.

Pictorial circles are usually drawn as ellipses or ovals. Fig. 13-7. Make these with a quick, easy movement of the pencil. Orthographic circles are sketched more easily if they are first laid out with points marking the circle diameter. Fig. 13-8.

With practice and patience, you can soon become a skillful sketcher.

QUESTIONS AND ACTIVITIES

1. What tools are needed for sketching?
2. Tell how to sketch a straight line.
3. What is an orthographic sketch?

4. Make sketches of three or four items found around the home such as a glass jar, a toy, a model car, or a hand tool.

There are several ways to show the shape of an object with drawings. One way is described in this chapter. For example, Fig. 14-1 shows the top, front, and end views of a coffee table. This type of drawing is called a *multiview* or an *orthographic projection*. *Orthographic* means placing the views at right angles to one another. *Projection* means to show the parts of an object on flat surfaces, or planes. To make an orthographic projection you must place the views of an object on planes lying at right angles to one another.

Every object has six views: a top and a bottom, a front and a back, a right side and a left side. To make a multiview drawing of most objects it is necessary to show only three views—usually the front, top, and right side. The front view is generally the most important or most descriptive view.

It is easy to see how orthographic projection works if you imagine that the object is inside a glass box. If you were to look straight into the front of the box and trace what you saw with a wax pencil, you would have

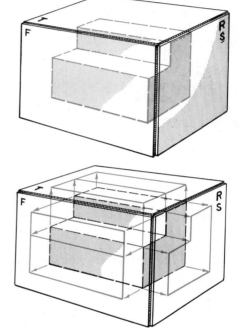

14-1. The top, front, and end views of a coffee table frame. A top of wood, glass, etc., can be placed on the frame.

14-2. Imagine an object enclosed in a clear box. Tracing what you see would transfer the views onto the planes (surfaces) of the box. The T indicates top, F is front, and RS, right side.

TOP

54

1½ 1½

FRONT

18

RIGHT SIDE
OR
END

16

1½ 1½

sketched the front view. Then if you did the same for the top and right-side views, you would have sketches of all three views. Fig. 14-2. Next, if you could fold the top and right-side views forward, you would have an orthographic projection showing three views of the object. Fig. 14-3. Orthographic projection is as simple as that.

Not all objects need three views. For example, cylindrical shapes such as a hockey puck or

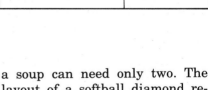

14-3. *When you unfold the box, you have an orthographic drawing of the object showing its three main dimensions—height, width, and depth.*

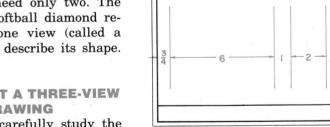

14-4. *Simple layouts such as this softball field can be made with one view.*

for dimensions and notes. Notice how the sizes of the views, the distances from the borders, and the distances between the views are added to determine the scale and paper size needed.

Draw the outline of each view. Add details such as holes. The arcs and circles should be drawn in first with object lines. It is easier to join two arcs with a line than to add two arcs to a line. Try it.

Add dimensions and notes to complete the drawing. Fig. 14-5d. Follow the same procedures for one- and two-view drawings.

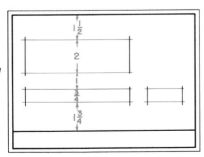

14-5a. *When laying out a three-view drawing, first study the object to be drawn.*

a soup can need only two. The layout of a softball diamond requires only one view (called a *plan view*) to describe its shape. Fig. 14-4.

LAYING OUT A THREE-VIEW DRAWING

You must carefully study the object to be drawn before you can begin. For example, look at Fig. 14-5a. You should select those views of the object which best describe it. It is important to select those views which show the object in its natural position. For example, it would be wrong to show a drawing of an automobile upside down.

To draw the object in Fig. 14-5a, block out the views with construction lines. Figs. 14-5b & c. Be sure to center your work on the paper and leave enough space around and between the views

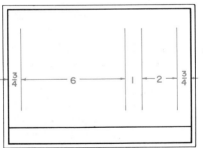

14-5b. *Lay out the horizontal overall dimensions.*

14-5c. *Lay out the vertical overall dimensions.*

14-5d. *Locate the circles, darken lines, and add dimensions.*

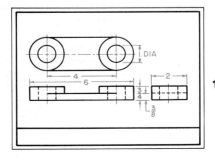

QUESTIONS AND ACTIVITIES

1. How many views does every object have? Name them.
2. Which three views are commonly shown?
3. Sketch some objects that need only one or two views to describe their shape.

4. Describe the steps for laying out a three-view drawing.

CHAPTER 15

Sectional and Auxiliary Views

It is very important that multiview drawings be as clear and understandable as possible. Sometimes special views, such as sections and auxiliaries, can help to make them clearer. *Sectional* views show interior or hidden details of objects. For example, it would be difficult to make an automobile carburetor without knowing what the inside looked like. The method of showing in-

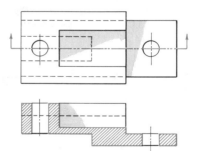

15-1b. Cutting-plane lines on one of the regular views show where the section is made. Here the cutting-plane line is on the top view. The sectional view has been drawn below it.

terior views in a drawing is seen in Fig. 15-1. Note how the cutting plane marks the place for the sectional view.

Auxiliary views show true views of surfaces which are set at angles to the planes of the trans-

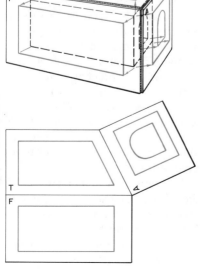

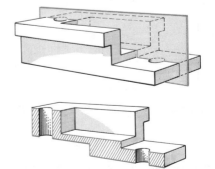

15-1a. Sections are made by passing an imaginary cutting plane through an object. By "removing" the front half of the object, you see its interior.

15-2. Auxiliaries are made by drawing one side of the transparent box parallel with the oblique (angled) surface of the object. Folding out the sides of the box produces an auxiliary view (A) of the oblique surface. The T indicates top and the F, front. Auxiliary views show the true shape and size of an oblique view.

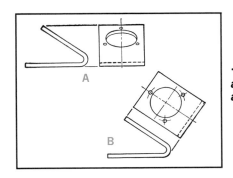

15-3. *Comparing orthographic (A) and auxiliary (B) views. Note that the auxiliary view is true size and shape.*

parent box. Fig. 15-2. Orthographic views of these inclined or angular surfaces would appear foreshortened (shorter than true size). An auxiliary drawing provides a view which is true in size and shape. Fig. 15-3.

QUESTIONS AND ACTIVITIES

1. What is a sectional view?
2. What is an auxiliary view?
3. What is the purpose of a cutting-plane line?
4. Sketch an example of an auxiliary view.
5. Sketch an example of a sectional view.

CHAPTER 16
Pictorial Drawings

When you hold a tool box in front of you, what you see is a perspective view of it. You can see that it has height, width, and depth and that some parts of the box are closer to you than others. If you were to make a drawing of what you saw, you would have a perspective drawing of the box. Fig. 16-1. Note how it resembles a photograph of the box. Such drawings are called *pictorials* because they are much like a picture. In a pictorial drawing, objects are drawn the way they appear to the eye. Besides the perspective, two other forms of pictorial drawings are used to describe the shapes of objects. These are the isometric and oblique drawings. Figs. 16-2 & 16-3.

PERSPECTIVE DRAWINGS
When you look down a railroad track or a long, straight highway, the track or road seems to get narrower and come together

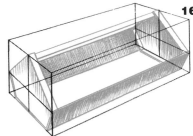

16-1. *A perspective drawing of a tool box.*

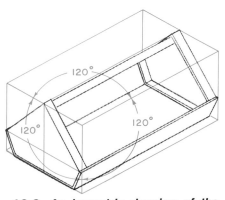

16-2. *An isometric drawing of the tool box. In isometric drawings, one corner appears closest. The rest of the object slants away along three axes that are 120 degrees apart.*

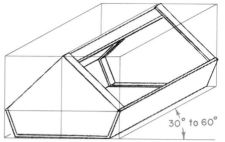

16-3. *An oblique drawing of the tool box. In oblique drawings the front of the object appears closest. The top and side slant away, usually at a 30- to 60-degree angle.*

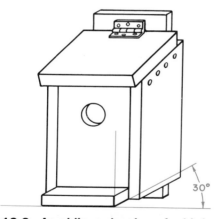

16-5. *An isometric drawing is constructed around three lines (axes) drawn 120 degrees apart (A). The completed drawing is shown at B. The letters X, Y, and Z represent the dimensions of the cabinet.*

at some distant point on the horizon. In perspective drawings, lines also meet at a point on the horizon. This horizon point is called a *vanishing point*. A one-point, or parallel, perspective drawing has one vanishing point. The two-point, or angular, perspective drawing has two vanishing points. Fig. 16-4. Architects and designers often use a perspective drawing to show realistic houses or furniture.

ISOMETRIC DRAWINGS

Isometric drawings are so called because they contain three views, or planes, with axes that lie 120 degrees apart. Fig. 16-5. *Isometric* means "equal measure." To make an isometric drawing, first draw a box that will enclose the object you want to draw. This box must also be isometric. Fig. 16-5, A. Note that the base lines are made with the

30° triangle. Then take the dimensions of the object you are drawing, transfer them to the proper view, and connect the lines. Fig. 16-5, B. All lines parallel with the 120-degree axes are called isometric lines and are true length. Lines not parallel to these axes are called nonisometric lines and are not true length.

Circles in isometric are drawn as ellipses. Use an ellipse template to draw these circles.

OBLIQUE DRAWINGS

The oblique drawing is made from the front view (plane) of an object, with the top and side

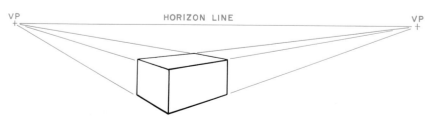

16-6. *An oblique drawing of a birdhouse.*

views lying back at any angle other than 90 degrees. The most commonly used angles are 30 degrees and 45 degrees. Fig. 16-6. It is the easiest of the three pictorial methods to use. Its most important advantage is that the front view is of a true shape. This is helpful if that view contains many arcs and curves, because

16-4. *This angular perspective drawing of a box has two vanishing points.*

arcs and curves in isometric or perspective are more difficult to draw.

There are two kinds of oblique drawings: cavalier and cabinet. In cavalier oblique, the side and top of the object are drawn their true lengths. This makes the object look distorted. Fig. 16-7. In a cabinet drawing, the depth is shortened by one-half so that it won't look out of proportion. Fig. 16-8.

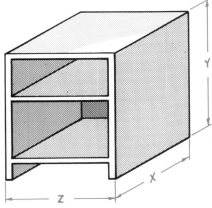

16-7. *Cavalier oblique.*

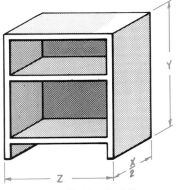

16-8. *Cabinet oblique*

QUESTIONS AND ACTIVITIES

1. List three types of pictorial drawings.
2. What is a vanishing point? What type of pictorial drawing is it used in?
3. What type of drawing contains three views with axes that lie 120° apart? How do you make a circle in this type of drawing?
4. What is the easiest of the three pictorial methods to make and use? What is the most important advantage of this method?
5. What type of pictorial drawing uses cavalier and cabinet styles? Describe each style.

CHAPTER 17
Dimensioning and Lettering

DIMENSIONING PRACTICES

The chapters dealing with multiviews, pictorials, sections, and auxiliaries explain ways of shape description. Shape description is one important part of drawing, but it is not enough to only describe the shape of an object. The drawing must also show its size. Size description involves placing numbers and letters on the drawing which show the object's width, height, depth, size of holes, and other information. Fig.

17-1. Putting the size description on a drawing is called *dimensioning*. Figure 17-2 shows a dimensioned drawing. It is complete because it accurately describes the shape and size of the snack tray. This drawing, as well as others in this book, shows how dimensions are properly placed on a drawing.

There are many things to learn about dimensioning so that you

17-1. *This three-view drawing of a block shows the placement of dimensions for width, height, and depth.*

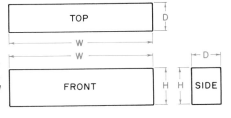

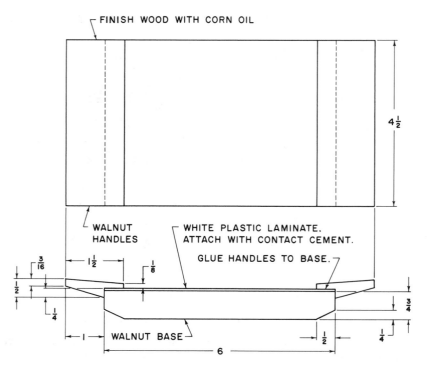

FINISH WOOD WITH CORN OIL

WALNUT HANDLES

WHITE PLASTIC LAMINATE. ATTACH WITH CONTACT CEMENT.

GLUE HANDLES TO BASE.

WALNUT BASE

$4\frac{1}{2}$

$\frac{3}{16}$ $1\frac{1}{2}$ $\frac{1}{8}$

$\frac{1}{2}$ $\frac{1}{4}$ 1 6 $\frac{1}{2}$ $\frac{1}{4}$ $\frac{3}{4}$

SNACK SERVING BOARD

17-2. *Drawing of a snack tray, showing the shape and size. Dimensions and lettering are neat and accurate.*

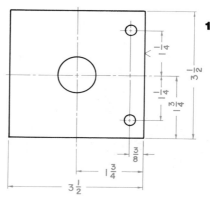

17-3. *Aligned dimensions.*

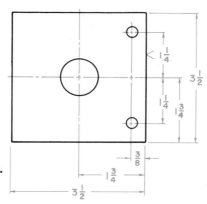

17-4. *Unidirectional dimensions.*

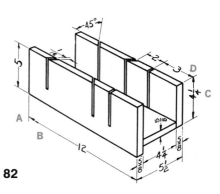

17-5. *Sample drawing showing: (A) Extension line. (B) Dimension line. (C) Size dimension. (D) Location dimension.*

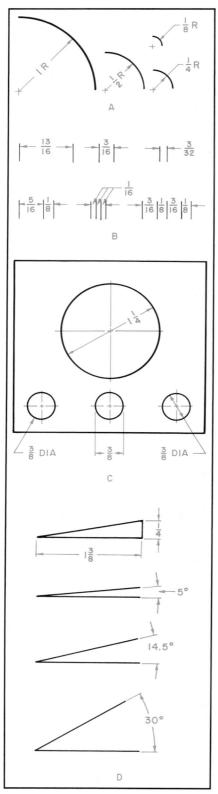

17-6. *Methods for dimensioning: (A) Arcs. (B) Narrow spaces. (C) Holes. (D) Angles.*

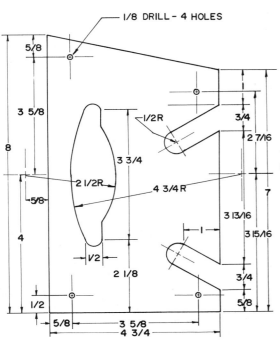

1/8 DRILL - 4 HOLES

17-7. *Crowded dimensions are difficult to read.*

mensions. They should be accurately placed, neat, and easy to read. Fig. 17-6.

4. Study an object carefully before you dimension it. There may be a better way of dimensioning to make it less cluttered and easier to read. Study the two drawings in Figs. 17-7 & 17-8. Which of the two is easier to read?

5. Avoid placing dimensions on the views; place them between views whenever possible. Stagger dimensions so that they can be easily read.

6. Dimension in inches up to and including 72 inches. Use feet and inches above this size. (See Chapter 18 for metric dimensioning.)

LETTERING PRACTICES

The style generally used for the numbers and letters on a drawing is *single-stroke Gothic*. Each letter is separate; that is, it is not connected to another letter

can describe the size of an object properly. Some of these important facts are:

1. There are two systems for placing dimensions on a drawing:

• In the aligned system, dimensions are placed so as to be read from the bottom and right side of the page. Fig. 17-3.

• In the unidirectional system, all numbers are placed so that they can be read from the bottom of the page. Fig. 17-4. Select either of these systems. Do not mix them on the same drawing.

2. Some dimensions are used to show the overall size of an object. These are called *size* dimensions. *Location* dimensions are used to locate holes, slots, etc., in relation to other features. Fig. 17-5. In this drawing note the difference between extension and dimension lines. Extension lines should not touch the object.

The arrowheads on dimension lines should just touch the extension lines. Arrowheads should be neat and narrow and should end in a sharp point. They should be about ⅛″ long. The width should be about ⅓ of the length. Do not

measure each arrowhead. Practice making them until you can do so quickly and neatly.

3. There are a number of ways to dimension narrow spaces, arcs, circles, and angles. The method you use depends upon how much space you have. Do not crowd di-

17-8. *This is Fig. 17-7 drawn on grid paper. See how much easier it is to read.*

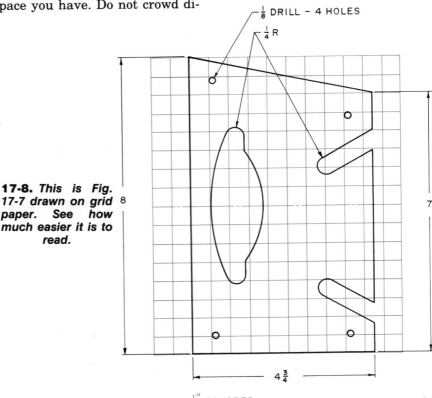

⅛ DRILL - 4 HOLES

¼ R

8

7

4¾

$\frac{1}{2}$" SQUARES

17-9. *Forming single-stroke Gothic vertical letters. This style of lettering is most often used in mechanical drawings.*

and *W* for example. Draw some horizontal guidelines on a piece of paper about ¼" apart. Practice forming letters. Make firm, straight strokes and regular curves. Continue to practice those letters which give you trouble.

Figure 17-9 shows the vertical style single-stroke Gothic letters. In this style the letters are straight up and down and block-like. However, because of the common use of slant in everyday writing, it is natural to form letters with a slant. Slant letters and numbers can be made with ease and speed. In slant lettering all vertical lines slant at 68°. Circles are made as ellipses with the major axis at 68°. The direction of the strokes and the order in which each is made are similar to the techniques for vertical upper-case lettering.

The spacing between letters is about one-fourth the width of any normal letter. The spacing between words is equal to the width of a normal letter, and the spacing between sentences is twice

as in script handwriting. This separation makes it unnecessary to trace back on the strokes of each letter. Therefore each stroke of a letter is a single line. Letters can be either lowercase or uppercase (small or capital). Uppercase letters are more often used.

Lettering should be done with a medium hardness pencil such as an H or 2H. Hold the pencil comfortably but firmly without tiring the hand. Keep the point sharp and rotate the pencil after forming a few letters. This will prevent the point from flattening and the strokes from getting too wide.

Study Fig. 17-9 to learn how to form letters. Note that some letters are wider than others, the *M*

MORE OR LESS THAN HEIGHT OF LETTERS
WIDTH OF AVERAGE LETTER
SELF CHECK YOUR LETTERING

AND LINE WORK. PRACTICE IS
¼ WIDTH OF AVERAGE LETTER → TWICE HEIGHT OF LETTERS

17-10. *Use correct spacing when lettering.*

$3\frac{7}{8}$ TWICE THE HEIGHT OF THE WHOLE NUMBER

17-11. *Tips for forming fractions.*

$7\frac{5}{16}$ division line in line with center of whole number. fractional numbers do not touch the line.

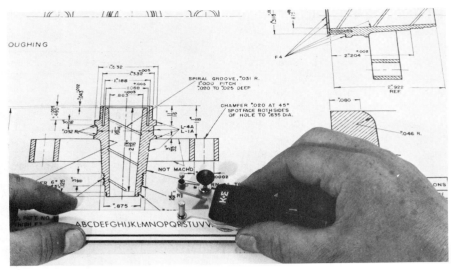

17-12. *Lettering with a lettering set.*

the height of a letter. Spacing between lines is equal to the height of a letter. Fig. 17-10. When lettering fractions, follow the suggestions shown in Fig. 17-11.

Lettering sets or instruments are often used to make neat, clear letters and numbers. These are precision instruments and must be used carefully. Fig. 17-12.

QUESTIONS AND ACTIVITIES

1. Name and describe the two systems used for placing dimensions on a drawing.

2. What is the difference between a size dimension and a location dimension?

3. Get a clean sheet of paper and practice making arrowheads. Follow the examples in the textbook for proper techniques.

4. Get a piece of paper that has ¼″ spaced lines on it. Use this as a practice sheet and begin to practice lettering. Follow the examples in the textbook to learn proper techniques.

CHAPTER 18
Metric Dimensioning

There are a number of things to consider in the change from customary to metric drafting. The most important of these are drawing scales and metric dimensioning. Metric scales were discussed in Chapter 12. Metric dimensioning, information on rounding off numbers, and de-signing in metrics are covered in this chapter.

METRIC DIMENSIONING METHODS

In the United States today a number of industries are switching to the metric system. Drawings are being made which show both metric and customary dimensions. This makes it possible to read a drawing of a part in metrics, but still use customary tools and machines to make the part. Or the drawing can be read with customary dimensions, but the part made with metric tools and machines. The practice of showing both customary and metric dimensions in the same drawing is called *dual dimensioning*. Fig. 18-1. In such combination dimensioning systems, all inch-millimetre conversions are done by the engineers and drafters. This makes it unnecessary for others who use the drawing to make these conversions and possibly make mistakes.

There are three common dual dimensioning methods:
- The bracket method.
- The position method.
- The conversion or equivalency chart method.

While there are others, these are the three used most frequently.

The *bracket* method is a size description system in which inch or millimetre equivalents are shown in brackets next to each dimension. (To obtain a metric

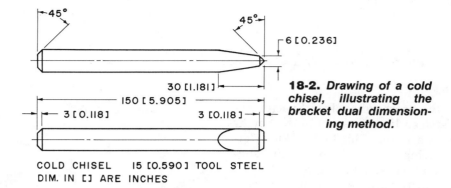

18-2. *Drawing of a cold chisel, illustrating the bracket dual dimensioning method.*

equivalent, multiply the inch dimensions by 25.4; one inch equals exactly 25.4 millimetres. To convert millimetres to inches, divide by 25.4.)

Shown in Fig. 18-2 is a drawing of a cold chisel. Note that the dimensions are both metric and customary. The millimetre dimensions are placed on the dimension line, and the equivalent inch dimensions are placed next to them in brackets. On this drawing, metric is the primary system. Of course, you can also put the inch dimensions first and the millimetre equivalents in brackets. Which system is primary depends on who designed the project or part and in which country the drawing will be used. In order to avoid any confusion, always add this note to the drawing: "DIM. IN [ ] ARE MILLIMETRES"; or "DIM. IN [ ] ARE INCHES."

The *position* method is shown in Fig. 18-3. The millimetre dimensions are placed above the dimension line and the equivalent inch dimensions are placed below it. You can also put the inch dimensions above the line and the millimetre equivalents below it, depending upon which is the primary dimension. A note should also be added to this drawing to avoid confusion:

$$\frac{mm}{inches} ; \text{or } \frac{inches}{mm} .$$

The *conversion chart* method is shown in Fig. 18-4, which is a metric drawing of the cold chisel. A conversion chart gives inch equivalents for each of the millimetre sizes. This method results in a much neater, less cluttered drawing, especially if the draw-

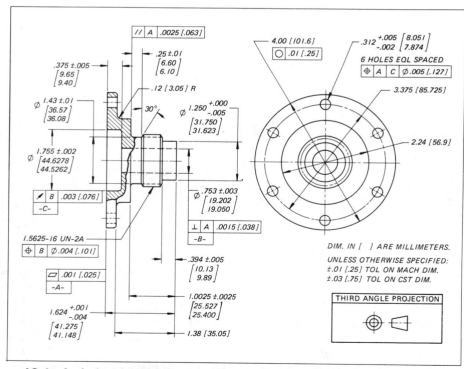

18-1. *An industrial drawing of a metal part showing the use of dual dimensions. The dimensions in brackets are millimetres. The dimensions on this drawing include the tolerances (range of acceptable sizes) for each part.*

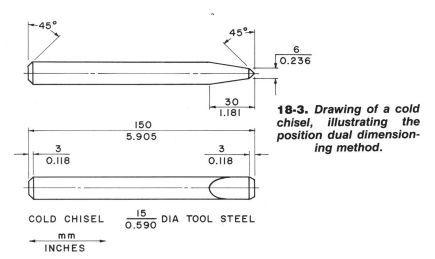

18-3. *Drawing of a cold chisel, illustrating the position dual dimensioning method.*

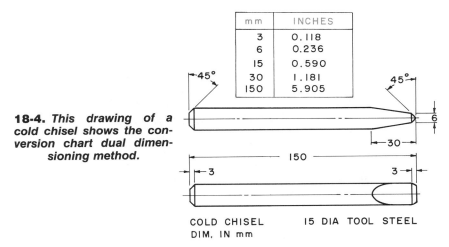

18-4. *This drawing of a cold chisel shows the conversion chart dual dimensioning method.*

tion shows the top view below the front view. The left-side view is shown to the right of the front view.

In order to avoid confusion among the users of drawings, the angle of projection must be shown by one of the symbols in Fig. 18-5 on all drawings to be used in foreign countries. Procedures for making drawings according to the first angle of projection can be found in any advanced drafting textbook.

To further avoid mistakes and confusion, paper with the word METRIC printed in large letters can be used for drawings. If this paper is not available, the word *metric* can be shown in the title block or the words "ALL DIM. IN mm" lettered on the drawing. The important point here is that the person who is making a part from the drawing must know what the dimensions mean in order to know whether to use customary or metric measuring instruments.

In metric drawings, all dimensions will be in millimetres or metres except for maps, where the kilometre is used. On engi-

ing has many dimensions and notes. It also makes it possible to use a computer to figure the equivalencies. The computer printout can be attached to the drawing, eliminating much hand lettering. These chart dimensions should be placed in ascending or descending order (from smallest to largest or vice versa) in order to make the chart easier to use. The ordering can be done automatically with a computer. These drawings should also carry the note "DIM. IN mm" or "DIM. IN INCHES" to let people know which system is used on the drawing itself. Conversion charts are becoming very popular in in-

dustry because of their convenient application to the computer.

METRIC DIMENSIONING RULES

All drawings showing both customary and metric dimensions must also indicate which angle of projection is used. The United States and Canada use the familiar third-angle projection in making multiview drawings. Third-angle projection shows the top view above the front view. The right-side view is shown to the right of the front view. Most other nations use the first angle of projection. First-angle projec-

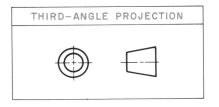

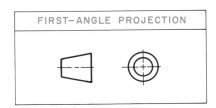

18-5. *First- or third-angle projection symbols should be placed on all drawings to be used in foreign countries.*

neering drawings, surface roughness is given in micrometres. Centimetres and decimetres should not be used, in order to avoid confusion among metric units. The mm and m are related by 1000 (1000 mm equal 1 m). Be careful that you do not mix units on a drawing. It is best and clearest if a drawing contains either mm or m dimensions, not both.

In fact, most drawings will be in mm. Even building construction drawings can be easily understood if they are so dimensioned. For example, a wall that is 15 900 mm long can easily be read as 15.900 m, and the carpenter will use a 30 m measuring tape to lay it out. The use of metres on a drawing is usually limited to very large construction, shipbuilding, etc. In these cases, dimensions in metres and decimal parts of metres are used. For example, the dimension 15 m and 900 mm is shown as 15.900 m.

In dimensioning, remember that a millimetre is very small:

1 mm = 1/25.4 inch or 0.039 370, about four-hundredths (0.04) inch

0.1 mm = 0.0039, less than four-thousandths inch

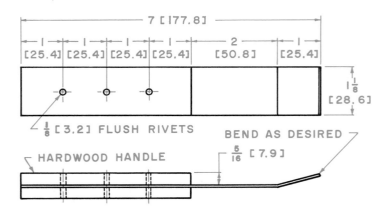

PUTTY KNIFE
0.020 [0.50] STAINLESS STEEL BLADE
DIM. IN [] ARE mm

18-7. *Dual dimension drawing of a putty knife.*

0.01 mm = 0.000 39, less than 4/10 000 of an inch

You can see that dimensions to more than two decimal places in millimetres are generally not needed.

Rounding Off

Decimal fractions are often rounded off to two decimal places. The rules to follow are:

● When the third digit after the decimal is greater than 5, the second digit is increased by one.

Example: 11.346 is rounded off to 11.35.

● When the third digit is less than 5, the second digit remains as it is. Example: 21.234 99 is rounded off to 21.23.

● When the third digit is 5, look at the numbers following. If they are greater than 0, round off to the next highest number. Example: 16.235 3 is rounded off to 16.24. If the 5 is followed only by zeros, the second digit is left unchanged if it is even, and increased by one if it is odd. Example: 3.045 00 becomes 3.04, and 3.035 00 becomes 3.04 also.

DESIGNING IN METRICS

Good drawing practice includes designing in metric units and not just converting inch values to millimetres. See Fig. 18-6. In this drawing of a serving tray, the millimetre is the primary unit of measurement.

If you study the plans for the putty knife in Fig. 18-7, you will see that you can use either inch or millimetre rules to measure the parts as you make the knife. You will recall that this is an example of bracket dual dimensioning. But look carefully at the

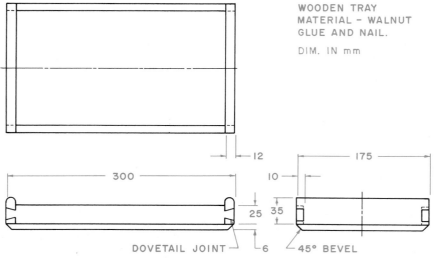

WOODEN TRAY
MATERIAL – WALNUT
GLUE AND NAIL.

DIM. IN mm

18-6. *Metric drawing of a wooden serving tray.*

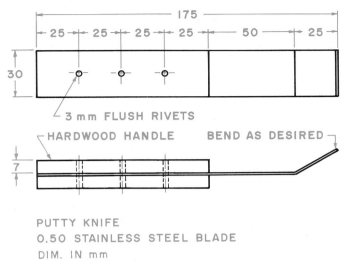

3 mm FLUSH RIVETS

HARDWOOD HANDLE   BEND AS DESIRED

PUTTY KNIFE
0.50 STAINLESS STEEL BLADE
DIM. IN mm

18-8. *A metric drawing of a putty knife.*

metric measurements. Notice that they are set at the nearest tenth of a millimetre. Now take a metric rule or scale in hand and study it. You can see that a millimetre is approximately the thickness of a dime. While it is easy to locate full millimetre dimensions, it is harder to find a decimal fraction such as 25.4 mm. Wouldn't it be easier to find a size of 25 mm? To make metric measuring simpler, you can round off numbers to the nearest whole metric size. Avoid using decimal fractions of millimetres when designing a project or a part unless the accuracy of the piece demands it. This is what designing in metrics is all about —making it simpler to make

projects or products based on metric dimensions.

Now study the drawing in Fig. 18-8. It too is a putty knife, but it is drawn in millimetres only. It is about the same size as the one in Fig. 18-7, but some important changes have been made. For example, the length of the handle is

100 mm instead of 4 inches. Note that 4 inches converts to 101.6 mm exactly. Obviously, the 101.6 mm can be rounded to 100 mm so that the handle is easier to make. Use this kind of logic as you design in metrics. Also, remember the standard material sizes, such as 1″ lumber as 25 mm and ½″ lumber as 12 mm. A chart of preferred numbers is shown in Fig. 18-9.

To summarize, conveniently rounded-off numbers should always be used as a *first* choice unless other considerations make it impossible. For example, use 65 millimetres or 70 millimetres in preference to 67. Chosen sizes should always be consistent with the available sizes of raw materials, fasteners, etc. Study the chapters in this book dealing with wood, metal, and plastic materials for metric sizes. You will soon learn to design metric projects which are much simpler and more logical to draw and make.

First-Choice Sizes

		Rising by 5		Rising by 10		Rising by 20	Rising by 50		Then	
1	5	20	55	80	150	200	300	650	800	3 000
1.2	6	25	60	90	160	220	350	700	900	4 000
1.6	8	30	65	100	170	240	400	750	1 000	
2	10	35	70	110	180	260	450	800	1 200	
2.5	12	40	75	120	190	280	500		1 500	
3	16	45	80	130	200	300	550		2 000	
4	20	50		140			600		2 500	

18-9. *When you design a metric project, try to use these metric sizes.*

QUESTIONS AND ACTIVITIES

1. List and describe three types of common metric dual dimensioning methods.

2. Round off the following numbers to two decimal places. Follow the rules given in the textbook: A. 12.345 B. 42.3249 C. 65.849 D. 2.797.

3. Why is it important to state the angle of projection on drawings that will be used in other countries?

4. Sketch a simple object such as an eraser or a pencil. Dual dimension it according to one of the methods described in this book.

CHAPTER 19

Detail and Assembly Drawings

You have already learned that drawings must have all the information needed to build a product. Such drawings are called *working drawings*. A working drawing for a fireplace tool set is shown in Fig. 19-1. This drawing shows all the dimensions and notes necessary to make the pieces. No other information is needed. The drawing also shows how the parts are assembled, or fit together. This is a common type of working drawing you will make and use in the shop.

There are other types of working drawings used in industry. A product such as an engine carburetor is made up of many parts. Each part is made separately, and for each part one must have a detail drawing. Still another drawing is needed to show how the parts fit together. This is called an assembly drawing.

Detail drawings are used to show how each part is made. It is necessary to make separate drawings of these parts because, in mass production, different people make different parts of the product. Such drawings include all dimensions plus notes about what materials to use and how to finish them. A typical shop detail drawing of a C-clamp is shown in Fig. 19-2. Note the information given in this drawing. All these parts are grouped together on one sheet for easy study. In industry each part would be drawn on a separate sheet.

Assembly drawings show the various parts fastened together to make the final product. Figure 19-3 shows the assembly of the parts shown in Fig. 19-2.

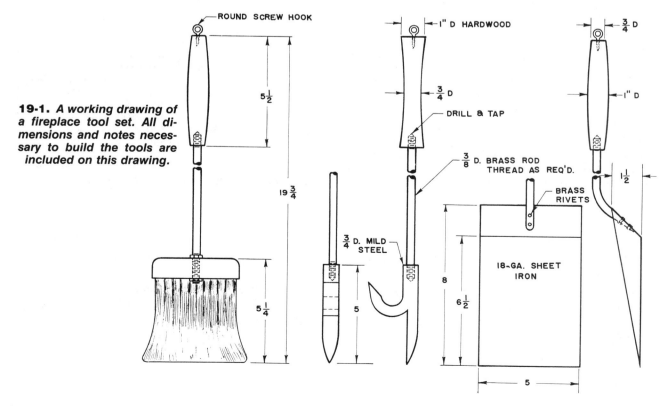

19-1. *A working drawing of a fireplace tool set. All dimensions and notes necessary to build the tools are included on this drawing.*

19-2. *Detail drawings of the parts of a C-clamp.*

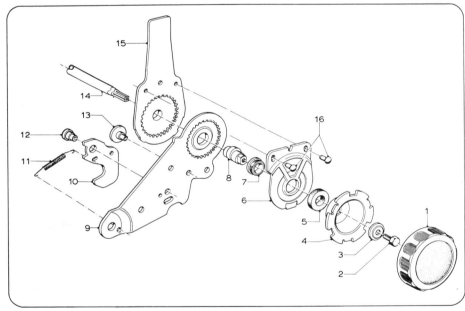

FILLETS AND ROUNDS $\frac{1}{8}$ R

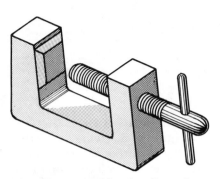

19-3. *An assembly drawing showing how the parts of the C-clamp fit together.*

Keiper USA

Industry also uses exploded assembly drawings. These are pictorial drawings that show the object "pulled apart." Fig. 19-4.

Here are some instructions to follow when preparing detail and assembly drawings.

1. Use only the views that are necessary. Sometimes two views are enough. Some objects require three views.

2. Select the views which show the object best.

3. For simple objects with few parts, place both the detail and assembly drawings on one sheet. More complex objects require drawings on separate sheets.

4. Make certain that the drawing contains all the information necessary to make the object.

19-4. *An exploded assembly drawing of an automobile seat adjustor.*

QUESTIONS AND ACTIVITIES

1. What is a detail drawing?
2. Why are separate detail drawings needed in industry?

3. What is an assembly drawing?
4. List four important instructions to follow in preparing detail and assembly drawings.

CHAPTER 20

Building Construction Drawing

Teledyne Post, Des Plaines, IL

20-1. *Architectural drawing is an important part of the drafting profession. This student is lettering a drawing.*

PLANNING A BUILDING

Designing a home or other building requires much thought and planning before the drawings can be made. In home design, the designer must know the family and how it lives in order to make the home suit the family's wants and needs. How large is the family? What activities do they enjoy? Should there be spaces for hobbies, playing musical instruments, and for study and reading? Where is the home to be built? How can the designer take advantage of the natural surroundings? These and many other questions must be answered. Whether designing a house or a hardware store, the designer must know how the building is to be used.

As ideas for rooms begin to appear, the designer sketches plans for room spaces and layouts. For example, he or she may think about the shapes of kitchens and how the work areas can be planned. Fig. 20-4. In this way the house begins to take shape.

Homes, offices, schools, and other buildings require complete plans before they are built. These plans or drawings contain all the information needed by the builders to finish the construction. This type of drawing is called "building construction drawing" or "architectural drawing." Fig. 20-1.

People who design homes and commercial buildings and draw plans for them are called *architects*. Architects often have architectural drafters working for them to help in drawing plans. When an architect designs a house, he or she often makes pictorial sketches first to show what the completed structure might look like. Fig. 20-2. The architect also prepares plans to show what interiors will look like. Fig. 20-3. As a student architectural drafter, you would have experiences in drawing house plans or plans for other kinds of buildings.

Forest Products Laboratory
20-2. *This is a pictorial sketch of an attractive, low-cost summer home.*

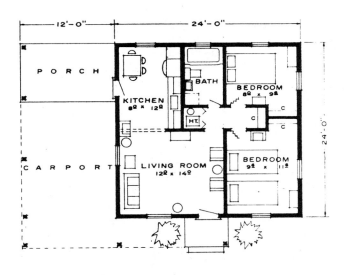

AREA - 576 sq. ft.

Forest Products Laboratory

20-3a. *A simple floor plan of the summer home. A floor plan shows the arrangement of the rooms as they would be seen from above.*

Forest Products Laboratory

20-3b. *A sketch of the home's kitchen. Compare this to the floor plan in 20-3a.*

STEPS IN DRAWING HOUSE PLANS

House plans are prepared in a number of steps. The drafter lays out the basic shape and size of the house. Next, the living, working, playing, sleeping, and eating areas are grouped together. These interior details are studied and refined. Changes and improvements are made. There are many problems to be solved: where to locate the chimney and fireplace, how to group plumbing fixtures, how to provide enough storage, where to place stairways and hallways, and where to locate doors and windows. As these problems are solved, the final floor plan takes shape.

The final floor plan includes basic information for the builder. Other plans are often drawn with information on plumbing, heating, and electricity. Sometimes all this information is put on one plan, if it will not look too cluttered.

In addition to the floor plans, others are needed. These are foundation plans, elevations, and construction details. *Elevations* are views of the outside of the building. There are generally four elevations, one for each side of the house. Fig. 20-5. Construc-

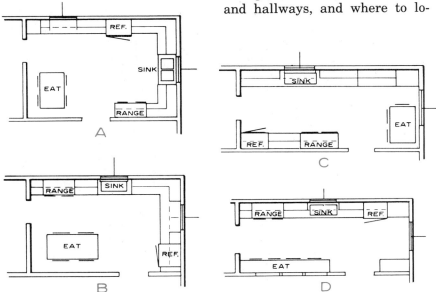

20-4. *Typical kitchen arrangements: (A) U-shaped. (B) L-shaped. (C) Corridor. (D) Straight-wall.*

RIGHT SIDE

20-5. *Right-side elevation.*

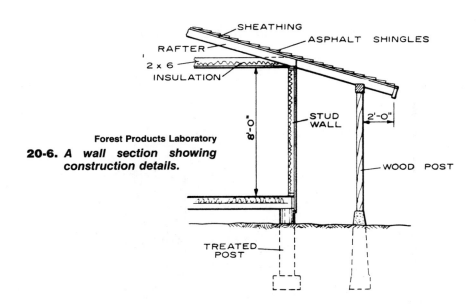

SHEATHING
ASPHALT SHINGLES
RAFTER
2 x 6
INSULATION
STUD WALL
8'-0"
2'-0"
WOOD POST

Forest Products Laboratory
20-6. A wall section showing construction details.

TREATED POST

tion details show selected parts of the house. They are drawn to a larger scale than the plans that show the whole house. Fig. 20-6.

When drawing house and building plans, select the proper scales so that the plans will fit on the paper. Plans and elevations are often drawn to a scale of $\frac{1}{4}'' = 1' \ 0''$. Construction details are usually to a scale of $\frac{3}{4}'' = 1' \ 0''$.

Architectural symbols are used to make drawings easier to read. Some examples are shown in Fig. 20-7.

An example of a metric house plan appears in Fig. 20-8. Note that the dimensions are in millimetres. Otherwise it looks like a customary house drawing.

As you begin your building construction drawing activities, remember to start with something small and simple. A backyard clubhouse, a small cabin, or an outdoor tool storage shed are some good projects for the beginner.

OUTSIDE DOOR
INSIDE DOOR
SWINGING DOOR
DOUBLE HUNG WINDOW

TUB
BATH
TOILET
SINK

FIREPLACE

20-7. Some symbols used in architectural drawing.

DN
UP
STAIRS

WINTER SPORT CABIN
ALL DIMENSIONS IN mm

1600
700
LIVING & SLEEPING 3300 X 2000
STORAGE
DOUBLE BUNKS
FLOOR FURNACE
WOOD-GAS RANGE
KITCHEN 1900 X 1800
TABLE
BATH 1400 X 1200
SHOWER
900
3600
1150
1500
1900
4200

20-8. Metric drawing of a winter cabin.

QUESTIONS AND ACTIVITIES

1. Define *architect*.
2. What is a floor plan?
3. What is an elevation drawing? How many are needed for a building such as a house?

4. What scale is generally used for plans and elevations? For construction details?
5. Draw a floor plan for a winter sports cottage or a summer beach cottage.

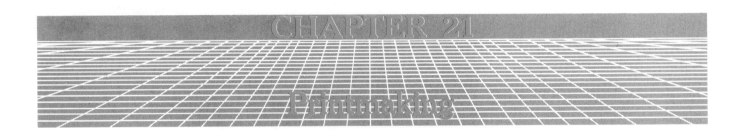

CHAPTER 21

Printmaking

Many shop drawings are made on ordinary opaque drawing paper. However, in industry and construction, many different people may use a drawing. To provide enough drawings for everyone, and to keep the original from becoming soiled and torn, copies are made. These copies are called prints.

To make a print, you must first make a tracing of a drawing. This is done by fastening a sheet of thin paper over a drawing and then tracing all lines and lettering. Most drawings in industry are made directly on tracing paper or vellum to avoid having to make a separate tracing.

The tracing is then used to make a print, much as you would make a photographic print from a negative. Three types of prints are generally used.

Dry-diazo prints have black or blue lines on a white background. The process is explained in Fig. 21-1. The dry-diazo process is convenient because no messy liquids are used. It is often called the Ozalid process.

The *moist-diazo* process is much like the dry-diazo process. The difference is that while the dry-diazo uses an ammonia fume developer, the moist-diazo uses a liquid developer applied with a roller. Fig. 21-2. This is also a fast, convenient printing method, often called the Bruning process. Fig. 21-3. In both diazo processes, the color of the lines depends upon the type of paper used.

Ordinary *office copiers* can also be used to make prints of small drawings. (Very large drawings won't fit on the copier.) These copies can be made from the original drawing. You don't need to make a tracing first.

As you learned in Chapter 3, drawings can be done on a computer. Paper printouts ("hard copies") of such drawings are produced by a printer designed for this purpose.

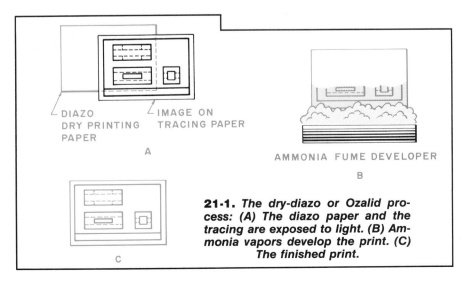

21-1. The dry-diazo or Ozalid process: (A) The diazo paper and the tracing are exposed to light. (B) Ammonia vapors develop the print. (C) The finished print.

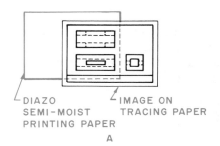

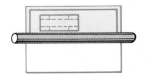

DIAZO
SEMI-MOIST
PRINTING PAPER

IMAGE ON
TRACING PAPER

A

CHEMICAL SOLUTION ROLLER
DEVELOPER

B

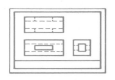

C

21-2. *The moist-diazo or Bruning process: (A) Exposure. (B) Development. (C) Finished print.*

Bruning Division,
Addressograph-Multigraph Corporation

21-3. *Using the moist-diazo machine.*

QUESTIONS AND ACTIVITIES

1. Why is it necessary to make prints of drawings?

2. Describe the dry-diazo process.

3. Describe the moist-diazo process.

4. What determines the color of the lines on a diazo print?

PART 3

Graphic Communications

CHAPTER 22

Introduction to Graphic Arts

Graphic arts deals with the placing of printed images onto a solid material. This solid material is usually paper. However, other solid materials such as cloth, metal, glass, or plastic may also have designs printed on them by one of the graphic arts processes. What we call the graphic arts industry also includes papermaking, ink making, and bookbinding, as well as printing.

The purpose of graphic arts is to communicate. The messages that are communicated, and the form of communication, vary widely. They include billboards and Bibles, textbooks and soup can labels, novels and wedding invitations. We look at road maps to check our location during a trip. We read the comics in the Sunday paper to be amused and entertained, and we read the newspaper to be informed. Every day products of the graphic arts industry influence our lives in some way.

PRESSES

Graphic arts is a very old craft. Thousands of years ago the Chinese printed from wooden blocks. The ancient Koreans printed from type made from porcelain. However, these inventions did not find their way to Europe, so for hundreds of years all European books were copied by hand. Therefore they were scarce and costly. This began to change when movable type was invented. *Movable type* consisted of individual letters which could be assembled into words, used for printing, and then disassembled and later reused. Johann Gutenberg first used movable type in Europe about 1439. After this date, reading material was no longer a luxury only the rich could afford.

Early printing presses were made of wood and operated by hand. Many were converted wine presses. The *typeform* (the assembled letters) was locked into the horizontal bed of the press. Then the typeform was inked and a sheet of paper was placed over it.

The platen of the press was lowered onto the paper until enough pressure was exerted for the inked letters to make an impression on the paper. Fig. 22-1. The platen was raised, the paper removed, and the entire process was repeated for the next sheet. It was slow, hard work. Only a few hundred sheets a day could be printed.

Today's presses are highly automated and very fast. There are sheet-fed presses which print one sheet at a time and web-fed presses which print from large rolls of paper. In the following

Paul M. Schrock

22-1. *Early printing presses were operated by hand. Note the two daubers on the stool. These were used to apply ink to the type.*

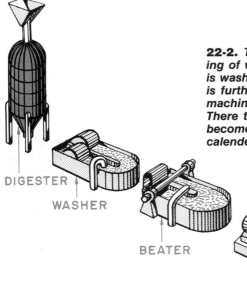

22-2. *The process of making paper begins with the cooking of wood chips in the digester. After cooking, the pulp is washed, then twisted and frayed in the beater. The pulp is further refined in the jordan. It then goes to the paper machine. (The fourdrinier was named after its inventors.) There the pulp is formed on a continuous wire screen to become paper. After drying, the paper is smoothed in the calender rolls and wound onto big spools to await further processing.*

DIGESTER

WASHER

BEATER

JORDAN

FOURDRINIER

FINISHED PAPER

DRYING AND PRESSING

CALENDER

chapters you will learn about the major printing processes and the way words and pictures are prepared for printing.

PAPER

The art of papermaking is many centuries old. A material similar to the paper we use was invented in China about 105 A.D. The earliest paperlike material was developed in Egypt. This material, called papyrus, was made from reeds which grew along the Nile River. Today most of the paper is made from wood pulp or from recycled paper. Figure 22-2 shows a diagram of the papermaking process. In making paper, the cellulose fibers are separated from the lignin (the "glue" that binds the fibers together). This lignin is burned as a waste product—up to 20 million tons a year. Research is now going on to find ways of using the lignin. One possibility is to combine it with slag (steel-making waste) to make an asphalt substitute.

INK

The other material needed for the printing process is ink. Printing inks are similar to paint. Both ink and paint contain pigment for color, vehicle to carry the pigment, and dryer to help the vehicle dry.

Originally, in the time of Gutenberg, printing ink was made from lampblack and linseed oil. Today printing inks are made in a variety of colors and for many special purposes. Many tons of ink are needed each year to print newspapers, magazines, books, and advertisements.

OCCUPATIONS IN THE GRAPHIC ARTS INDUSTRY

The graphic arts industry is one of the largest in the country. Today over one million men and women are employed at some occupation connected with the graphic arts field. About one-third of these work in the printing crafts, while the rest are designers, artists, salespeople, managers, clerks, and laborers. Occupations in printing include composition (setting type), photography, platemaking (making printing plates), presswork, and binding.

QUESTIONS AND ACTIVITIES

1. What does graphic arts deal with and what is its purpose?
2. Describe movable type. How did the invention of movable type affect people?
3. Define typeform.
4. What was the earliest paperlike material made from? What is paper made from today?

5. What three ingredients must ink contain and what purpose does each of these ingredients serve?
6. Name three occupations in the printing industry.

The process of creating a printed message begins with planning. What will the message say? How long will it be? Who will be reading it? Will there be illustrations? What will the layout (the arrangement of type and illustrations) look like? Once these and similar questions about design have been answered, the message is set in type and the artwork prepared. The type and art are then arranged into pages, the printing plates are made, and finally the message is printed.

Later chapters in this section will discuss various methods for setting type, preparing artwork, and making plates. Because the methods chosen for a particular job depend at least in part on the printing process that will be used, we will first briefly review the major printing processes.

There are several methods by which an image is placed on material, but most printing is done on paper by one of four major processes:

- Relief printing.
- Lithography.
- Gravure.
- Stencil.

RELIEF PRINTING

Relief printing is done from raised surfaces. These may be pieces of type (Fig. 23-1), carved linoleum blocks, or various kinds of printing plates. Ink is applied to the raised portion of the surface, then transferred to the paper. This is the oldest form of printing. Probably the most familiar example is printing with a rubber stamp.

Letterpress printing, a very important form of relief printing, is usually done with one of three kinds of presses. These are the platen press, the cylinder press, and the rotary press. You might wonder why all letterpress printing cannot be done with one kind of press. Think of all the products that are printed. Printed jobs range in size from small labels for medicine bottles to thousand-page telephone directories. Each kind of press has been designed for a certain type of work.

If you have a hand-lever press in your shop, you have a form of *platen press*. It brings the type and paper together to make a print. Fig. 23-2. Early platen presses were described in Chapter 22. Platen presses used in in-

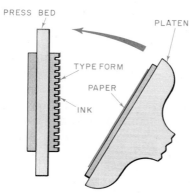

23-2. The platen press. Many small printing jobs are done on this kind of press.

dustry today are usually run by electric motors. They may be either automatic or fed by hand. Such presses are used for printing business cards, stationery, tickets, and similar items.

In the *cylinder press* the paper is wrapped around a cylinder and rolled over the type to transfer the ink. Fig. 23-3. The typeform moves horizontally back and forth and the paper revolves over it. This kind of press is used for printing such things as booklets, folders, catalogs, and labels. Some cylinder presses have two printing units so that more than one color may be printed as the sheet of paper moves through the press.

The *rotary press* is the fastest press for relief printing. The type material for this press is made into curved plates that fit on a cylinder. The paper goes between the curved plates and another cylinder, called the impression

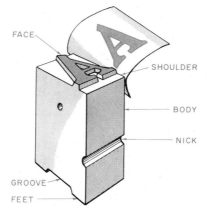

23-1. Relief printing is done from raised surfaces like this piece of type. Note that each part of a piece of type has a name.

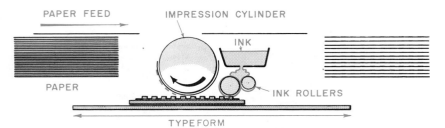

23-3. *The cylinder press prints from flat plates or type. The paper is printed as it rolls over the type.*

cylinder. Fig. 23-4. The most familiar use of rotary presses is for printing newspapers and magazines from rolls of paper. Many high-speed rotary presses can print and fold thousands of newspapers every hour. Some rotary presses print sheets rather than continuous rolls of paper.

LITHOGRAPHY

In lithography the printing is done from a flat surface, not a raised one as in letterpress. Lithography is based on the principle that grease and water do not mix. The image areas on the flat lithographic plate—that is, the parts which are to print—are covered with a greasy substance that will repel water. The plate is then coated with a water solution which sticks to the plate where there is no grease. In the areas where the greasy image has been applied, the water does not stick, but the oil-base printing ink will.

The image can be transferred directly from the plate to the paper. However, in *offset lithography* the image is transferred (or "offset") from the plate to a rubber-covered cylinder, called a blanket, then from this cylinder to the paper. Fig. 23-5.

Originally, stones were used for the lithographic process. The images were placed on the stone by drawing with a grease pencil or wax crayon. This process is still used by artists for making lithographic prints. Industry uses

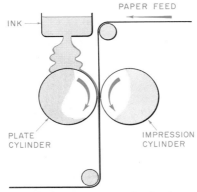

23-4. *The rotary press is the fastest kind of letterpress. It uses curved printing plates.*

a much faster and more precise method called *photo-offset lithography*. This process is also called *offset lithography* or simply *offset*. There are various ways to produce a printing surface for offset lithography. One common way is described here.

The material to be printed is first produced as images (words and pictures) on film. The images may be positive or negative. The film is placed on top of a sensitized metal lithographic plate. Then the image is transferred to the plate in much the same way as a photographic print is made. After processing, the areas to be printed attract ink but repel water just as if the image had been placed on the plate with a wax crayon.

Offset lithography has become a very important process. Many items that used to be printed by

the letterpress process are now printed by lithography. This textbook was printed by offset lithography. Many large mail order catalogs are printed by this process. So are many bulletins, bank checks, stock certificates, and some decals. Large printing firms commonly have complete offset-lithographic departments.

GRAVURE PRINTING

Gravure printing, also called intaglio, is similar to engraving. The printing is done from recessed or sunken images. Fig. 23-6. This process is the exact opposite of relief printing, in which the printing surfaces are raised. In the *rotogravure* process the printing is done by transferring ink to the paper from cells. These cells are holes of varying depths etched into a copper cylinder.

The first step in the process is to photograph the original material to be printed. Film positives are then made. These are used to transfer the design to the copper cylinder, with the help of a special carbon tissue material. When the carbon tissue is applied to the copper cylinder and developed with warm water, a copy of the image is left on the

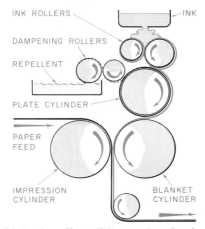

23-5. *In offset lithography the image is printed on a rubber blanket, then on paper.*

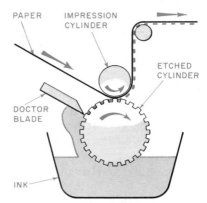

23-6. *Gravure printing is done from sunken or etched surfaces of a copper cylinder or plate.*

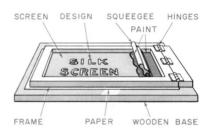

23-7. *The squeegee forces the ink or paint through the screen onto the paper in screen process printing.*

cylinder. The entire cylinder is then etched with acid, forming the cells in the shape of the image.

The copper cylinder is placed on a gravure press for printing. The cylinder turns in a trough of ink and the image cells are filled. The rest of the cylinder is inked too, so it is wiped off with a doctor blade. When the paper comes in contact with the cylinder, the ink in the cells is transferred to the paper.

Modern high-speed rotogravure presses can print continuous sheets of paper as wide as eight or nine feet. The copper cylinders can print millions of copies before wearing out.

The gravure process is used for printing such items as paper money, wallpaper, wrapping paper, box labels, packaging materials, and postage stamps.

Design: Getting the Message Across

The purpose of a printed message is to communicate information, whether it's a birthday greeting or instructions for building a satellite. That purpose can be helped or hindered by the design of the message.

The design requirements discussed in Chapter 7 apply to printed matter as well as to other products. The printed message must be functional, it must be made of the right materials, and it must be pleasing to look at.

To be functional, the message must be easy to read and understand. Otherwise the information will not be communicated. Suppose you are designing a poster for a new store. You would want to make the poster and the words on it large enough to be easily read from a distance of a few feet. The poster must convey the necessary information (name of store, location, etc.) in a clear, easily understood manner.

The right materials must be used. The poster must be printed on heavy paper that will not wrinkle or tear easily. It should have good printability so that the ink will not smear or run. At the same time, it should not cost more than your client can afford.

To get people's attention, the poster must be pleasing to look at. The design principles discussed in Chapter 7—unity, variety, proportion, and balance—apply to printed matter as well as to other products. These principles should be kept in mind when choosing typefaces and

artwork and when arranging these elements into a layout.

A functional, attractive product printed on quality materials will enhance the message it contains. It will get people's attention so that they read the message, and it will make an impression so that they remember the message.

SCREEN PROCESS PRINTING

Screen process printing is really stencil printing. A stencil is simply a pattern with certain parts cut out. It is placed over the surface to be printed. Ink or paint is brushed through the cut-out areas, forming words or a design on the surface beneath.

In the screen process the stencil is attached to a screen. The type of screen used depends on the design being printed. A coarse mesh is used for large,

heavy designs. For detailed work, a fine mesh is used. A squeegee forces ink or paint through the stencil and the screen, onto the object being printed. Fig. 23-7.

Screen process printing can be done on curved and irregular shapes as well as on flat surfaces. It may be used on paper, cloth, glass, plastic, wood, metal, and many other materials. For these reasons screen process printing is often used on packaging materials and textiles.

QUESTIONS AND ACTIVITIES

1. Briefly describe relief printing.

2. Briefly describe how each of the three kinds of presses used in letterpress printing operates.

3. How is lithography different from letterpress printing? On what principle is lithography based? Briefly describe basic lithography.

4. What is used to transfer images in gravure printing? Name three things that are printed using the gravure process.

5. Briefly describe screen process printing.

CHAPTER 24
Composition and Photography

Once the content and design of a message have been determined, the words are set into type and the artwork is prepared.

COMPOSING TYPE

Composition is the assembling of type characters (symbols such as letters, numbers, and punctuation marks) into words, lines, and paragraphs. For many years after the invention of movable type, all type was composed, or set, by hand. Today most type is set by machine. Hand-set type is still used sometimes for small, specialized jobs such as wedding invitations, business cards, and some advertisements.

There are many methods for composing type, but they can all be divided into two general categories. *Hot composition* methods use type which was cast in a mold from molten metal. The type can be assembled by hand or by machine. Typesetting done without the use of hot metal is called *cold composition*. Cold composition can be mechanical (using a typewriter, for instance) or electronic.

Type comes in many designs, called *typefaces*. The text of this book was set in a typeface called Century Schoolbook. The captions were set Helvetica Black. Type is one of the most important materials used by a printer. You should learn the various kinds of type available in your shop.

Printers use a special measuring system in their work. The standard unit of this system is the *pica*, which equals about ⅙ of an inch. The pica is divided into twelve equal parts called points. A *point* is equal to about 1/72 of an inch.

The printer's measuring tool is the line gauge. Fig. 24-1. It is divided into inches and picas. Type is measured in points with the line gauge. For example, the type being measured in Fig. 24-2 is 24 points.

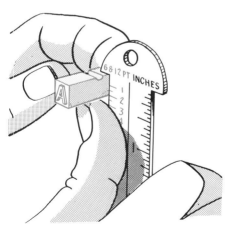

24-2. *Measuring the size of a piece of type using a line gauge.*

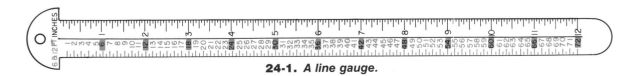

24-1. *A line gauge.*

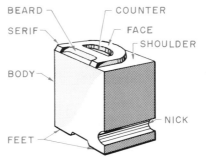

24-3. *A piece of foundry type with the parts labeled.*

BEARD — COUNTER
SERIF — FACE
SHOULDER
BODY
FEET
NICK

24-5. *Type is stored in a type bank. This type bank also stores leads and slugs.*

Hot Composition

COMPOSING TYPE BY HAND

Metal type is made from an alloy of lead, tin, and antimony. Type used for hand composition is called *foundry type* because it is made in foundries (factories which cast metal). Each piece of type has one letter, punctuation mark, or other symbol on it. Fig. 24-3.

Type is stored in drawers which are divided into compartments for each letter, number, and space. This drawer is called a type case. One such case is the California job case. Fig. 24-4.

Type cases are stored in a cabinet called a type bank. Fig. 24-5. Type banks also provide room

to store spacing material and a top surface to hold the type case when type is being set.

Leads (pronounced "leds"), 2 points thick, and *slugs,* 6 points thick, are used to make spaces between lines. Fig. 24-6. To provide space at the beginning of paragraphs and between words, several sizes of spaces are needed. The basic space is the *em quad,* which is the square (length times width) of the type's body size. Thus an em in one typeface may not be the same as an em in another typeface. Figure 24-7 shows size relationships.

Type is set, or composed, in a *composing stick.* Fig. 24-8. To set type the composing stick must

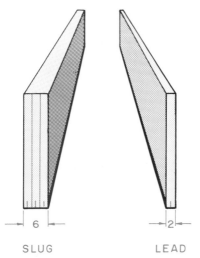

6

2

SLUG   LEAD

24-6. *Line spacing materials are called leads and slugs. Slugs are 6 points in thickness, and leads are 2 points thick.*

24-4. *The California job case.*

ffi	fl	3-EM	4-EM	'	k		l	2	3	4	5	6	7	8		$						
j	b	c	d	e		i	s	f	g	ff	9		A	B	C	D	E	F	G
?										fi	0								
!	l	m	n	h	o	y	p	w	,	EN QUAD	EM QUAD		H	I	K	L	M	N	O
z																			
x			3-EM SPACES				;	:	2-EM		P	Q	R	S	T	V	W
q	v	u	t	a	r		.	–	3-EM QUAD		X	Y	Z	J	U	&	ffl

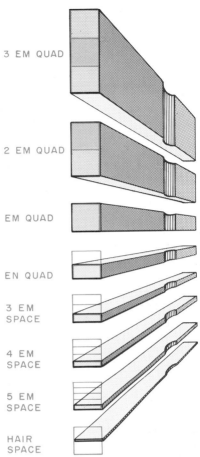

24-7. *Relative sizes of type spacing material.*

3 EM QUAD

2 EM QUAD

EM QUAD

EN QUAD

3 EM SPACE

4 EM SPACE

5 EM SPACE

HAIR SPACE

a time and place them in the composing stick from left to right with the nick on the letter toward the open side of the stick. You can read the words you are setting from left to right, but the letters are upside down and backwards. **CAUTION:** *Type metal contains lead. Never place type in your mouth.*

After the line is set, it must be spaced out to the proper length. This is called *justification*. The spaces and quads shown in Fig. 24-7 are used for this purpose.

When the stick is ½ to ¾ full, it should be transferred to a three-sided flat metal tray called a *galley*. With the open end of the galley to your left, place the composing stick into the galley. Grip the typeform firmly and slide it onto the galley. Place the type in the lower left-hand corner of the galley with the nicks toward the

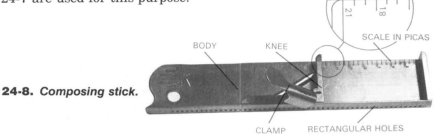

24-8. *Composing stick.*

BODY

KNEE

SCALE IN PICAS

CLAMP

RECTANGULAR HOLES

first be set to the desired length of line. Raise the clamp and slide the knee to the desired setting. Press the clamp back down to lock the knee. Insert a piece of line spacing material in the composing stick. Usually a slug is used at the beginning of a job.

Select the type you wish to use and place the case on top of the type bank.

Be sure to grasp the case firmly to avoid spilling. Stand in front of the type case and hold the composing stick in the left hand as shown in Fig. 24-9. Hold the stick at a slight angle with the thumb against the type. This keeps the type from falling over. Pick up the pieces of type one at

24-9. *Correct way to hold the composing stick. Note how the thumb holds the type in the stick.*

open end. Slide the typeform. Do not pick up the type.

Cut a piece of string that is long enough to go around the form four or five times. Start the string at one corner and go around the form clockwise. The string should overlap the start so that it can be pulled tight. Wind the string around the form several times, then tuck the end of the string under the windings. This will hold the type tightly together.

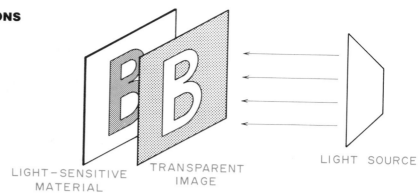

LIGHT-SENSITIVE MATERIAL   TRANSPARENT IMAGE   LIGHT SOURCE

24-10. *A beam of light travels through the transparent area and strikes light-sensitive material behind it, forming an image.*

Composing Type by Machine

Typesetting machines for hot composition are much faster than hand typesetting. The machine stores typeface molds rather than pieces of type. An operator sits at a keyboard and selects the desired characters and spacing. The machine casts the characters from molten type. Some machines (such as the Monotype) cast individual pieces of type. Others (like the Linotype) cast entire lines of type. After use, the type is remelted so that it can be cast again.

Cold Composition

As mentioned earlier, there are mechanical and electronic methods of cold composition. In both methods, the end result is an image on film or paper. In mechanical cold composition, the type is composed and then it is photographed. Electronic methods (called photocomposition) compose type directly onto film or paper.

Mechanical Methods

Some methods of mechanical composition are described here:

• Existing printed work may be used as camera copy (copy that is photographed). Preprinted display lines, borders, decorations, etc., can be purchased for this purpose. The type must be clean and of good quality so that it will reproduce well on film.

• Hand lettering can be used as camera copy.

• Strike-on composition uses the principle of forming an image by striking paper with a key containing a character. Typewriters, either standard or specially designed models, are used for much composition work. The specially designed typewriters have provisions for *justifying* lines (making all lines the same length). Some IBM models have interchangeable typing elements; each element has a different typeface.

• Linoleum block carving was developed to compose images that could not be cast by hot composition methods. Originally the image was drawn on a wood block that was the same height as the type characters. The areas that were not to print were carved away, leaving a raised design. The woodcut was then used for relief printing. Because wood is difficult to carve, a thin layer of linoleum is now added to a wood block and made type high. The design is then carved into the linoleum.

As you can see, mechanical cold composition is more versatile than hot composition. Since the image is photographed rather than assembled from metal type, you are not restricted to the characters available on metal type.

You can use various typewriter elements or even create your own typeface, hand letter it, and photograph it.

Photocomposition

In photocomposition, characters are generated by a phototypesetter which sends a beam of light through the transparent image of a character onto light-sensitive film or paper. Fig. 24-10. The transparent images (characters) may be stored on rotating disks, film strips, rotating turrets, or stationary grids. These *master character sets* vary in the number and kinds of typefaces and sizes which they contain.

Some phototypesetters generate characters using a cathode ray tube (CRT). The characters are generated in much the same way as images on a television screen. The characters may be stored on master character sets as with conventional phototypesetters, or the character storage may be digital. In digital storage, each character is described by a number code which gives the position of the character's various parts. The code is read by a computer, and the computer then activates the character generation system. The image is then exposed to film or paper.

Still newer phototypesetters use lasers to expose light-sensi-

24-11. *Computers with word processing capabilities are being used by some publishers to compose type.*

tive film or paper. A computer controls the movement of the laser beam. As the beam moves quickly back and forth across the film or paper, the type characters are formed, dot by dot. Such machines can set over 500 characters per second.

Phototypesetting machinery may be operated manually using a keyboard, by instructions coded on punched paper tape, or by a computer tape or disk. Whatever the input source, its purpose is to tell the phototypesetting machine which characters to set, what the typeface and size should be, and what the line length and spacing should be.

Many units have a video display terminal (VDT) which shows the type that has been composed. Such a feature enables the operator to examine composed lines and make corrections or other changes.

In today's electronic composition systems, the functions of composing type and setting type may be performed by two different machines. For example, the words for this book were composed on a computer with a word processing program. Fig. 24-11. Special codes were also entered to specify typeface, spacing, etc. The computer stored all the information digitally on floppy disks. These disks were then used with phototypesetting equipment to generate the words onto film.

The entire composing-typesetting operation is becoming more automated. Today there are scanners which can "read" properly prepared typed copy and convert it into code for a typesetting machine. The typesetting field is very complex, and it is changing rapidly. As with so many other fields, the computer has played a key role in changing the way type is composed. Fig. 24-12.

24-12. *This diagram shows how one large newspaper sets copy automatically with computerized equipment.*

Courtesy *Los Angeles Times*

24-13. *A process camera. These cameras are usually built into a wall so that the rear of the camera, which is the part where the film is placed, opens into the darkroom.*

24-15. *An example of continuous tone art.*

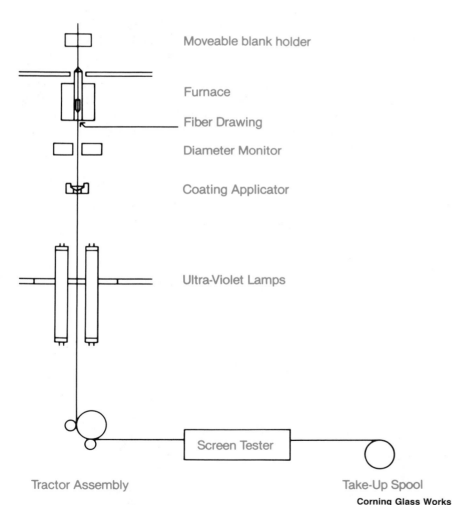

Moveable blank holder

Furnace

Fiber Drawing

Diameter Monitor

Coating Applicator

Ultra-Violet Lamps

Screen Tester

Tractor Assembly

Take-Up Spool

24-14. *This schematic showing the production of optical fibers is an example of line art.*

PHOTOGRAPHY

So far we have discussed how type is composed. But printed items often include pictures as well as words. One way to convert pictures into a form that can be printed is to carve them into a linoleum block. They can then be used for relief printing. Another way is to make a lithograph as discussed in Chapter 23. Obviously these methods are slow and impractical for today's high-volume publishing. Today pictures are prepared for printing using photographic techniques. The camera used is a *process camera*. This camera has special lenses for photographing flat copy to make sharp negatives. By using this camera, the image may also be either enlarged or reduced to the desired size. Fig. 24-13. Industry uses computers to determine exposure times and get the best possible reproduction.

There two kinds of pictures: line art and continuous tone art. *Line art* is made up of solid lines. Fig. 24-14. There is no shading, no gradation of tone. Mechanical drawings are line art. So is hand lettering. Type copy and pre-printed display lines and borders can also be treated as line art.

Continuous tone art has gradations of tone. A black and white photo, for example, has tones ranging from black to white with varying shades of gray in between. In printing, however, there can be no tonal gradations.

Either there is ink on the paper or there isn't. Therefore continuous tone art must be *screened*. This means a screen is placed between the film in the camera and the art to be photographed. The screen breaks up the image into dots which can be printed. A picture which has been screened is called a *halftone*. If you look closely at the pictures in a news-

paper, you will see that they are halftones. The dots are larger and closer together in the darker parts of the pictures and farther apart in the lighter areas. This creates the illusion of shading. The photographs in this book have been screened, too, but it is not as easy to see the dots because a finer screen was used. Fig. 24-15.

QUESTIONS AND ACTIVITIES

1. Define composition. Briefly describe the two general categories of composition.
2. What is a pica? A point?
3. Describe foundry type and tell what kind of composition it is used in.
4. What is the difference between a type bank and a type case?
5. Describe leads and slugs and tell what they are used for.
6. What does one em equal? What is an em quad and what is it used for?
7. Basically, what is done in mechanical cold composition? Explain the advantage of using mechanical cold composition instead of hot composition.
8. Describe three methods of mechanical cold composition.
9. What does "justifying" lines mean?
10. Briefly describe photocomposition.
11. What is the difference between line and continuous tone art? Give an example of each.
12. Describe how halftones are made.

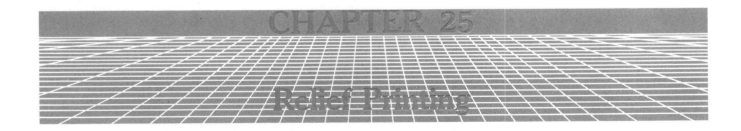

CHAPTER 25

Relief Printing

In the last chapter, you learned about composition. The next step in the printing process is to prepare the images (type and illustrations) so that they will carry ink to the paper or other surface. In relief printing, such as letterpress, the ink-carrying surface must be raised. Images composed from hand-set foundry type, machine-set hot metal type, or linoleum-block carving are already

raised. The flat images produced by cold composition methods can also be used for relief printing. However, they must first be converted to plates that have raised images.

USING HAND- OR MACHINE-SET TYPE

Hand- or machine-set type must be tied (see Chapter 24) and locked in a special frame called a

chase before it can be put on the press. Before locking up the type-form, however, a *proof* (sample copy) is made ("pulled") to check whether everything is spelled correctly and positioned properly.

Making a Proof

To pull a proof, place the galley on the bed of the proof press. Fig. 25-1. The open end of the galley should be toward the cylinder. Be

25-1. *One type of proof press.*

Vandercook Co.

sure the ends of the string are not under the form.

Place a small amount of ink on the ink plate. Distribute the ink evenly over the plate using a brayer. Then ink the typeform by rolling the brayer lightly over the form as shown in Fig. 25-2.

Place a sheet of paper carefully on the typeform. Roll the cylinder over the form to transfer the ink to the paper. Do not roll the cylinder more than once over the form because the print may be blurred.

Before removing the galley from the proof press, use a cloth and solvent to clean the ink from the face of the type. **CAUTION:** *Be sure to place the used cloth in a safety can.*

Remove the galley with the form and place it back on the type bank. Read the proof and replace wrong or damaged type.

Tie up the form and make another proof. If it is clean (no mistakes or broken type), the typeform is made ready for the press.

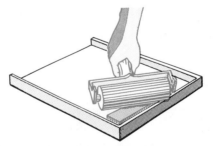

25-2. *Inking the typeform with a brayer.*

This process is called imposing the form. During the imposing process, the typeform is locked up in a *chase,* a metal frame which holds the type on the press.

Locking Up the Form

To lock up a form for the press, follow these steps:

1. Clean the top of the imposing surface (a smooth metal or stone top table).

2. Slide the typeform from the galley to the imposing surface.

3. Place the chase over the typeform.

4. Move the form to the proper position within the chase. The long side of the form should be parallel with the long side of the chase. The head of the form should be to the bottom or left of the chase. Fig. 25-3.

5. Place furniture around the form. *Furniture* is the larger pieces of spacing material, made of wood. Each piece of furniture should overlap, or "chase," the other piece around the form. Fig. 25-4.

6. Place quoins at the top and right of the form. *Quoins* are blocks that can be expanded with a key. Fig. 25-5.

7. Fill in the rest of the chase with enough furniture to hold the

25-3. *Position of the type in the chase. The head is at the wide end. Therefore it is placed towards the bottom of the chase.*

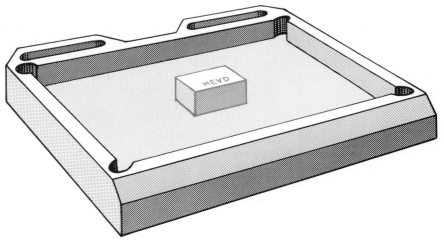

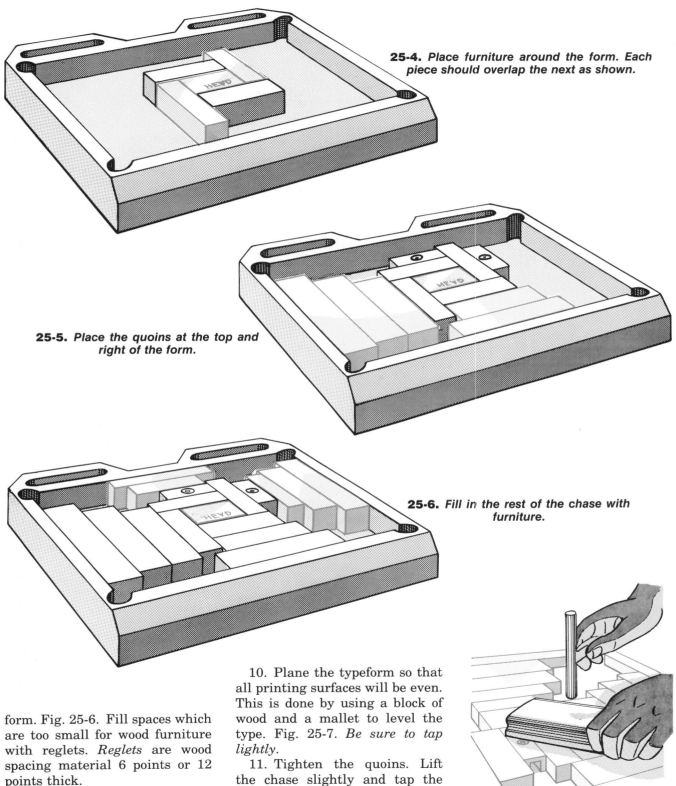

25-4. *Place furniture around the form. Each piece should overlap the next as shown.*

25-5. *Place the quoins at the top and right of the form.*

25-6. *Fill in the rest of the chase with furniture.*

25-7. *Plane the typeform with a wooden block and a mallet.*

form. Fig. 25-6. Fill spaces which are too small for wood furniture with reglets. *Reglets* are wood spacing material 6 points or 12 points thick.

8. Remove the string from the form.

9. Lightly tighten the quoins.

10. Plane the typeform so that all printing surfaces will be even. This is done by using a block of wood and a mallet to level the type. Fig. 25-7. *Be sure to tap lightly*.

11. Tighten the quoins. Lift the chase slightly and tap the type with your fingers. If there is no loose type, the locked form is ready for the press.

The Platen Press

As stated earlier, letterpress printing is done from raised surfaces like the typeform you are now ready to print. The platen press is one kind of press for printing typeforms. There are two types of platen presses: the hand lever, Fig. 25-8, and the power press. Some power platen presses have automatic feeders which feed the printing paper to the press.

FOR YOUR SAFETY . . .

The platen press can cause serious injury if it is not operated correctly and carefully. Always follow your teacher's instructions and these safety rules:

• Before starting, tuck in loose clothing, roll up long sleeves, and remove jewelry. These can get caught in the press, causing you to be pulled in, too.

• Never reach into a moving press to apply ink, make adjustments, or remove sheets.

• Only one person at a time should operate the press.

• Cloths soaked with solvent or ink are flammable. They must be placed in a metal safety can.

Preparing the Press

The first step in printing your typeform is to *dress* the platen. This is the operation of covering the platen with packing and tympan paper, a tough, oily paper of uniform thickness.

First, remove the old packing, which is held on the platen by two bales, one at the top and the other at the bottom of the platen. Next, cover the platen using two sheets of book paper and one pressboard as packing. These should be cut slightly smaller than the platen. The amount of packing will vary from press to

25-8. Hand lever press.

press; so check with your instructor about how much packing should be used. Cut the tympan paper so that it is as wide as the platen and long enough to extend under each of the bales. Clamp the lower bale over the tympan sheet. Pull the tympan tight and clamp the upper bale. Fig. 25-9. Be sure the pressboard is not clamped under the bale.

Next, ink the press. This is done by placing a small amount of the desired color ink on the left side of the ink disk. Turn the press and allow the ink to distribute evenly. NOTE: If the form is placed on the press before the ink is well distributed, the letters will fill up with ink.

After the press is inked, lift the chase with the typeform into the bed of the press. The quoins should be up and to the right. Fig. 25-10. Let the ink rollers go over the form a few times to ink it.

Make sure the grippers are to the edge of the platen. Take a trial impression on the tympan. Using a line gauge, measure and mark where the gauge pins are to be placed. Place two gauge pins at the bottom of the sheet and one on the left side. Fig. 25-11. Be sure the gauge pins' points come back through the tympan.

Now place a piece of paper against the gauge pins. Turn the press and print a copy. Check the position of the gauge pins by measuring the print on the sheet. Fig. 25-12. If necessary, move the

25-9. Pull the tympan tight and clamp the upper bale.

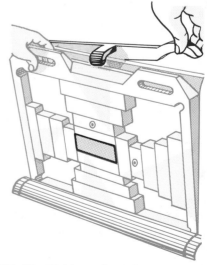

25-10. Placing the chase in the press. Lift the chase up and over the ink roller. Be sure the chase is clamped tightly.

gauge pins until the printing is in the proper position on the sheet. Remove the ink from the tympan paper so it does not transfer to the back of the papers as they are being printed. Moisten a cloth with solvent and wipe off the ink, then apply talcum powder or chalk dust to absorb any moisture.

Printing

Before starting production, some final adjustments in the packing may be necessary to make the entire impression even. This process is called *makeready*. Make an impression on a sheet of paper to be printed. Check the printing for low places. If the impression is too light, additional packing may be required. If the impression is too heavy, remove some packing. Also check the amount of ink. Too much ink means an easily smudged print. Once all the makeready is completed, the job is ready to run. Now the desired number of copies may be printed. Follow your teacher's instructions for operating the press.

After the copies have been printed, remove the chase from the press and clean the ink from

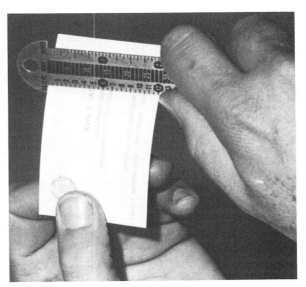

25-12. *Checking with a line gauge to make sure the printing is straight.*

the type. Clean the press by first cleaning the ink disk. Next, clean the ink from the rollers. Unlock the typeform and return all lock-up material to the proper place. Now sort the type back into the proper case. Be sure all tools and materials are returned to their storage places.

LINOLEUM BLOCK PRINTING

Another form of relief printing uses linoleum block in place of type. With the linoleum block printing process, you can make

interesting designs on paper or cloth. The first step is to make a *layout*—a drawing of the design to be carved in the block. Since this is a relief printing process, the design on the linoleum block must be drawn in reverse. To do this, draw the layout on tracing paper. This way the design can be seen from the back of the paper.

To transfer the design to the block, first tape a piece of carbon paper to the face of the block. Then tape the layout face down on the carbon paper. Now trace around the design; the carbon will transfer to the block. Fig. 25-13. The designs for linoleum blocks should not have fine lines because these break easily and are hard to carve.

Remove the layout and the carbon. The block is ready to carve. Fig. 25-14. Place the block in a bench hook. Be sure to keep your fingers behind the block when carving. With the V-shaped block-cutting tool, outline the design. After the design has been outlined, use the larger tool—the gouge—to remove the unwanted linoleum.

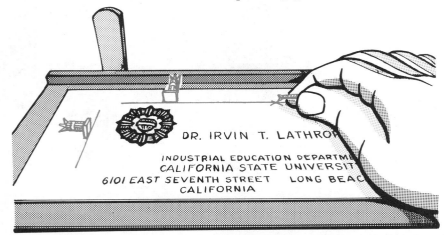

25-11. *Place two gauge pins at the bottom and one at the left side. Insert the gauge pins to the line.*

25-13. *Tracing the design taped over the linoleum block.*

mounted to type-height on a block and proofed on a proof press.

The photoengraving can be used for printing, but usually a duplicate plate is made.

DUPLICATE PLATES

Original typeforms and photoengravings are expensive. Most printers do not use original plates on the press. Instead they use duplicates. Molds are made from the original plates, and the duplicate plates are cast from these. Three types of duplicate plates are stereotypes, electrotypes, and rubber plates. *Stereotypes* are produced from paper molds. They don't reproduce fine detail. Stereotypes are often used in newspaper printing. *Electrotypes* are high-quality duplicate plates made from metal molds. They are used for letterpress printing of books and magazines. Rubber duplicate plates are used in most printing on paper bags, cellophane, and plastic. One plate is good for a million or more uses. Printing from rubber plates is called *flexography*.

To print, place paper over the inked block and apply pressure. The best way is to place the block in a platen press and apply pressure. Another way is to apply the ink, place the block on a piece of paper on the floor, and stand on the block. Still another way is to place the inked block on a piece of paper and press with a rolling pin.

light-sensitive coating. Thus the image on the negative is reproduced on the plate. The plate is developed to fix (set) the image and to wash the coating off the unexposed areas. Next the plate is put in an etching bath. The etching solution eats away the upper portions of the unexposed areas. The image areas remain as a raised surface. The plate is

PHOTOENGRAVINGS

Letterpress plates are either original or duplicate plates. You already know about some kinds of original plates—the hand-set or machine-set typeforms and linoleum blocks. Another type of original letterpress plate is the *photoengraving*.

The photoengraving process converts pictures into raised printing surfaces. First, the original drawing or photo is photographed with a process camera. The resulting negative is placed face down on a sensitized metal or plastic plate and exposed to a strong light. The light passes through the transparent areas of the negative onto the plate's

25-14. *Cutting a linoleum block.*

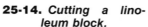

QUESTIONS AND ACTIVITIES

1. What are proofs and why are they needed?
2. What is done during the imposing process?
3. Define each of the following printing terms: images, chase, quoin, furniture.
4. What are three safety rules to remember when using the platen press?
5. What is "dressing" the platen?
6. What is done in the process called make-ready?
7. Briefly describe the steps you would follow to make your own linoleum block print.
8. What is the purpose of photoengraving?
9. Very simply, how are duplicate plates made? Name the three types of duplicate plates and tell what each is made from and what each is used to print.

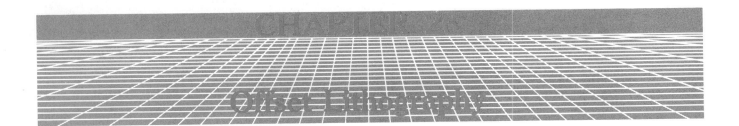

Offset Lithography

As you remember, lithography is printing from a flat surface rather than a raised one. In offset lithography, the image is transferred from a plate to a rubber blanket on a cylinder and then to the paper. The image is placed on the plate in much the same way a photograph is made. Before the image can be put on the plate, film negative (or, in some cases, a film positive) must be prepared.

PREPARING COPY

As explained in Chapter 24, there are two kinds of copy (material to be printed): line copy and continuous tone copy. The two types of copy must be prepared separately for printing.

Line Copy

There are several ways to prepare line copy for offset printing. One way to prepare the type is to hand set it, take a proof (called a *reproduction proof),* and then photograph the proof. Another method is to set type with a machine such as a Linotype, proof it, and then photograph that.

By far the most copy set for offset printing is produced by cold composition. Cold composition may include copy made by using hand lettering or line drawings, preprinted type or drawings, or a typewriter. Today almost all copy for offset printing is set by some form of phototypesetting. The end result of all these cold composition methods is line copy on paper or film. If it is on paper, the copy is photographed to obtain a negative or positive for platemaking.

Continuous Tone Copy

All continuous tone copy must be screened. Otherwise, it would not reproduce well in print. Copy which is screened is called a halftone. Because continuous tone copy is screened, it is photographed separately from line copy.

After film of line copy and continuous tone copy has been made, the two sets of film are put together in making the flat.

MAKING THE FLAT

The negatives (or positives) are positioned and taped onto a special masking paper. This process is called *stripping.* Fig. 26-1. The completed assembly of film and paper is called a *flat.*

One type of paper used in making the flat is goldenrod. Goldenrod is an opaque yellow paper that has guidelines on it to help in positioning the film. The arrangement of the type and illustrations follows the layout that was prepared earlier, usually during the planning of the job.

The taping of the negatives is done on a light table. A light table consists of a large piece of frosted glass with a light underneath. The light table provides a smooth, flat surface on which to work. Fig. 26-2.

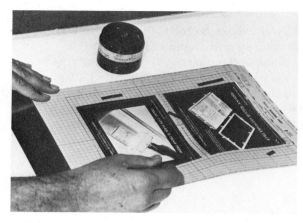

26-1. *Stripping negatives onto masking sheet. Note the bottle of opaquing fluid.*

26-4. *Developing an offset plate.*

After the negatives have been taped in place, the paper is turned over and cut so as to leave the printing area of the negative clear. All other areas remain covered by the paper. No light must go through any place but the image area. Unwanted holes and scratches are blacked out with opaquing solution. The flat is now ready to use for making the offset plate.

PREPARING THE OFFSET PLATE

Several kinds of plates are used for offset lithography. The *presensitized* plate is a metal, paper, or plastic plate that has been coated to make it sensitive to light. The flat is placed on the plate and the two are put in a platemaker. Fig. 26-3. The platemaker has a vacuum frame to hold the flat and plate in tight contact. It also has a bright light. The plate and flat are exposed to the light. The light goes through the transparent areas of the flat and strikes the light-sensitive coating of the plate.

After exposure the plate is developed using the proper solution for the type of plate. Fig. 26-4. Most offset solutions include an etch and a developer. If the plate is not to be used right away, it is covered with a solution of gum arabic to protect it.

PRINTING

The actual printing procedure may vary somewhat depending on the press that is used. Fig. 26-5. Here are the general steps. To operate the press in your school

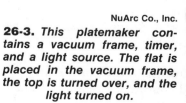

NuArc Co., Inc.
26-3. *This platemaker contains a vacuum frame, timer, and a light source. The flat is placed in the vacuum frame, the top is turned over, and the light turned on.*

NuArc Co., Inc.
26-2. *The light beneath the work surface makes it easier to see the image areas of the film.*

AM International

26-5. *A modern offset press. For small jobs, offset duplicators are used. They are similar to offset presses, but smaller and simpler.*

shop, follow your teacher's instructions.

1. Place ink into the ink fountain.

2. Put fountain solution in the water reservoir. At this point you might run the press to distribute the ink on the ink rollers and to moisten the fountain rollers.

3. Place paper in the paper feeders and adjust so that the paper feeds properly into the press.

4. Fasten the plate to the press. Begin by attaching the lead edge of the plate to the clamp and turning the press by hand. Fig. 26-6.

Fasten the tail of the plate to the clamp, as shown in Fig. 26-7.

Tighten the clamp until the plate is held tightly. Do not tighten too much, or the holes in the plate may be torn.

Remove the gum arabic from the plate with a small sponge that has been moistened with fountain solution.

5. Adjust the delivery end of the press so that the paper comes out of the press and stacks neatly.

6. Proceed to print the desired number of copies. After all of the copies have been printed, the plate must be removed and the press cleaned. Both the ink rollers and the fountain should be cleaned, following the directions for the press being used.

If the plate is to be stored for use again, it should be coated with gum arabic to preserve the surface.

Addressograph-Multigraph

26-6. *Attach the lead edge of the plate to the plate clamps. Turn the hand wheel to roll the plate around the cylinder.*

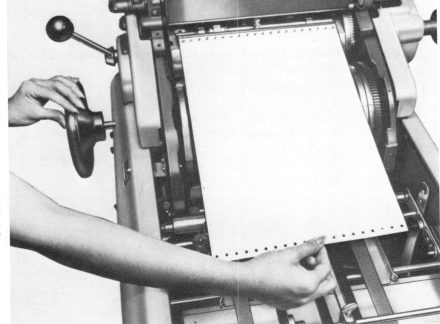

FOR YOUR SAFETY . . .

Observe these safety rules when operating the offset press:

- Tuck in loose clothing, roll up long sleeves, and remove jewelry.
- Except for normal operating functions, the press should *not* be adjusted while it is running.
- Do not run the press at a faster speed than you can handle comfortably. Use a speed that keeps you in full control.
- Cloths that have solvent or ink on them must be put in metal safety cans.
- Do not operate the press until you have received instructions from your teacher and are sure you know what to do.

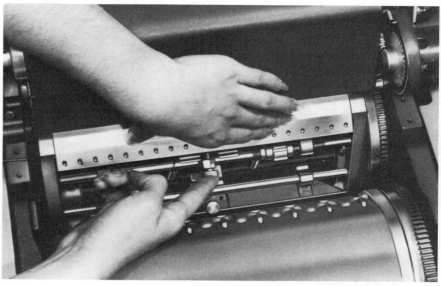

Addressograph-Multigraph

26-7. *Fasten the tail of the plate with the plate clamp tightening screw.*

COLOR PRINTING

Color printing requires a separate plate for each of the four process colors. *Process colors* are the primary colors used in printing: magenta, cyan, yellow, and black. Four negatives are required to make the plates. Each negative contains a dot pattern for one color. The full-color original is photographed four times through different filters and with different halftone screen angles to produce the negatives. This process of making screened negatives from the full-color, continuous tone original is called *color separation.*

Plates are made from the negatives. These plates are then printed, one after the other, on the same sheet of paper. The right combination of yellow, magenta, cyan, and black inks can produce any color of the spectrum.

QUESTIONS AND ACTIVITIES

1. Define copy. Describe the two kinds of copy.
2. Describe the process called stripping.
3. What is a flat?
4. When and why is opaquing solution used?
5. What are three safety rules to remember when using the offset press?
6. What are process colors? Name them.
7. Define color separation and briefly describe how it is done.

As you remember from Chapter 23, screen process printing is really printing by forcing ink or paint through a stencil which is attached to a screen that has been stretched over a wooden frame. The equipment needed for screen printing is not expensive. In fact, with a little skill in woodworking, much of it can be homemade. There are also supply houses which furnish complete kits ready for printing.

EQUIPMENT

Printing Frame

The printing frame may be made in the school shop or at home, or it may be bought. The size of the frame depends on the size of the designs to be printed. Most schools and commercial shops have many sizes of frames for printing many different designs. If you build a frame, it should be about six to eight inches larger than the largest stencil to be used. The extra space is needed as a reservoir for the ink when printing. For a school, the 8″ × 10″ and 10″ × 15″ frames would probably be best. These would be large enough to print posters, but not so large as to waste stencil material or ink. Some schools have one or two very large frames for making large posters and many smaller frames for student use.

The printing frame consists of four pieces of wood joined together at the corners. These form an open rectangle over which the screen is stretched. Usually the frame is made of pine or basswood strips about 1½ inches wide. Whatever wood is used, the frame must be rigid and strong enough to withstand the stretching of the screen. The frame is attached with hinges to a flat base.

Screen

The screen serves as a base for mounting the stencil and as a means of controlling the ink flow. The larger the openings in the screen, the more ink will reach the printing surface.

Although silk is most often used, the screen may also be made of organdy, nylon, or a metal such as stainless steel. The mesh openings vary in size, and the size used depends on the design that is being printed. Coarse screens are used for large, heavy designs. Fine screens are used for designs that are small and detailed.

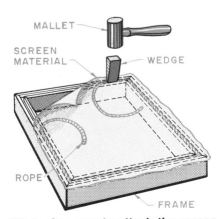

27-1. *One way to attach the screen to the frame is with a cord (rope).*

MALLET

SCREEN MATERIAL

WEDGE

ROPE

FRAME

To attach a screen to the frame, remove the frame from the base and turn the frame over on a table. Spread a piece of silk (or other material) over the frame so that the threads run parallel to the sides of the frame. Cut the silk so that it is about one inch larger than the frame on all four sides.

There are two common ways to fasten the screen to the frame. One way is to use tacks or staples. If the frame has a groove around it, the silk may be attached with a cord wedged into the groove over the silk. Lay the silk on the frame and start forcing one end of the cord into the groove with your fingers. Start at a corner. Pull the silk fairly tight as you force the cord into the groove. When the cord is in the groove on all four sides of the screen, force it further down into the groove using a mallet and a blunt wooden wedge. This method will hold the silk tighter than the tacks. Fig. 27-1.

After the screen is fastened to the frame, it should be washed in warm water with a detergent to remove any sizing it may contain. The frame is then attached again to the base.

Squeegee

A squeegee is used to push the ink through the open mesh of the screen onto the printing surface. It has a rubber blade and a wooden handle. It should be about two inches shorter than the inside width of the frame.

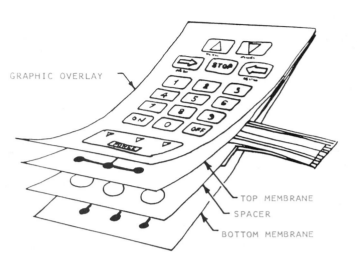

GRAPHIC OVERLAY

TOP MEMBRANE

SPACER

BOTTOM MEMBRANE

Membrane Switches

Screen printing is an old process which is finding new applications in today's technology. One example is membrane switches. These finger-touch switches are used on the control panels of many electronic devices in place of conventional mechanical switches.

To produce a membrane switch, two pieces of polyester film are screen printed with electrically conductive ink to form the circuit for the switch. A polyester spacer, die-cut at contact points, is sandwiched between the two printed circuit films. The die-cut holes allow a touch of a finger to close the contacts, and this in turn completes the circuit. When the finger is lifted, the circuit is broken. A graphic layer is added to the top to allow the user to identify the different switches.

Some of the advantages that membrane switches have over ordinary mechanical switches are:

—They are easier to manufacture and therefore less expensive.

—There are no mechanical, moving parts to wear out.

—They are sealed to keep out dust, dirt, and moisture.

—They are easier to clean because there are no projecting parts.

—They consume little electrical energy.

—They are bright, attractive, and easy to use.

—They are flat and compact and easier to apply to electric or electronic equipment.

Squeegees can be bought at art supply stores.

Stencil

Stencils for screen process printing may be made from many materials. The easiest and least expensive is paper. A simple stencil may be made from a good grade of bond paper.

The stencil paper should be large enough to cover the entire screen. Trace the design onto the paper from the layout. With this type of stencil material, designs with loose pieces—for instance, the letter O—should be avoided.

Cut the design in the paper,

27-2. *Cutting a paper stencil.*

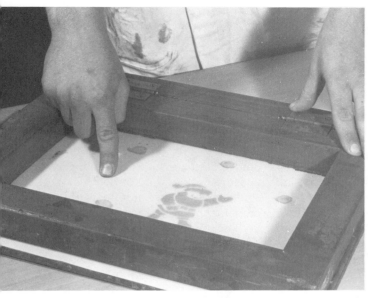

27-3. *Glue the stencil to the screen with water-soluble glue.*

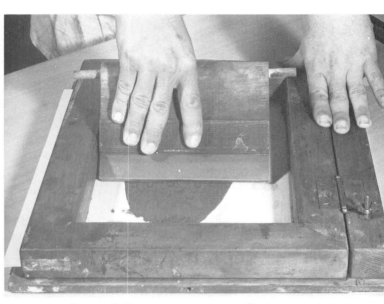

27-4. *To print, pull the squeegee across the screen at about a 45-degree angle.*

using a sharp knife. Be sure to use a piece of cardboard or other material under the stencil to avoid cutting the tabletop. Cut all the way through the stencil material. Be careful not to cut yourself. Fig. 27-2.

There are other ways to make stencils. In schools hand-cut lacquer film is often used rather than paper. There are also photographic methods of making stencils.

MAKING A SCREEN PROCESS PRINT

After the stencil is prepared, it is attached to the bottom of the screen with water-soluble glue or with tape. Fig. 27-3. If the stencil material is not large enough to cover the entire screen, the areas around the stencil are covered with a blockout material such as glue or masking paper. This stops the paint from going through the screen at any place but the design.

To print the design, place a sheet of paper on the base of the frame. Put a small amount of ink into the screen frame and pull the squeegee across the screen. Fig. 27-4. This forces the ink through the cut-out parts of the stencil and onto the paper.

Screen process printing has several advantages. First, a fairly heavy layer of ink is pushed through the screen. This means a light color can be printed over a dark color. Also it is possible to print on almost any material or object that can be placed directly under the screen. The screen process is used for printing on such things as bottles, sweat shirts, and T-shirts and for printing panel markings for radios and automobiles, posters, drinking glasses, and decals. It is a small but important part of the graphic arts.

QUESTIONS AND ACTIVITIES

1. Describe the equipment needed for screen process printing and tell the function of each.

2. What is the screen made of? What type of screen is used for large, heavy designs? For small, detailed designs?

3. Describe how to make a screen process print once all the equipment is prepared.

4. What is a membrane switch? What are three advantages of using these switches over conventional mechanical switches?

CHAPTER 28

Taking and Developing Pictures

In the preceding chapters you learned about photography for the graphic arts. You read how process cameras are used to photograph flat line or continuous tone copy in preparation for making plates. This chapter will discuss a more familiar type of photography—taking pictures of three-dimensional subjects.

Photographs are made by the reflection of light from the subject being photographed to a light-sensitive film. Bright areas reflect more light than dark areas. The film is therefore exposed in direct relation to the amount of light. The camera which holds the film is basically a lightproof box with a small, glass-covered opening to admit light and a shutter to close the opening after the film has been exposed.

TYPES OF CAMERAS

There are many types of cameras in common use today. The simplest is the box camera. The box camera does not have a way to control exposure or change focus. The shutter speed is slow, but under the right conditions the box camera will take acceptable pictures. The highly popular pocket-size cameras are examples of modern box cameras. Fig. 28-1.

Two other cameras in popular use today are the thirty-five millimetre and the twin lens reflex camera. The 35 mm cameras are popular because of their small size and great depth of field. (Objects will be in focus from close up to far away.) Thirty-five millimetre cameras are basically of two types: the kind with a rangefinder and the single lens reflex. Those cameras with a rangefinder use separate lens systems for viewing and for taking the picture. Fig. 28-2. The single lens reflex (SLR) camera uses a mirror and lens system so that the viewing is done through the same lens that the picture is taken with. Fig. 28-3. The mirror

moves out of the way during exposure. Thus you cannot see the subject you are photographing during the actual exposure time.

The twin lens reflex camera has two lens systems: one to take the picture and one to view the subject. When the image is sharp in the viewfinder, it will be just as sharp on the film. Fig. 28-4.

TAKING PICTURES

No matter how expensive or complex a camera is, it cannot take good pictures unless certain procedures are followed.

● The most expensive camera in the world will not give you a good picture unless the camera is held steady. When you are taking pictures, squeeze the shutter release gently so that you don't move the camera. For long exposures, place the camera on a tripod or other nonmoving base.

● To avoid light streaks and fogged negatives, the camera must be correctly loaded away from bright light. Each type of camera loads differently; so be sure to read and follow the directions for the camera you are using.

● There is no substitute for correctly exposed film. On very simple cameras, the exposure controls are preset at the factory. Pictures can be taken only in fairly bright light. On more complex cameras, it is possible to set the exposure for existing light conditions. Some cameras give you a choice of automatic or manual settings.

Eastman Kodak Company

28-1. Disc cameras are one type of modern box camera. Focus and exposure are preset at the factory. The photographer simply snaps the shutter.

GAF

28-2. *Example of a 35 mm camera with a separate rangefinder.*

28-3. *Common 35 mm SLR camera.*

The f-stop and the shutter speed control exposure. The *f-stop* refers to the size of the lens opening. Most of the more complex cameras have an f-stop range of 16, 11, 8, 5.6, 4, 2.8, 2, and 1.4. The higher the number, the smaller the lens opening. Therefore the higher numbers are used when taking pictures in bright light. Higher f-stops also give greater depth of field.

Shutter speed is the amount of time that light is allowed to pass through the lens and strike the film. The shutter speeds on most of the more complex cameras are $1/1000$, $1/500$, $1/250$, $1/125$, $1/60$, $1/30$, $1/15$, $1/8$, $1/4$, $1/2$, and 1 second. Some cameras also have a "B" setting. This keeps the shutter open for as long as the shutter release is held down.

The more light, the faster the shutter speed can be. Faster shutter speeds are also needed when taking pictures of a moving subject. Otherwise, the pictures will turn out blurred.

The combination of f-stop and shutter speed that will be used depends on the amount of light that is reflected from the subject being photographed and on the type of film being used.

In very bright light, you would use a high f-stop number and a high shutter speed. In low light,

you would use a low f-stop number and/or a slow shutter speed. Remember, though, as the f-stop numbers decrease, so does the depth of field. If you want things both close up and far away to be in focus, you must use a higher f-stop. To compensate for the smaller lens opening, you can decrease shutter speed. But longer exposures can result in blurred pictures if the camera or the subject moves. To get the best exposure for a particular picture, you must know what combination of f-stop and shutter speed to use. Modern cameras usually have built-in light meters which help you determine the correct settings.

● It is important to use the right film. First, of course, you must buy the film that will fit your camera. Simple cameras use "126" or "110" film. The 35 mm

28-4. *A twin lens reflex camera.*

cameras use 35 mm film. Twin lens reflex cameras use a slightly larger film size. Film has an ISO rating. This rating (formerly called the ASA rating) indicates the film's sensitivity to light. The higher the number, the more sensitive the film. Films with higher numbers are said to be "faster" because they require shorter exposure times.

● Another factor to consider when taking a picture is focusing. Focusing is the process of selecting the correct subject-to-camera distance. On some cameras the focus is preset. Only subjects within a certain range, say 5 to 15 feet, will be in focus. On more expensive cameras there are ways to adjust the camera for various distances. This is done with the help of rangefinders or ground-glass focusing.

There are two types of rangefinders. The *split image* rangefinder splits the image in half. You look through the viewfinder and turn the camera's focusing ring until the two halves align. The *coincidental* rangefinder uses a superimposed image. When the subject is out of focus, the same

28-5. *Common type of development tank.*

image appears twice. When the subject is in focus, only one image appears in the viewfinder.

With ground-glass focusing, you simply turn the focusing ring until the image in the viewfinder looks sharp.

DEVELOPING FILM

After you have exposed a roll of film in your camera, what is known as a "latent image" has been formed on the film. The image is not visible. The film must be processed, or *developed,* to make the image appear.

The most common method of developing film in homes and schools is the tank method. The tank method is popular because once the film is in the tank, the rest of the procedure may be done under ordinary room light. Figure 28-5 shows a development tank. There are three types of tanks in common use today. These are the plastic apron, the plastic reel, and the metal reel. The plastic reel is the easiest to load and the metal reel the most difficult. Whichever type of tank you intend to use, practice loading unexposed film in the light before trying to load your film in the darkroom. Fig. 28-6.

Correct development is a must for obtaining a usable negative.

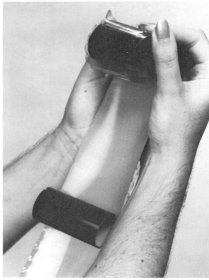

28-6a. *Loading for a plastic apron development tank.*

The time and temperature requirements vary with the type of film and the developer. Be sure to follow the manufacturer's instructions, listed on most film packages.

Even after the film has been developed and the image is visible, it is not permanent until the unexposed and undeveloped silver salts have been removed from the film. This is done with an acid fix solution. However, because the developer is alkaline and the fixer is acid, the film must first be washed in a solution called a stop bath. Otherwise, the fixer would be contaminated with developer.

After the film has been fixed, all of the chemicals must be washed from the film with running water. To avoid water spots,

the film is dipped in a wetting agent, then hung up to dry in a dust-free place. After the film is dry, it should be stored in a protective cover such as a glassine envelope.

Black and White

The following step-by-step procedure is recommended for developing black and white film by the tank method.

1. Place the reel, the tank, and the lid in a convenient position.
2. Turn the room light off.
3. Carefully remove the paper

28-6b. *Loading a plastic reel.*

28-6c. *Loading a metal reel.*

backing from the film. Be sure to touch the film only by the edges. There is also about one inch at the beginning and at the end of the film which may be safely handled without danger of touching the image area.

4. After the reel is loaded, place it in the tank and put the light-tight cover on the tank.

5. Turn on the room lights.

6. Select the correct developer and check its temperature. The temperatures of all three solutions—developer, rinse, and fixer—should be within a few degrees of each other. Refer to a time and temperature chart for the film being developed to find the correct temperature. Make sure you have more than enough developer to fill the tank completely.

With certain types of tanks it might be best to fill the tank with cool tap water. This prewets the film and helps to reduce air bubbles, called "air bells," which may cause spots on the negative. If a rapid developer is to be used, prewetting the film will decrease development time.

7. Set the timer to the correct development time as given by the time and temperature chart.

8. Pour the water from the tank.

9. Pour the developer into the tank as fast as possible. Fig. 28-7. Start the timer as soon as the tank is full.

10. As soon as the timer has been started, rap the tank smartly against the bottom of the sink in order to dislodge any air bells. Agitate (shake) the tank according to the film manufacturer's directions. Improper agitation will cause air bells and streaked negatives.

11. When development is complete, pour out all of the developer. Be sure to pour the developer back into the right

28-7. *Pouring developer into a tank. Fill the tank quickly to ensure even development.*

container. If at any time you think you may have contaminated a solution, throw it away. Or, if you used a developer in a diluted form, be sure to discard it.

12. Pour enough stop bath into the tank to completely cover the film. If you do not have stop bath, use a tap water rinse for 60 seconds after the development. Be sure to agitate while the film is in the stop bath.

13. Pour out the stop bath or rinse water and pour in enough fixing solution to completely cover the film. Again, it is important to shake the tank, especially for the first few seconds. The film should then be agitated at least once every 90 seconds. It should stay in a normal acid fixing solution from 7 to 10 minutes. Then return the fixer to the proper container.

14. Wash the film in the tank for at least 20 minutes using running tap water. After the film has been washed, it should be immersed for at least one minute in a wetting agent. This helps to eliminate water spots and streaks during drying.

15. Hang the film in a dust-free room to dry. Place a weighted clip on the bottom of the film to prevent it from curling. One typical drying arrangement is shown in Fig. 28-8.

16. When the negatives are dry, place them in protective envelopes until you are ready to print them.

Color

Developing color film is similar to developing black and white film, but more steps are involved. The film must be developed, bleached, washed, fixed, washed again, stabilized, and dried. Several manufacturers sell color processing kits which include all the necessary chemicals plus detailed instructions for use.

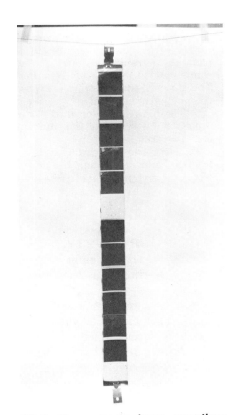

28-8. *One way to hang negatives for drying. Be sure to place a weight on the bottom to keep the negatives from curling.*

Taking Pictures without a Camera

A *hologram* is a three-dimensional picture made on photographic film or plate. Instead of using a conventional camera, holography forms images with the intersecting beams of a laser.

Basically, a hologram is made as follows:

1. The light from a laser is split into two beams: an object beam and a reference beam.

2. The object beam is directed through lenses, which spread it out. Then it is bounced off a mirror onto the subject. The beam reflects off the subject and onto the photographic film.

3. The reference beam is also sent through lenses which spread it, but this beam does not strike the subject. Instead it is sent directly to the photographic film.

4. As you'll recall from Chapter 6, light travels in waves. When the two beams converge at the film, their waves create interference patterns. Where the crests and troughs of the waves are in step, they reinforce each other and create stronger waves. Where the waves are out of phase, the troughs of one tend to cancel out the waves of the other. The interference patterns expose the photographic film.

5. The film is developed to create a permanent record of the interference patterns.

To view the hologram, a laser light is beamed through the developed film from behind. The hologram bends the laser light, re-creating the original waves that were reflected from the subject. Because light waves from every part of the subject were recorded everywhere on the film, the hologram is seen as a three-dimensional object, floating in the air.

When laser light passes through a hologram, it bends in a single direction because laser light is made of only one wavelength. Ordinary white light is made up of many wavelengths. Thus when white light passes through a hologram, each wavelength is bent at a different angle. The different views recorded on the hologram are seen as a blur of colors. To make a hologram viewable in ordinary white light, it is processed to reduce the number of views.

Although holography was invented in 1947, most of its applications have been developed in recent years. You have probably seen holographic stickers. Some credit card companies have begun putting holograms on their cards to make counterfeiting more difficult. Holography is being combined with computer graphics to create three-dimensional images of nonexistent objects. This technology has applications in such fields as industrial design.

Some holograms bend light without creating an image. These are called holographic optical elements. This type of hologram is used to combine laser beams of various wavelengths so that numerous telephone calls can be transmitted at the same time over a single optical fiber. Another application of holographic optical elements is the holographic combiner. This is a transparent plate which is mounted in front of the windshield in an airplane and displays flight path information. Because it is transparent, the pilot can look out of the airplane while at the same time monitoring the flight-path information. Holographic optical elements are also being used in supermarket UPC scanners. They bend the scanner's laser beam to various angles, improving the scanner's ability to read the UPC codes on oddly-shaped packages.

The future promises even more remarkable applications for holography, such as genuine 3D movies and television shows, X-ray holograms of molecules, and holographic data storage for computers. Robots may be equipped with holographic memories. Working on an assembly line, the robots would be able to pick out objects that were identical to the shapes in their memories. These are just a few of the many possibilities of this exciting technology.

FOR YOUR SAFETY. . .

Observe these safety rules when working in the darkroom:

● The darkness itself can cause problems. Let your eyes become accustomed to the darkness before you begin work. To prevent falls, make sure the floor is dry and free of obstacles. Clean up any spills immediately.

● To avoid shock, do not handle electrical equipment when your hands are wet or when you are standing near or on a wet surface.

● Handle chemicals with care. Make sure they are stored in closed containers and properly labeled. Plastic containers are preferred because they don't break. Pour out the chemicals carefully to avoid splashing.

● If you will be mixing chemicals, read and follow instructions carefully so that you will mix the right chemicals in the right proportions. Glacial acetic acid, which is used for stop baths, is particularly dangerous. Always pour acid into water. Water should not be poured into acid because it can cause the mixture to bubble and spatter.

● If you are handling dry chemicals, avoid breathing the dust they create.

● Some people are allergic to the chemicals used in developing film and in making prints. Avoid putting your hands in the chemicals. When doing tray processing, it is a good idea to use tongs. There should be a set of tongs for each tray.

● Wash your hands thoroughly after handling chemicals.

● Dispose of chemicals properly. Follow your teacher's instructions.

● The darkroom should be well ventilated.

Taking Pictures without Film

Work is underway to perfect the filmless camera. Instead of using film, this camera stores images electronically on a disk similar to those used in computers. (NOTE: These cameras are not the same as the pocket-sized disc cameras which use disc-shaped film.) With the use of a playback unit, the images can be transmitted over telephone lines or viewed on a television screen.

The disks are erasable and reusable. One advantage: speed. The camera can take several dozen pictures per second. One disadvantage: the picture quality is not as good as that of a conventional color print. Research is being conducted to improve the quality of the picture and to develop ways to print out permanent copies.

QUESTIONS AND ACTIVITIES

1. Describe the two basic types of 35 mm cameras.

2. What five procedures should you remember to follow in order to take good pictures?

3. What does the f-stop of a camera refer to? What type of f-stop settings are used in bright light?

4. Define shutter speed. When should you use the faster shutter speeds?

5. What does a film's ISO rating indicate?

6. Define focusing. Describe the three types of focusing on the more expensive cameras.

7. Why is an acid fix solution needed in the developing process? Why is a stop bath needed?

8. What are four safety rules to remember when working in the darkroom?

9. What is a hologram? What forms holographic images?

CHAPTER 29

Printing Pictures

After a negative is developed, it is further processed to make the images positive. The images may be on film (slides) or on paper (prints). This chapter will discuss making prints. A print is made by passing light through a negative onto a piece of sensitized paper by either the contact or projection method. By either of these methods, any number of copies can be made from one negative.

Contact printing produces positives of the same size as the negative. *Projection printing* allows you to choose the size of the print. Some projection prints are large enough to cover a wall. These are called "photomurals."

CONTACT PRINTING

A contact print is made by placing a piece of light-sensitive material, usually a printing paper, and a negative in direct contact with each other, emulsion to emulsion. Printing papers come with different kinds of light-sensitive emulsions (coatings). The type of emulsion to choose depends on the method of printing that will be used.

To make a contact print, a device to hold the negative and the printing paper tightly together is required so that a sharp image will be produced. One such device is the printing frame shown in

29-1. *Contact printing frame. The printing paper and the negative are placed in the frame and exposed to light.*

Fig. 29-1. In this frame, the negative and the paper are kept tightly together between a glass plate and a spring-loaded back. The negative is placed emulsion side upward on the printing glass. Then the paper is placed emulsion side downward on the negative, and the back is locked securely in place. The paper is exposed to white light for a set time. After exposure, the paper is processed using developer, stop bath, fixer, and a water rinse.

Making contact prints is quite simple. First set up three trays, as shown in Fig. 29-2. NOTE: Observe the safety rules for working in the darkroom. These rules are listed in Chapter 28.

1. To remove dust, brush the negative carefully on both sides with a negative brush. Place the negative on the printer glass with the emulsion side up. Slide the border mask into position over the edges of the negative if white borders are desired on the print. Be sure that the printer glass is clean and free of lint and dust.

2. Turn on the safelight and turn off the room lights. (A safelight is a lamp with a filter to screen out rays that would harm the light-sensitive paper.)

3. Grade 2, normal contrast paper must be used. With clean, dry hands, open the package of paper and take out one sheet. Cut this sheet into test strips about one inch wide. Then return all test strips but one to the envelope. You will use these test strips to determine the exposure time.

4. Place one of the test strips diagonally across the negative with the emulsion side down.

5. Close the cover of the printer and expose the test strip. The exposure time will depend on the density of the negative and the wattage of the bulb in the printer. There is no standard printing time, but a good starting point is to expose the paper for ten seconds.

6. Slide the test strip into the tray of developer. Agitate the developer by rocking the tray constantly during the development time. Develop the test strip for a full two minutes. This will assure you of the fullest range of tones that the paper can produce.

7. When the two minutes are over, pick the test strip up with tongs and drop it into the stop bath. *Do not put the tongs from the developer into the stop bath.* Agitate the test strip for at least thirty seconds.

8. Using a second set of tongs, remove the test strip from the stop bath and place it in the fixer. Agitate the strip constantly for at least the first ten seconds and periodically after that. After two minutes the test strip may be inspected under room light. If the strip appears too light or too dark, the printing time (exposure time) will have to be lengthened

29-2. *Arrangement of solution trays for print processing.*

or shortened. A longer printing time makes the print darker; less printing time will lighten the print. You may need to do several tests to determine the best exposure time.

9. Repeat the printing process with a full sheet of printing paper. Expose it for the amount of time determined from inspecting the test strip. Process the print in the same manner as the test strip, but leave it in the fixer longer. Prints should be left in the fixer for eight to ten minutes. If prints are left in fresh fixer longer than ten minutes, they will tend to bleach.

10. After the prints have been fixed for the proper time, they must be washed to remove all traces of the processing chemicals, particularly the fixer. All prints should be washed in running water for about one hour.

For this, some form of print washer or a tray siphon is recommended to make sure that the prints are thoroughly washed. Hypo clearing agent may be used to shorten the washing time if desired. Follow the manufacturer's instructions for using hypo clearing agents.

11. Air dry the prints on a photo blotter.

PROJECTION PRINTING

One of the advantages in making your own prints is that you can be creative in the use of light and materials. A print made by projection allows you to change the tone relationships as well as the size of the finished print. By using the many print control methods available, you can create unique and pleasing pictures. Only after you have practiced with the various projection print-

ing control methods will you fully realize the potential of the enlarging process. Fig. 29-3.

Procedure for making projection prints:

1. Set up the three trays for printing as you did for contact printing. The chemical mixing and temperature requirements are also the same.

2. Carefully dust the negative to be enlarged and place it in the negative carrier (Fig. 29-3) emulsion side down. When placing or removing negatives in the carrier, do not pull the negative between the pressure plates or it will be scratched.

3. If you are using an adjustable easel, set the masking guides for the desired border size.

4. Turn on the safelight and turn off the room light.

5. Turn on the enlarger light.

6. Focus the enlarger with the lens wide open until the image is sharp.

7. If the image on the easel is too small, raise the enlarger head and refocus. If the image is too large, lower the enlarger head and refocus.

8. After adjusting the image to the correct size, make sure that you have the sharpest possible focus with the lens wide open.

9. Close the lens down until the fine detail just begins to disappear, but is still plainly visible.

10. Select a sheet of No. 2 paper or place the No. 2 filter in the enlarger. (Remember, a No. 2 paper or No. 2 filter must be used with a normal negative.) Take a sheet of paper from the package and cut it into one-inch test strips and return all but one of these strips to the package.

11. Place one of the test strips on the easel, emulsion side up, in a position to sample the widest possible changes in tone.

12. Cover three-fourths of the

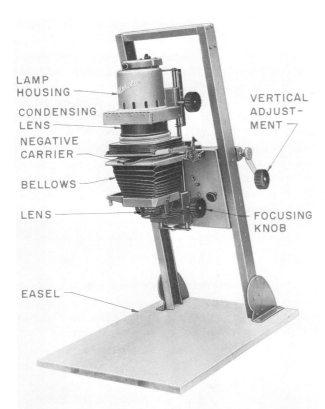

LAMP HOUSING
CONDENSING LENS
NEGATIVE CARRIER
BELLOWS
LENS
EASEL
VERTICAL ADJUSTMENT
FOCUSING KNOB

29-3. *An enlarger for making projection prints.*

29-4. *A test strip.*

29-5. *A finished print.*

test strip with a piece of cardboard. Expose the strip for five seconds. Move the cardboard cover so that about one-half of the test strip is uncovered. Expose it for five seconds. Move the cardboard again to uncover three-fourths of the test strip and expose for five seconds. Uncover the entire test strip and expose for another five seconds. This will fill a strip with exposures of 20, 15, 10, and 5 seconds. Fig. 29-4.

13. Process this test strip in exactly the same way as outlined for contact printing.

14. Inspect the test strip under room light and select the length of exposure which produced the best result. You may find that a time between two of the blocks would yield the best print. If this is the case, make another test strip, exposing for the time that will produce the best print. When you have a satisfactory strip, you are ready to make a full-size print. If the contrast is not satisfactory, a harder or softer grade of paper or another filter should be used.

15. Place a full sheet of paper in the easel and expose for the proper time. NOTE: Before handling any paper, make sure your hands are clean and dry. Process in the same way used for processing the test sheet. Fig. 29-5.

16. The prints may be washed and dried as described for contact printing.

Color Printing

Color printing is similar to black and white printing. You will need an enlarger that can hold color filters or one with a color head (built-in filters). You will also need paper made for color printing and, if the enlarger does not have a color head, you will need filters. After exposure, the paper is developed in seven steps: prewet, develop, stop bath, wash, bleach-fix, wash, and dry. Specific procedures and times vary, so follow the instructions that come with the processing kit.

QUESTIONS AND ACTIVITIES

1. What is the main difference between contact and projection printing?

2. Briefly describe how to determine the proper exposure time for negatives.

3. What is a safelight?

4. What are some advantages of using projection printing instead of contact printing?

Many offices need copies of letters, memos, etc., quickly and inexpensively. Many schools need rapid, inexpensive ways to duplicate lessons and tests. Office duplicating processes and machines fulfill these needs. The most commonly used methods of duplicating are:

- Electrostatic.
- Spirit.
- Stencil, or mimeograph.

ELECTROSTATIC DUPLICATING

Electrostatic duplicating produces copies directly from the original. It uses an electrostatic charge and a dry powder called "toner." The process is often called xerography. An example of a xerographic copying machine is shown in Fig. 30-1.

To make a copy, the original is placed face down on a glass called a platen. Fig. 30-2, #1. Reflected light from the original is measured by a photocell which signals the lens (#2) to open or

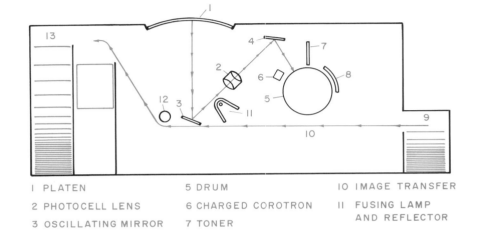

1 PLATEN	5 DRUM	10 IMAGE TRANSFER
2 PHOTOCELL LENS	6 CHARGED COROTRON	11 FUSING LAMP AND REFLECTOR
3 OSCILLATING MIRROR	7 TONER	
4 FIXED MIRROR	8 ELECTRODE	12 COPY BRUSH
	9 PAPER INPUT UNIT	13 OUTPUT TRAY

30-2. *Path of paper through a copier.*

close depending upon the light. The oscillating mirror (#3) scans the original and reflects the image through the lens to the fixed mirror (#4). The fixed mirror reflects the image to the selenium drum (#5) which receives a posi-

tive charge from the corotron (#6).

The positive charge on the drum is in the same pattern as the reflected image. In other words, the positive charge is held on the dark areas of the image, but not on the light areas. Toner (#7), which has a negative charge, is then poured between the drum and the electrode (#8). Because opposite charges attract, the toner is drawn to the positively charged image areas on the drum. The darker the image, the more toner clings to it.

The paper input unit (#9) stores the copy paper, which is always kept at the proper feed level by an automatic elevator. As the positively charged copy paper passes under the drum, the toner is drawn to it. Thus the im-

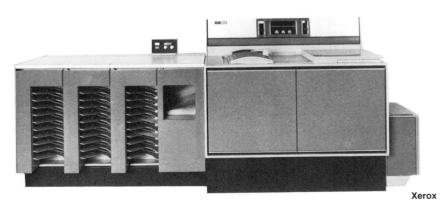

Xerox

30-1. *An electrostatic copying machine.*

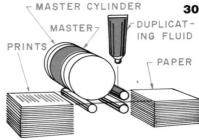

MASTER CYLINDER
MASTER
PRINTS
DUPLICAT-
ING FLUID
PAPER

30-3. *The spirit duplicating process.*

age is transferred from the drum to the paper (#10). The paper is then moved to the fusing lamp (#11) where the toner is heated and fused to the paper. Any excess toner is brushed away by the copy brush (#12). Finished copies are stacked in the output tray (#13).

All this is done within a few seconds. Many of today's machines can also enlarge or reduce the size of the image. Other features include duplexing (printing on both sides of a sheet), collating (stacking pages in proper order), and printing in color.

SPIRIT DUPLICATING

In spirit duplicating, a master which has a carbon image is placed on a cylinder. As paper is fed into the machine, it is moistened by duplicating fluid. When the moistened paper comes in contact with the master, a small amount of carbon is transferred to the paper. Fig. 30-3.

The master may be drawn, handwritten, or typed. A sheet of carbon paper faces the back of the master. Thus, whatever is drawn, written, or typed on the front of the master will show up in reverse on the back of the master. One master may be used to reproduce several colors at one time by using different colors of carbon paper while preparing the master.

When the master is completed, it is placed on a spirit duplicator such as the one in Fig. 30-4, and the desired number of copies is made.

MIMEOGRAPH

The mimeograph process uses a stencil through which ink is forced onto paper. The stencil is placed on the outside of the mimeograph cylinder. As paper is fed through the machine, the impression roller presses it against the stencil on the cylinder. The ink, which is stored inside the cylinder, flows through the openings in the stencil to make the image impression on the paper. Fig. 30-5.

Mimeograph stencils may be typed, handwritten, or drawn. The sheets are made of a fibrous tissue which is coated on both sides. When the image is prepared on the stencil sheet, the coating is pushed aside by the pressure of typewriter keys or stylus, leaving the fibrous base tissue exposed. During the duplicating process, the ink passes through the fibrous tissue wherever the coating has been pushed aside.

ELECTRONIC PRINTING

Many businesses do a large amount of printing—mass mailings, company reports, newslet-

ters, and so on. In recent years some of these companies have begun using electronic printers in place of office duplicating machines or printing presses. In electronic printing the graphic image is generated electronically and sent to an imaging device which prints the image on paper. No photography, film, or printing plates are needed. Various types of electronic printing are now available or soon will be. A few are described here.

● Some kinds of *laser typesetters* (see Chapter 24) qualify as electronic printers since they can form images on nonphotographic materials.

● In *ink jet* printing, a cluster of very small nozzles squirts droplets of ink onto paper. A computer controls this printing unit, directing it to spray ink in various patterns and densities to form symbols and characters.

Ink jet printing can be used to address envelopes for mass mailings, eliminating the need to print separate labels and glue them to the envelopes. A computer data base provides the names and addresses to the jet printer, which sprays them directly onto the envelopes at speeds of over 1000 characters per second.

Ink jet technology is also being applied to computer printers. The high speed and the ability to

Bell & Howell
30-4. *A spirit duplicator.*

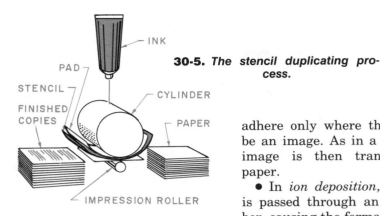

30-5. *The stencil duplicating process.*

print graphics as well as words make ink jet printing useful for CAD/CAM printouts, business charts, and word processing printouts.

• *Laser electrostatic* printers are similar to office copiers. The difference is that no "original" is used. Instead, a computer-controlled laser beam discharges the copying drum so that toner will

adhere only where there should be an image. As in a copier, the image is then transferred to paper.

• In *ion deposition,* electricity is passed through an air chamber, causing the formation of ions (electrically charged particles). At one end of the air chamber is a screen. The holes in the screen are opened and closed by a computer to create a graphic pattern. The ions that pass through the screen form this same graphic pattern on a drum. The drum picks up toner, and the remainder of the process is similar to copying.

• In *magnetography,* a print

head is magnetized by computer control. Toner is attracted to the magnetized areas, forming the desired images which are then transferred to paper.

One drawback of electronic printing is the rather low resolution. (The higher the resolution, the sharper the images will be.) Electronic printers typically compose images at a resolution of 300 lines per inch. Compare this with computerized typesetters which have a resolution of 800 lines per inch or better. Another drawback of electronic printers is their inability to print in full color. Some ink jet printers can use inks in colors other than black, but even these cannot print more than one color at a time. As these problems are resolved, electronic printing will become more widely used.

QUESTIONS AND ACTIVITIES

1. Which duplicating process depends in part on magnetic principles?

2. Name three things that some xerographic copying machines can do besides make duplicates of the original.

3. Which duplicating process uses both a master and its carbon image on a cylinder?

4. What is another name for the stencil duplicating process? Briefly describe this process.

5. Briefly describe ink jet printing.

Part 4

Production: Manufacturing and Construction

CHAPTER 31

Introduction to Metalworking

The metalworking industries make metals from raw ores and then build usable products from them. These industries use various mechanical means of doing this, such as turning, milling, extruding, drawing, and casting. In a broader sense, metalworking may be said to include any method of changing the shape, size, and physical form of metals, such as casting, powder metallurgy, and heat treatment. Important processes of the metals industries will be explained in the following chapters.

The demand for better materials and improved manufacturing methods has led to many innovations. Robots and other high-tech devices are widely used in the metalworking industries. Fig. 31-1.

It is almost impossible to think of any object that does not contain metal, or that does not require metal for its manufacture or production. Tools and ma-chines, from typewriters to gasoline engines, are made chiefly of metals. All forms of transportation—automobiles, ships, aircraft, trains—are moving masses of metal and metal parts. Almost everything depends on metal, from the modern skyscraper to the production and distribution of electricity. With these facts in mind, it is not difficult to see that the metalworking industries are important to every other industry—farming, transportation, manufacturing, construction, and power being among the major ones. As a necessity for life and a wealth producer, metalworking is second only to farming.

Metalworking has a great deal to offer as a career. The field is very broad and includes a wide range of career opportunities. The chart in Fig. 31-2 shows the different types of metalworking industries.

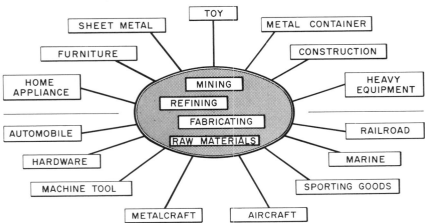

METALWORKING INDUSTRIES

TOY • SHEET METAL • METAL CONTAINER • FURNITURE • CONSTRUCTION • HOME APPLIANCE • HEAVY EQUIPMENT • AUTOMOBILE • RAILROAD • HARDWARE • MARINE • MACHINE TOOL • SPORTING GOODS • METALCRAFT • AIRCRAFT

MINING • REFINING • FABRICATING • RAW MATERIALS

31-2. Metals are used by many industries in the production of goods, as shown in this chart.

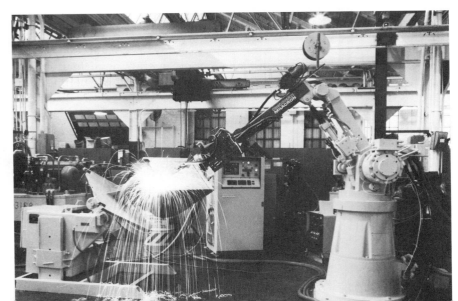

Cincinnati Milacron

31-1. This robot (a metal product) is welding a seam on a backhoe bucket (another metal product).

QUESTIONS AND ACTIVITIES

1. List three of the mechanical methods used to work metal into usable products.

2. List four of the important metalworking industries.

3. Why is the metalworking industry important to other industries?

4. Robots are becoming more and more common in the metalworking industries. What do you think are the advantages and disadvantages of having robots do factory work?

CHAPTER 32

Metal Materials

Metals are among the most widely used materials. They are useful to us because of their physical and mechanical properties, which are very different from those of wood, concrete, or plastic. For instance, metals are strong enough to withstand high loads at high temperatures, yet flexible enough to be worked easily. Fig. 32-1.

PROPERTIES OF METALS

The many kinds of metals used in schools and industry have different properties. Some metals are very hard. This means they resist scratching. Others are strong, yet tough and ductile. *Ductile* means capable of being drawn out or hammered thin. Metals are also good conductors

of electricity and heat. Because of the many properties of the various kinds of metals, almost every product need can be met.

Gemaco

32-1. One property of metals is that they can withstand high temperatures. That makes metal a good choice for fireplace tools.

The chart in Fig. 32-2 describes some of the most important and useful properties and the way these are tested. The test most commonly used in the school shop is the hardness test. Figs. 32-3 & 32-4.

The properties of metals are due to their structure; that is, the special way their atoms are grouped together to make up a solid material. Pure metals have simple structures. Alloys have more complex structures.

Alloys

Most metals used in product manufacturing are alloys, not pure metals. Pure metals are seldom used except in the laboratory. *Alloys* are composed of two or more metals or of metals and small amounts of nonmetallic elements. By carefully alloying, new metals with desirable properties are created. Some of the results are surprising. For example, nickel can be combined with copper to produce an alloy that is stronger than either nickel or copper. A great number of alloys can and have been made because of the ease with which metal atoms combine with other atoms.

One of the most important alloys used in industry is steel, which is an alloy of iron and carbon. The carbon content of a metal is shown as a percentage or as "points." One percent (1.00%) carbon is called 100 carbon, 100 point carbon, or point 100 carbon. You can tell the difference among steels by grinding a piece and studying the sparks. High-carbon steel gives off a bomblike cluster of sparks, while low-carbon steel gives off a long, spread-out pattern. Fig. 32-5.

Metallurgy is the science of making metals and alloys for practical use. For thousands of years it was an art where results came from hard experience and little understanding. Now it is a science using the basic principles of the microstructure (microscopic structure) of metals and alloys and how that structure affects their behavior and properties. Scientists continue to experiment with metals, to create the new and improve the old.

KINDS OF METALS

Our supply of metals comes from the earth in the form of metal *ores*. That is, the metals are combined with other materials, such as rock. The ores are refined and then made into forms usable by the manufacturing industry. Many kinds of metal—steel, copper, nickel, lead, aluminum, gold, silver, and others—are used to make the products we need.

The following paragraphs describe metals you will probably find in the shop and how they are refined. Pay special attention to the properties and common uses of each so that you will be able to select wisely for any projects you make. *Ferrous metals* are mostly iron in content. *Nonferrous metals* contain little or no iron.

Ferrous Metals

The main materials in this group are iron and steel. The chart in Fig. 32-6 describes some of the more common ferrous metals and their uses.

IRON

Iron is one of the basic elements used in making cast iron, wrought iron, and steel. The great iron ore producing areas in this country are in Minnesota, Michigan, and Wisconsin. In these areas there are also large amounts of taconite, a lower grade ore which is balled into pellets rather than shipped as raw ore. The steelmaking centers are in Indiana, Ohio, and Pennsylvania.

STEEL

Steel is one of our most important metals. It is tough, durable, plentiful, and is used in every industry. Well over 500 000 people in this country are involved in the mining, refining, and fabricating of steel.

Shown in Fig. 32-7 (p. 140) is a chart of the steelmaking process. Note that the first important step is to take out the iron from the ore. This is done in a blast furnace. Coke, iron ore, and limestone make up the charge for the furnace. The coke supplies the

32-2.
Metal Properties and Tests

Property	Description	Test
Hardness	Resists penetration, wear, cutting, and scratching; difficult to form and machine.	Brinell, Rockwell, Vickers, and microhardness tests. A metal ball or diamond-shaped point is pressed against a test specimen.
Tensile strength	Resists stretching.	Tensile tests. A specimen is pulled under a tension load to break it in two.
Ductility	Easy to form without tearing or rupturing.	A test specimen is stretched or bent without breaking.
Toughness	Resists shock and impact; tough materials are not brittle.	Charpy or Izod impact tests—a notched specimen is struck to fracture it.
Elasticity	Resists bending, returns to original shape after bending; springy metal.	A test of stiffness or springiness—a deflection or vibration test where the load is applied and then removed.
Compressive strength	Resists squeezing.	Specimen is compressed (pushed together) until it fractures.
Fatigue strength	Resists fracturing under repeated loads.	Specimen is subjected to one or a combination of repeated loads, such as bending or twisting.

Hardness Testing

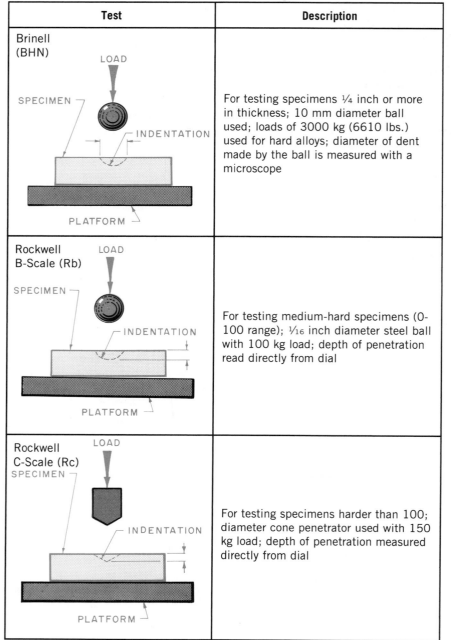

Test	Description
Brinell (BHN) LOAD SPECIMEN INDENTATION PLATFORM	For testing specimens ¼ inch or more in thickness; 10 mm diameter ball used; loads of 3000 kg (6610 lbs.) used for hard alloys; diameter of dent made by the ball is measured with a microscope
Rockwell B-Scale (Rb) LOAD SPECIMEN INDENTATION PLATFORM	For testing medium-hard specimens (0-100 range); ¹⁄₁₆ inch diameter steel ball with 100 kg load; depth of penetration read directly from dial
Rockwell C-Scale (Rc) LOAD SPECIMEN INDENTATION PLATFORM	For testing specimens harder than 100; diameter cone penetrator used with 150 kg load; depth of penetration measured directly from dial

32-3. *Three kinds of hardness tests.*

Wilson Mechanical Instrument Division
32-4. *A Rockwell hardness tester.*

SPARK IDENTIFICATION PATTERNS
– STEEL –

STAINLESS

LOW CARBON

HIGH CARBON

CAST IRON

32-5. *Steel can be identified by the spark pattern given off when grinding.*

heat (about 3500 degrees F), and the limestone serves as a flux; that is, it combines with the impurities to draw them away from the iron. The iron is then made into steel by one of three processes: open-hearth furnace, electric furnace, or basic-oxygen furnace.

In each of these three systems, raw iron, which is fairly soft, is made into steel by adding other metals and chemicals to it to form alloys. The alloys are varied to make special steels for drawing into pipe, for casting, for machining, and for many other uses.

After the steel is made, it is

32-6.
Common Ferrous Metals

Name	Characteristics	Uses
Cast iron	2-6 percent carbon; white and gray types are very brittle; malleable cast has been annealed	Main material used in iron castings
Wrought iron	Almost pure iron; little or no carbon in it; easily worked	Ornamental work
Low-carbon steel	Known as mild or machine steel; 3 to 30 point carbon; cannot be hardened; easily welded, machined, and formed; cold-rolled steel	Rivets, chains, machine parts, forged work, school projects
Medium-carbon steel	Stronger than low-carbon steel; 30 to 60 point carbon; can be hardened	Machine parts; bolts, shafts, axles, hammer heads
High-carbon steel	Known as tool steel; 60 to 170 point carbon; best steel for heat-treating and hardening; hard to cut and bend	Tools such as drills, taps, chisels, etc.
Alloy steel	Alloying elements are added to steels to produce steels with special properties: *nickel* increases strength and corrosion resistance; *chromium* adds toughness and resistance to wear; *tungsten* adds strength and heat resistance; *vanadium* adds shock resistance	wire cables, rails gears, axles, bearings cutting tools, armor springs, gears, splines
Sheet steels	Hot- or cold-rolled sheets, generally coated for protection: a *zinc* coat gives a rustproof finish (galvanized sheet); a *tin* coat gives a bright, corrosion-proof finish suitable for food tins (tinplate)	Sheet metal products, heating and air conditioning ducts, automobile bodies

poured into ingot molds. Later it is milled into bars, sheets, plates, and other forms to be used by industries.

Instead of reheating ingots and rolling them into blooms, slabs, or billets, these three shapes of steel can also be made by a new process called *continuous casting*. Fig. 32-8. Molten steel is poured continuously into a water-cooled mold that is open at the top and bottom. A starting bar temporarily closes the bottom. The steel gradually cools and begins to become solid in the mold. Then the starting bar is slowly pulled downward, drawing the steel with it. The rate at which molten steel is poured in at the top is matched with the rate at which the solid steel is pulled out at the bottom. In this way, a long, continuous piece is formed. It can then be cut into lengths as desired. Different shapes of molds are used, depending upon whether blooms, slabs, or billets are being made.

Some steels are hot rolled (squeezed between rollers while red hot). Such metals have a bluish scale on the surface and are called hot-rolled steel (HRS). Cold-rolled steels (CRS) are smoother and have a clean shiny surface finish.

Nonferrous Metals

These metals contain little or no iron and are soft and easily worked. They are good conductors of heat and electricity and are generally more durable than iron and steel.

COPPER

Copper is reddish brown and can be bought as sheets, tubes, or rods in a variety of sizes. Copper is perhaps best known for its uses in the electrical industry—it is second only to silver as a conductor of electricity. It is also used for working utensils, for roofing, and as an alloy for other metals. It is easily worked into bowls, but hammering makes it harder. (This is known as work hardening.) To soften the copper, it must be heated red hot and plunged into water. Copper tarnishes easily and does not machine easily.

Figure 32-9 describes the process of making copper. Large copper deposits are found in Chile, Canada, and the United States. Some of the mining centers in

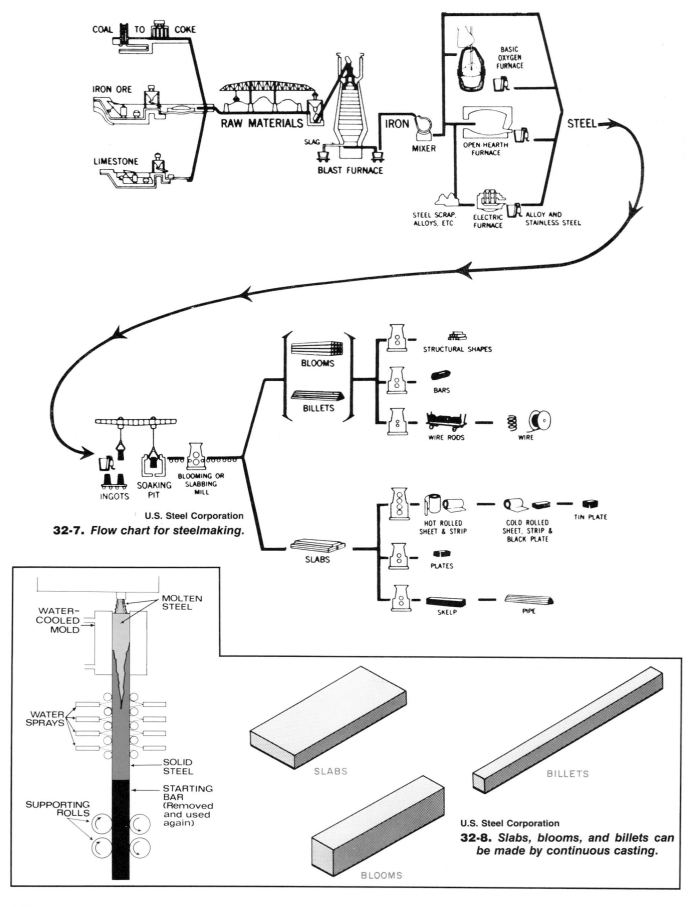

COAL TO COKE

IRON ORE

LIMESTONE

RAW MATERIALS

SLAG

BLAST FURNACE

IRON

MIXER

BASIC OXYGEN FURNACE

OPEN HEARTH FURNACE

STEEL SCRAP, ALLOYS, ETC

ELECTRIC FURNACE

ALLOY AND STAINLESS STEEL

STEEL

BLOOMS

BILLETS

STRUCTURAL SHAPES

BARS

WIRE RODS

WIRE

INGOTS

SOAKING PIT

BLOOMING OR SLABBING MILL

U.S. Steel Corporation

32-7. Flow chart for steelmaking.

SLABS

HOT ROLLED SHEET & STRIP

COLD ROLLED SHEET, STRIP & BLACK PLATE

TIN PLATE

PLATES

SKELP

PIPE

WATER-COOLED MOLD

MOLTEN STEEL

WATER SPRAYS

SOLID STEEL

STARTING BAR (Removed and used again)

SUPPORTING ROLLS

SLABS

BILLETS

BLOOMS

U.S. Steel Corporation

32-8. Slabs, blooms, and billets can be made by continuous casting.

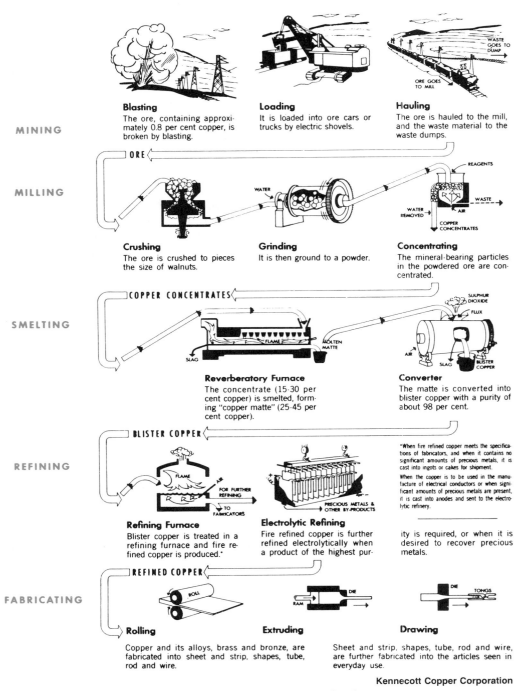

MINING

Blasting
The ore, containing approximately 0.8 per cent copper, is broken by blasting.

Loading
It is loaded into ore cars or trucks by electric shovels.

Hauling
The ore is hauled to the mill, and the waste material to the waste dumps.

MILLING

Crushing
The ore is crushed to pieces the size of walnuts.

Grinding
It is then ground to a powder.

Concentrating
The mineral-bearing particles in the powdered ore are concentrated.

SMELTING

Reverberatory Furnace
The concentrate (15-30 per cent copper) is smelted, forming "copper matte" (25-45 per cent copper).

Converter
The matte is converted into blister copper with a purity of about 98 per cent.

REFINING

Refining Furnace
Blister copper is treated in a refining furnace and fire refined copper is produced.*

Electrolytic Refining
Fire refined copper is further refined electrolytically when a product of the highest purity is required, or when it is desired to recover precious metals.

*When fire refined copper meets the specifications of fabricators, and when it contains no significant amounts of precious metals, it is cast into ingots or cakes for shipment.

When the copper is to be used in the manufacture of electrical conductors or when significant amounts of precious metals are present, it is cast into anodes and sent to the electrolytic refinery.

FABRICATING

Rolling
Copper and its alloys, brass and bronze, are fabricated into sheet and strip, shapes, tube, rod and wire.

Extruding

Drawing
Sheet and strip, shapes, tube, rod and wire, are further fabricated into the articles seen in everyday use.

Kennecott Copper Corporation

32-9. *Flow chart for copper production.*

this country are in Montana, Arizona, and Utah.

BRASS, BRONZE, AND PEWTER

Brass is an alloy of copper (90 percent) and zinc (10 percent). It is yellow-gold and more brittle than copper. Brass comes in forms similar to copper.

Bronze resembles brass in both color and working properties. It is an alloy of about 90 percent copper and 10 percent tin, with small amounts of nickel or aluminum.

Pewter is also an alloy containing about 10 percent copper and 90 percent tin. It is an excellent

MINING

Bauxite is mined, crushed, washed, and dried. The miner in the drawing is drilling a blasting hole.

REFINING

Processed bauxite is converted to aluminum hydrate crystals. These crystals are roasted to produce a white powder called alumina. The drawing shows a converter.

SMELTING

Alumina is reduced to metallic aluminum in carbon-lined pots. Electric current passes through the molten alumina to free the metal which is then siphoned off into pig molds.

FABRICATING

Ingots are formed into bars, sheets, wires, and tubes for industrial use.

32-10. *Aluminum production.*

ALUMINUM

Aluminum is silver-white, is remarkably light, and is easily worked. It work-hardens quickly and must be softened as follows: Cover the metal with chalk, heat until the chalk turns brown, and allow the metal to air-cool. It too is available in sheets, tubes, rods, and bars. Aluminum is very light, corrosion-resistant, and an excellent conductor of electricity.

Although it is soft, aluminum can be made strong by adding copper or zinc. Commercially pure aluminum is designated as No. 1100. Other alloy numbers go up to 7072 to indicate hardness and workability. Aluminum alloys are used for bridges, airplanes, house siding, kitchen utensils, and many other products.

Large deposits of aluminum ore (bauxite) are found in Arkansas. Other mining areas are in Africa, Australia, Brazil, and the

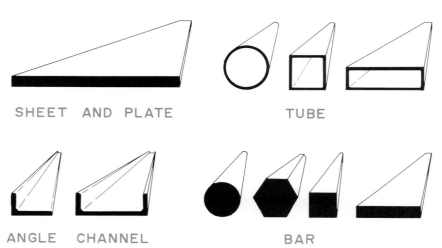

SHEET AND PLATE   TUBE

ANGLE   CHANNEL   BAR

32-11. *Metals come in many shapes and sizes. Here are some of the most common shapes. These are generally found in the school shop.*

material for making bowls and vases because it is so easily formed.

STERLING SILVER

This lovely, warm, lustrous metal is used for jewelry and fine tableware. An article marked "sterling" must contain not less than 0.925 parts of silver with 0.075 parts of copper or some other alloying metal added for hardness.

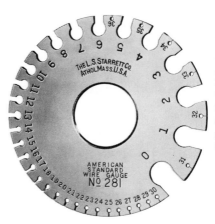

The L. S. Starrett Co.

32-12. *Disk-type gauge. Place the metal sheet into several openings to find the one that fits correctly. This will be the smallest one into which the metal fits easily. The number stamped near the opening tells the gauge of the metal. This measuring tool can be used for wire as well as sheet metal.*

32-13.
Metal Sheet Thicknesses

Gauge number	Brown & Sharpe or American Standard for nonferrous wire and sheet metals (inches)	United States Standard for ferrous sheet metals (inches)	ISO metric replacement sizes* (mm)
16	0.0508	0.0625	1.25
18	0.0403 (32 oz.)	0.0500	1.00
20	0.0320 (24 oz.)	0.0375	0.80
22	0.0253 (20 oz.)	0.0313	0.63
24	0.0201 (16 oz.)	0.0250	0.50
26	0.0159	0.0188	0.40
28	0.0126	0.0156	0.315
30	0.0100	0.0125	0.250
32	0.0079	—	0.200

*Based on ISO R 388, not yet adopted in USA. Gauge numbers will not be used; use mm sizes only.

U.S.S.R. Figure 32-10 shows the flow of raw material to finished aluminum. Today about 27 percent of aluminum is recycled. Using recycled aluminum to make new products is less costly than refining ore, and it helps reduce the problem of waste.

METAL SHAPES AND SIZES

The metals used in school and industry are available in a number of different shapes and sizes. Fig. 32-11. Bars, tubing, angles, etc., are available in lengths from 10 to 20 feet. Sheet metals are sold by weight or by the square foot.

The thickness of sheet metal is indicated by a gauge number.† A disk-type gauge is used to make this measurement. Fig. 32-12. Ferrous metals—those which contain a high percentage of iron —are measured with a United States Standard gauge. Galvanized iron and tinplate are ferrous metals. The Brown and Sharpe, or American Standard, gauge is

used for most nonferrous metals —metals containing little or no iron such as copper, aluminum, and brass. Fig. 32-13.

Other methods of measuring sheets are also used. Aluminum is measured in decimal units.

Copper is sometimes measured by the weight per square foot. For example, 20-gauge copper weighs 24 ounces per square foot. The charts in Fig. 32-14 describe the common sizes of metal bar and angle materials.

Materials Testing

In order to choose the best materials for a product—whether it's a bridge or a spaceship or a running shoe—it is necessary to know how materials will behave under certain conditions. This is the main purpose of materials testing: to determine how a given material will behave under specified conditions.

There are numerous kinds of tests and testing procedures. This chapter lists some of the tests done on metals; Chapters 58 and 65 describe tests done on plastics and wood. Some categories of tests are:

—Destructive and nondestructive. Destructive testing damages the test specimen so that it cannot be used again. It is done to determine the limits of a material's strength. For example, compression tests are destructive. A specimen is squeezed until it breaks. Nondestructive testing does not damage

the specimen. It is done to locate defects in materials and products. An example of a nondestructive test is X-ray radiography. This test is often used to inspect welds for defects.

—Dynamic and static. In dynamic testing, a load is applied to a specimen and immediately removed. Impact tests are an example of this. In static testing, the load remains on the specimen, and it is usually applied gradually rather than suddenly.

—High-, low-, and room temperature. Most tests are done at room temperature.

—Field and laboratory. Field tests are done under actual service conditions. Laboratory tests are done under simulated service conditions, usually in such a manner that results can be closely observed and precisely measured.

---

†Gauge is sometimes spelled gage. Either spelling is acceptable.

32-14.

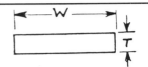

Common Flat Metal Bar Sizes

Thickness		Width	
inch	mm*	inch	mm*
1/16	1.6	3/8	10
1/8	3	1/2	12
3/16	5	5/8	16
1/4	6	3/4	20
5/16	8	1	25
3/8	10	1 1/4	30
7/16	11	1 1/2	40
1/2	12	1 3/4	45
9/16	14	2	50
5/8	16		
11/16	18		
3/4	20		
7/8	22		
1	25		
1 1/8	28		
1 1/4	30		
1 5/16	32		
1 3/8	35		
1 1/2	40		
1 3/4	45		
2	50		

*Metric replacement sizes based upon ANSI B32.3-1974.

Common Round, Square, and Hexagon Metal Bar Sizes

Diameters or Across Flats Distances	
inch	mm*
1/8	3
1/4	6
3/8	10
1/2	12
5/8	16
3/4	20
1	25
1 1/4	30
1 1/2	40
1 3/4	45
2	50

*Metric replacement sizes based upon ANSI B32.4-1974.

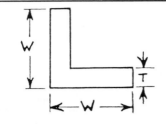

Common Steel Angle Sizes

Thickness	
inch	mm*
3/4	20
1	25
1 1/4	30
—	35
1 1/2	40
1 3/4	45
2	50

Width	
inch	mm*
1/16, 1/8, 3/16	3, 4, 5
1/8, 3/16, 1/4	3, 4, 5
1/8, 3/16, 1/4	3, 4, 5
—	3, 4, 5
1/8, 3/16, 1/4	3, 4, 5, 6
1/8, 3/16, 1/4	3, 4, 5, 6
1/8, 3/16, 1/4, 5/16, 3/8	3, 4, 5, 6, 7, 8

*Based on sizes recommended in ISO R657/1, not yet adopted in USA.

QUESTIONS AND ACTIVITIES

1. What is a ductile metal?

2. What are the characteristics of a hard metal? Name the three tests for metal hardness.

3. What is an alloy?

4. What is a ferrous metal? What is a nonferrous metal? Name two examples of each.

5. Name two alloying elements that can be added to steel and describe the special properties they would give to the steel.

6. What are the characteristics of machine (or mild) steel?

7. What three materials make up the charge for the blast furnace in steelmaking?

8. What is the best known use of copper and why is it so good for this purpose? What are some other good uses of copper?

9. What combinations of metals form brass? Bronze? Pewter?

10. What is aluminum ore called? Where can it be found in the United States?

11. What is alumina?

12. What would you use to measure the thickness of a piece of copper wire? A piece of tinplate?

CHAPTER 33

Metal Layout and Pattern Development

If you want to cut a certain shape out of metal, you first need to draw that shape on the metal. This process is called *layout*. Layout is a key step in making a project. Accuracy is important. Study your project plans carefully before laying out your patterns. Selection of proper materials is also important, and so is care to avoid waste.

LAYOUT TOOLS

You will soon become familiar with the tools used in layout work. They must be used constantly to check proper sizes and squareness. The basic tools you will use are shown in Fig. 33-1.

Squares are available in several sizes and styles. They are necessary for laying out rectangles and checking the squareness of stock. Solid or adjustable heads with 6″ or 12″ blades are most common. The combination set is a familiar and important tool in the shop. Fig. 33-2.

Circles are scribed (marked) in metal with *dividers*. These tools are also used for accurately marking out and transferring distances. Fig. 33-3.

The *scriber* is the metalworker's pencil. Its sharpened point produces a clean line on metal. This tool is also called a scratch awl. Fig. 33-4.

The *sheet and wire gauge* is used to measure the thickness of metal sheets and the diameter of wire.

The *ball-peen hammer* is used for general pounding and

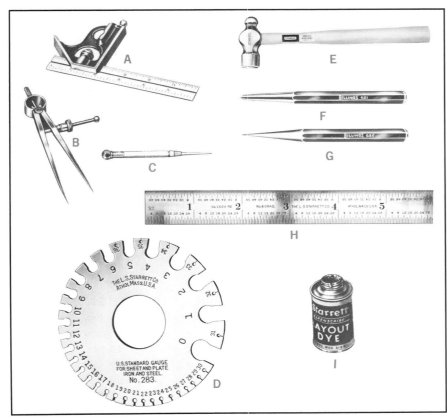

33-1. *Layout tools for metalwork: (A) Square. (B) Dividers. (C) Scriber. (D) Sheet and wire gauge. (E) Ball-peen hammer. (F) Center punch. (G) Prick punch. (H) Metal rule. (I) Layout dye.*

straightening. It is also used with center and prick punches to produce marks on metal. Ball-peen hammers range in size from four ounces to two pounds. A one-pound hammer is satisfactory for layout work. Smaller layout hammers are also used.

Center punches are used to mark the centers of holes before drilling. If you don't indent the metal slightly, the drill will skip around and mar the surface. The point of this punch is ground at 90 degrees.

Prick punches are similar to center punches, except that they are ground to a 30-degree point. They are used for making light indentations as location points for scribing straight lines, or for centers of circles to be made with dividers. Fig. 33-5.

Metal rules are important mea-

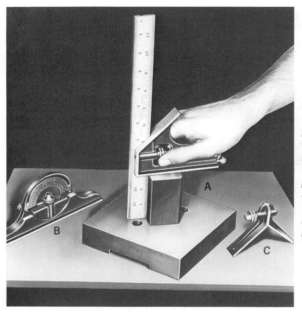

The L. S. Starrett Company

33-2. *The combination set. The squaring head (A) is used for general-purpose measuring and scribing of lines at 90 degrees and 45 degrees. The protractor head (B) produces accurate lines at any angle from 0 to 180 degrees. The centering head (C) is used for marking the centers of circles and arcs.*

The L. S. Starrett Company

33-4. *To scribe lines, hold the square in place and pull the scriber firmly along the edge of the blade.*

suring and layout instruments. Be careful not to dent or knick their edges. Rules come in many lengths; common shop sizes are 6″, 12″, and 36″. See Chapter 8 for information on reading the rule. Steel tapes are frequently used for measuring long pieces.

Layout dye makes it possible to see layout lines clearly. This fluid can be removed with turpentine or paint thinner. Common chalk may also be used in layout work.

MEASURING HEAVY STOCK

Metal bars, tubes, rods, and angles must be accurately measured before they are cut to length. The steel tape or metal rule is generally used for this measuring. Hold the end of the rule directly over the end of the stock and make a scriber mark at the desired length. Holes to be drilled are located in a similar manner and then marked with the center punch.

DEVELOPING SHEET METAL PATTERNS

Before sheet metal objects can be cut and formed, a pattern of the project must be made. For example, sheet-metal workers cut out patterns from flat pieces of metal and then bend them to form ductwork for heating, ventilating, and air conditioning systems. Figure 33-6 shows four

methods of pattern development. Patterns can be made on paper or directly on the metal.

Instructions for preparing these patterns, and uses of the various kinds of patterns, are described in the following paragraphs.

Angular Developments

Angular developments are used for boxes, trays, and other

The L. S. Starrett Company

33-5. *Locating points for scribing circles. A prick punch and a ball-peen hammer are used.*

33-3. *The dividers are used to scribe circles on metal.*

The L. S. Starrett Company

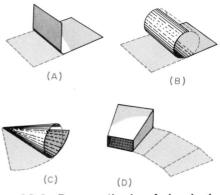

33-6. *Four methods of developing sheet metal patterns: (A) Angular. (B) Cylindrical. (C) Conical. (D) Triangulation. The huge metal duct was made from cylindrical and triangulation developments.*

eliminating excess material. A notch is cut at 45 degrees across the corners of the lap seams. Notch cuts ranging from 30 to 45 degrees may be used for single and double hems.

Cylindrical Developments

Cylindrical, or parallel-line, developments are used for cans or other cylindrical containers with vertical sides. Figure 33-8 is an example of this kind of pattern. It is made by drawing a top and side view of the bird feeder, and dividing the top view into equal parts. These numbered points are then projected (dropped down) into the side view.

A stretchout line is drawn next to the side view. On this line, step off the number of spaces shown on the top view. In this case, there are twelve equal spaces. Number these points and drop perpendicular lines from them.

rectangular items. Figure 33-7 is a layout for a utility box. This was made by drawing a rectangle equal to the size of the box bottom and adding the end and side pieces in position. Material for

hems and lap seams was added next. For most small boxes, ¼″ or ⅜″ hems and ⁵⁄₁₆″ lap seams are adequate. Note that notches are laid out on the pattern to provide for a better fit of the bends by

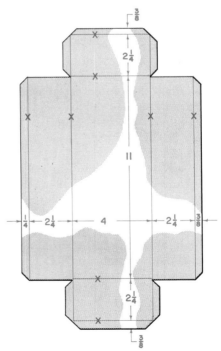

33-7. *A sheet-metal utility box laid out by the angular development method. Note the notch cuts made to remove unneeded material.*

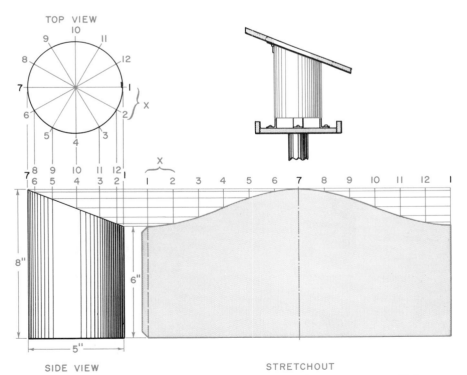

33-8. *This bird feeder was made from a parallel-line development. Make certain that all points are projected properly.*

TOP VIEW

SIDE VIEW

STRETCHOUT

Next, extend the numbered points of the side view over to the stretchout. Where like-numbered lines intersect, make a mark. Connect these marks with heavy lines to get the outline of the stretchout. Don't forget to add material for the grooved seam (three times the width of the seam).

Cylinders without the slanted line in the front view may be laid out by merely finding the circumference (the distance around the outside) of the cylinder. To find the circumference, multiply the diameter by 3.1416 (*pi*). A circumference rule may also be used.

Conical Developments

Conical, or radial-line, developments are used for making cone-shaped objects such as funnels, pails, covers, and shades. The patio-lamp shade in Fig. 33-

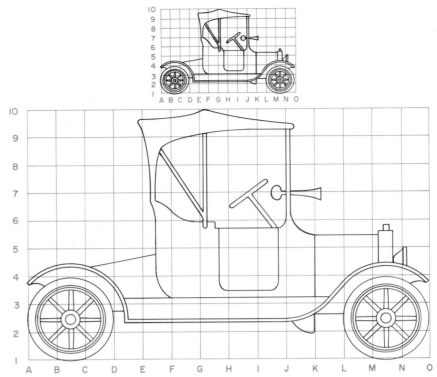

33-10. *Enlarging a pattern by using a grid system. Patterns may be reduced in a similar manner.*

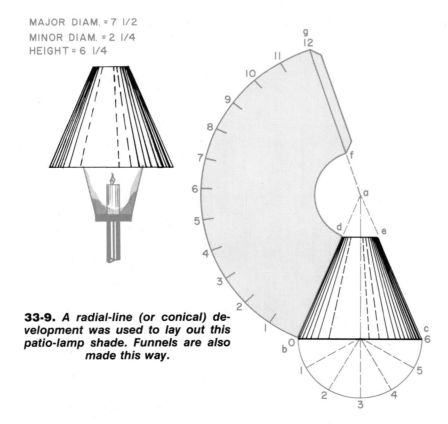

MAJOR DIAM. = 7 1/2
MINOR DIAM. = 2 1/4
HEIGHT = 6 1/4

33-9. *A radial-line (or conical) development was used to lay out this patio-lamp shade. Funnels are also made this way.*

9 is a good example of this. Cone patterns are made by drawing a full-size front view of the object. Extend the tapered (slanted) lines until they intersect at point "a." Draw semicircle "bc" and divide it into an even number of equal parts. Number these divisions. Using "a" as center, draw arc "bg" with radius "ab." Use the same center and draw arc "df" with radius "ad." On arc "bg" mark off twice the number of equal divisions found on the semicircle. Draw a line from the last division point (in Fig. 33-9 this is point 12) to "a" to form the layout "dbgf." Add ¼" for a plain lap, riveted seam.

Triangulation

Triangulation is used to develop transition pieces in sheet metal ductwork. Examples of this are square-to-round pieces and pieces that connect large rectangles to smaller ones. Very few triangulation pieces are made in the school shop.

ENLARGING PATTERNS

Sometimes you may want to develop a pattern from a picture or print in a book or magazine. These are often too small, but they can be enlarged by the use of grid systems. Fig. 33-10. Directly over the print draw a grid system with small squares. Letter the vertical lines and number the horizontal ones, as shown in the top part of Fig. 33-10.

Perhaps you are planning to triple the size of the pattern. If so, make another grid system with squares three times as large. Letter and number the lines as before. If a line on the pattern crosses the grid system where lines J and 7 come together, make a small dot at point J7 on the large grid. Transfer enough of these points to show the outline of the pattern. Connect the points with lines to complete the pattern.

QUESTIONS AND ACTIVITIES

1. Name four of the common layout tools and tell what they are used for.

2. What are the three heads of the combination square used for?

3. Name four methods of pattern development and tell what types of objects each might be used for.

4. How much metal must be allowed to form a grooved seam on a cylindrical container?

CHAPTER 34
Metal Cutting Principles

Cutting is very important in metalwork. Even metal which is to be worked by any of the other three methods—forming, fastening, and finishing—must first be cut. For example, a tool box could not be formed until the sheet had been cut to size.

To many people the term "cutting" means work done with a knife or shears. However, the metalworker knows that cutting is done in a variety of ways, some of which do not make use of any tool with a blade or cutting edge.

Cutting is done for two main purposes: separating and removing. Cutting a large piece of metal into two or more smaller pieces, as with shears, is *separating*. Cutting away small bits of metal, as with a file, is *removing*.

Drilling, turning, milling, shaping, and grinding (abrading) are common metal cutting processes, both for school and industrial purposes. Sawing, shearing, and electro-chemical cutting are other important kinds of cutting. These eight methods are illustrated and described in Fig. 34-1. Tools and machines listed are the ones you will probably see or use in school and industry.

Science and technology have developed many new ways of cutting metal such as by electric discharge machining, chemical mill-

ing, and numerical control. These newer cutting methods are, for the most part, merely refinements of the eight basic kinds listed in Fig. 34-1. The computer and laser are becoming very important in metal machining. (Laser cutting is described in Chapter 42.)

In Chapters 34-42, you will find descriptions and examples of each of the eight basic cutting methods. You will read about some of the common tools used and learn how to use them. It will be helpful for you to refer back to Fig. 34-1 often. You can see then how these cutting systems differ, and also how much they are alike.

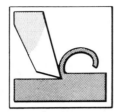

34-1.
Cutting

Cutting is the process of removing or separating pieces of material from a base material.

KIND OF CUTTING	DEFINITION	EXAMPLES
Sawing	Cutting with a tool having pointed teeth equally spaced along the edge of a blade.	Cutting with a hacksaw, metal band saw, jeweler's saw.
Shearing	Cutting, usually between two cutting edges crossing one another, or by forcing a single cutting edge through a workpiece.	Cutting with tin snips, bench shears, squaring shears, hollow punch, solid punch, cold chisel.
Abrading	Cutting by wearing away material, usually by the action of mineral particles.	Grinding, polishing, buffing, and operations using a sharpening stone or abrasive papers.
Shaping	Cutting by moving a single-edge tool across a fixed workpiece in a straight-line cutting path.	Operations on the metal shaper.

(Continued)

Cutting (Continued)

KIND OF CUTTING	DEFINITION	EXAMPLES
Drilling 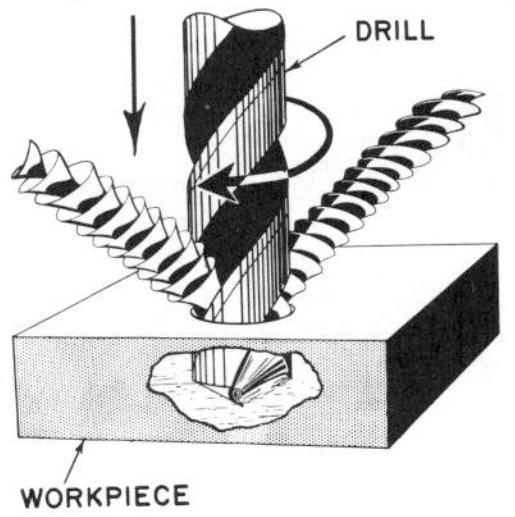	Cutting with a cylindrical tool usually having two spiral cutting edges.	Operations using the hand drill, electric hand drill, drill press, twist drill.
Milling 	Cutting with a tool having sharpened teeth equally spaced around a cylinder or along a flat surface.	Operations using the horizontal milling machine, vertical milling machine, hand file, needle file.
Turning 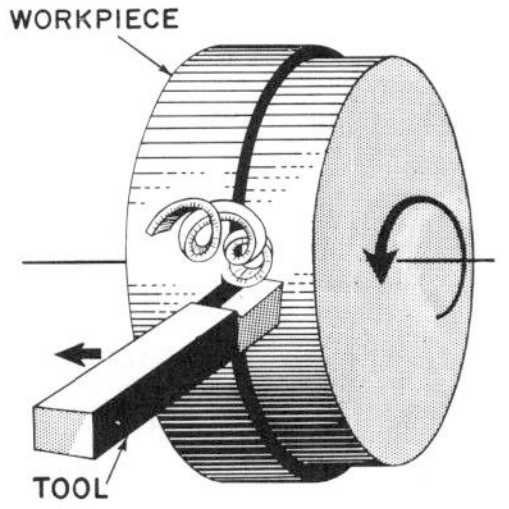	Cutting by revolving a workpiece against a fixed single-edge tool.	Operations on the metal lathe.
Cutting Electro-Chemical 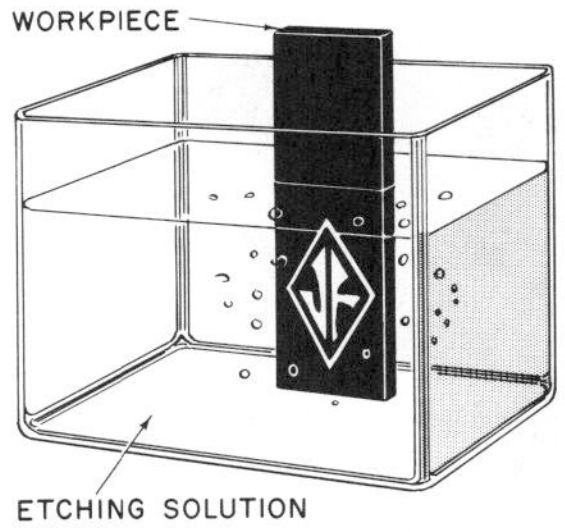	Cutting with heat or acids.	Etching, oxyacetylene and carbon-arc cutting, laser cutting.

QUESTIONS AND ACTIVITIES

1. Define metal cutting.
2. Cutting is done for two purposes. Name them. How do they differ?
3. List the eight kinds of metal cutting.
4. What kind of cutting method is buffing?

CHAPTER 35

Sawing Metal

Sawing is a way of separating material (such as cutting a short piece of stock from a longer piece) with a tool having pointed teeth equally spaced along the edge of a blade. Chips are produced in sawing, but not in shearing. Common metal sawing tools are the hacksaw, the metal-cutting band saw, the hole saw, and the jeweler's saw.

The *set* of a saw refers to how much the teeth are pushed out in opposite directions from the sides of the blade. The set of a blade helps to prevent the blade from binding or jamming as it cuts through metal. A few drops of machine oil on the blade will also make the cutting easier.

FOR YOUR SAFETY . . .

Sawing is a quick way to cut metal, but you need to be careful. Keep these safety rules in mind whenever you saw metal:
- The blade of the saw should be sharp. A dull blade is dangerous because it is more likely to slip and cut you.
- Keep your hands away from the sharp edges of the saw.
- Make sure the metal to be cut is held securely in a vise.
- Sawing produces rough areas and sharp ridges on the metal. These are called *burrs*. File off all burrs to avoid cuts and scratches.

HACKSAWS

Hacksaws are used to cut metal sheets, rods, bars, and

35-1. *Hacksaw with blade in place.*

pipes. There are two parts to a hacksaw: the frame and the blade. Fig. 35-1. Common hacksaws have either an adjustable or solid frame. Adjustable frames can be made to hold blades from 8 to 16 inches long, while those with solid frames take only the length blade for which they are made.

Hacksaw blades are made of high-grade tool steel, hardened and tempered. These blades are about one-half inch wide, have from 14 to 32 teeth per inch, and

are from 8 to 16 inches long. The blades have a hole at each end, which hooks to a pin in the frame.

Using the Hacksaw

Good work with a hacksaw depends not only upon the proper use of the saw but also upon the proper selection of the blades for the work to be done. Figure 35-2 shows which blade to use. Coarse blades with fewer teeth per inch cut faster and do not clog up with chips. However, the finer blades with more teeth per inch are necessary when thin sections are being cut.

To use the saw, first install the blade in the hacksaw frame so that the teeth point away from the handle of the hacksaw. Fig. 35-3. This is done because hand

35-2. *These are the blades recommended for use on different metals and metal shapes. The tooth count is the number of saw teeth per inch.*

14 Tooth – Soft materials, larger cross sections

18 Tooth – General use, same blade on several jobs

24 Tooth – Cross sections 1/16" to 1/4"

32 Tooth – Cross sections 1/16" or less

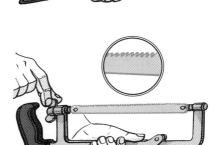
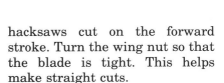

35-3. *Installing a hacksaw blade. The blade's teeth should point away from the handle.*

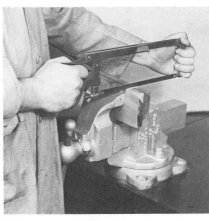

35-4. *Hold the hacksaw as shown here. Note that the metal has been sandwiched between pieces of scrap wood. This makes it easier to cut thin sheets of metal.*

hacksaws cut on the forward stroke. Turn the wing nut so that the blade is tight. This helps make straight cuts.

Place the material to be cut in a vise or hold it with clamps. Protect the workpiece by covering the vise jaws with pieces of copper sheet. The proper method of holding the hacksaw is shown in Fig. 35-4. Grasp the saw firmly with both hands as shown. Apply pressure on the forward stroke, which is the cutting stroke. Use long, slow, steady strokes.

Hacksaw Safety

The main danger in using hacksaws is injury to your hand if the blade breaks. The blade will break if too much pressure is applied, when the saw is twisted, when cutting too fast, or when the blade becomes loose in the frame. If the work is not tight in the vise, it will sometimes slip, twisting the blade enough to break it. Be careful when using the hacksaw.

METAL-CUTTING BAND SAW

The metal-cutting band saw cuts as a hacksaw when in the

down position. Fig. 35-5. It can also be used in an upright position as a conventional band saw for cutting arcs and circles. Make sure that the workpiece is securely fastened in the vise. Use cutting oil to lubricate and cool the blade. Wear safety glasses or goggles to protect your eyes from flying chips of metal.

The L. S. Starrett Company
35-6. *This hole saw (inset) is used to cut holes in ¼″ metal.*

HOLE SAW

The hole saw, Fig. 35-6, can be mounted in a drill press and used to cut holes in metal. The sizes of the holes range from ¾″ to 2″ or 3″. Be sure your instructor has demonstrated this tool before you use it.

JEWELER'S SAW

The jeweler's saw makes internal cuts. Fig. 35-7. Workpieces to be cut are held over a V-block fastened in a vise. Fig. 35-8. A hole must be drilled in the metal workpiece, and the blade fed through the hole. The teeth of the blade should face toward the handle. Cut carefully on the down stroke so as not to break the fragile blades. Blades come in sizes ranging from No. 6/0 (thinner than a thread) to No. 14 (about ⅛″ wide).

Rockwell-Delta Company
35-5. *Cutting a piece of pipe on a metal-cutting band saw.*

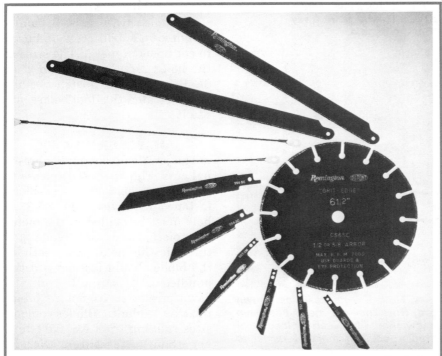

Remington

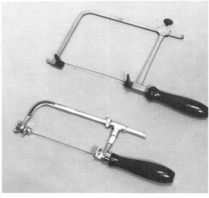

35-7. *Two styles of jeweler's saws.*

Toothless Saws

"Grit-Edge" blades are a fairly recent development in cutting tools. These blades have no teeth. Instead, the cutting edge is made up of thousands of tungsten carbide particles permanently bonded to alloy steel. Tungsten carbide is one of the hardest materials known.

"Grit-Edge" blades cut on both the forward and reverse strokes. Because they have no teeth, they outlast conventional saw blades.

Various kinds of blades are made for cutting different materials. For example, the round blade is used in circular saws for cutting fiber-reinforced cement, synthetic marble, plastic laminates, composites, fiberglass, sheet steel, plywood, and tempered hardboard. "Grit-Edge" blades should not be used on softer materials such as copper, aluminum, or ordinary wood.

35-8. *Using the jeweler's saw.*

QUESTIONS AND ACTIVITIES

1. Define sawing.
2. What is meant by the "set" of a saw blade? What is its purpose?
3. What are hacksaws used for? How should the blade be installed? On what stroke is the cut made?

4. What hacksaw blade would you use to cut a large piece of a soft metal? A piece of angle iron?
5. What is a jeweler's saw used for? What direction should the teeth of the blade face? On what stroke is the cut made?

Almost all cutting can be viewed as shearing. Even tearing a piece of paper in your hands is a kind of shearing. However, when people in industry speak of shearing, they usually have a certain kind of cutting in mind. Do you remember how the chart in Chapter 34 defined the term? It said that shearing is cutting between two edges that cross one another or cutting by forcing a single cutting edge through a workpiece.

Shearing is done chiefly for separating. It produces no chips; so it is different from sawing. Shearing is a fast, clean, accurate way to cut light and heavy metals. It is widely done in industry. Industrial shears are generally power driven and can cut metals up to ½" thick. This chapter describes shearing tools commonly found in the school shop.

FOR YOUR SAFETY . . .
● Keep shearing tools sharp. Dull tools are dangerous.
● Keep your fingers away from cutting edges.
● The ragged edges produced by shearing metal are as sharp as razor blades. Be careful; they cause very painful cuts. Wear gloves when hand-shearing large pieces.

TIN SNIPS

Sheet metals of gauge 20 or lighter are easily sheared with tin snips. These tools are also called hand snips or hand shears, and come in three common styles. Fig. 36-1.

Be careful how you hold tin snips so that you do not pinch yourself when cutting.

The *straight snips* are for making straight cuts and range in size from 6" to 14". To make straight cuts on larger pieces, place the sheet metal on a bench with the scribed guideline over the edge of the bench. Hold the sheet down with one hand. Be careful of sharp edges. With the other hand hold the snips so that the flat sides of the blades are at right angles to the surface of the work. If the blades are not at right angles to the surface of the work, the edges of the cut will be slightly bent and burred. Any of the hand snips may be used for straight cuts. When notches are too narrow to be cut out with a pair of snips, make the side cuts with the snips and cut the base of the notch with a cold chisel.

Aviation snips are double hinged for easier cutting. They are designed for right-hand or left-hand irregular cuts or for straight cutting. *Trojan shears* are for curved cuts.

To cut large holes in sheet metal, start the cut by punching a hole in the center of the area to be cut out. With aviation snips or some other narrow-bladed snips, make a spiral cut from the starting hole out toward the scribed circle and continue cutting until the scrap falls away.

SQUARING SHEARS

The squaring shears, Fig. 36-2, are for straight shearing of large

36-1. Tin snips are used for light shearing: (A) Straight snips. (B) Aviation snips. (C) Trojan snips.

Niagara Tools

36-2. The foot-operated squaring shears: (A) Blade. (B) Safety guard. (C) Slide gauge. (D) Housing. (E) Foot treadle. (F) Bed. Remember to keep your feet out from under the treadle and your fingers away from the safety guard. Do not cut rods, bars, and wire on this machine.

Niagara Tools

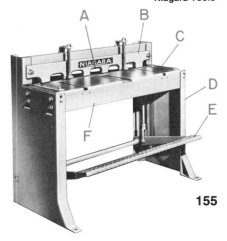

36-3. *Using the squaring shears. This is a hand-operated bench model.*

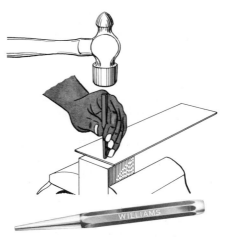

36-4. *The solid punch comes in several sizes for punching holes up to ⁷⁄₁₆″ in diameter.*

pieces of metal up to 18 gauge. (The gauge capacity is stamped on the machine. Do not exceed this.) The shears come in bench and floor models and in several sizes to take various widths of metal. Do not use this machine to cut rods, bars, or wire as these will damage the blade. Never permit anyone to stand near the squaring shears while you are using them. Be sure to keep your hands and feet away from the blade and treadle when shearing. Fig. 36-3.

PUNCHES

Holes are punched in light metals with the solid (hand) and hollow punches. To use them, place the metal over a block of wood and strike the punch sharply with a ball-peen hammer. Figs. 36-4 & 36-5.

CHISELS

Chisels are used for chipping or cutting metal that is softer than the materials of which the chisels are made. Cold chisels (chisels made for chipping or cutting cold metal) are classified according to the shape of their points, and the width of the cut-

ting edge denotes their size. The most common shapes of chisels are flat, cape, round nose, and diamond point. Fig. 36-6.

The type of chisel most commonly used is the flat cold chisel, which serves to cut rivets and thin metal sheets, to split nuts,

36-5. *The hollow punch is used for larger holes, ranging from ¼″ to 2″ in diameter.*

and to chip castings. The cape chisel is used for special jobs like cutting keyways (channels), narrow grooves, and square corners. Round-nose chisels make circular grooves and chip inside corners with a fillet (rounded, inside curve). Finally, the diamond-point is used for cutting V-grooves and sharp corners.

Using a Chisel

As with other tools, there is a correct technique for using a chisel. Select a chisel that is large enough for the job. Be sure to use a hammer that matches the chisel; that is, the larger the chisel, the heavier the hammer.

As a general rule, hold the chisel in the left hand with the thumb and first finger about 1 inch from the top. It should be held steadily but not tightly. The finger muscles should be relaxed; so if the hammer strikes your

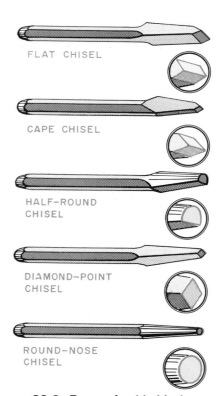

FLAT CHISEL

CAPE CHISEL

HALF-ROUND CHISEL

DIAMOND-POINT CHISEL

ROUND-NOSE CHISEL

36-6. *Types of cold chisels.*

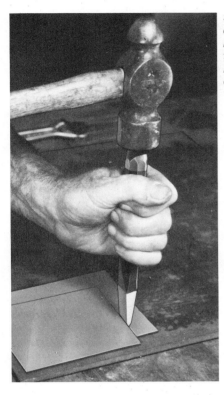

36-7. *Using a cold chisel. Hold the chisel firmly in your fist and strike its head squarely with a hammer. When the chisel head "mushrooms," dress it on the grinder. The chisel's cutting edge should be ground to an angle of 60 degrees.*

The Beverly Shear Mfg. Corp.

36-9. *Complex shapes can be cut on the throatless shears.*

hand, your hand will slide down the tool and lessen the effect of the blow. Keep your eyes on the cutting edge of the chisel, not on your hand, and swing the hammer in the same plane as the body of the chisel. When using a chisel, always wear goggles to protect your eyes. Fig. 36-7.

Hall Enterprises

36-8. *Slitting shears, also called bench shears.*

BENCH SHEARS

One method of cutting heavy metal is to use the slitting shears, also called bench shears. Fig. 36-8. This tool comes in many sizes and capacities. The capacity is stamped on the machine. A common shear can cut ¼″ × 2″ bar stock and ⅝″ rod. Hold the workpiece firmly and keep your fingers away from the blade. For large pieces, have a friend help you hold the workpiece.

A similar machine is the throatless shears. Fig. 36-9. It is used for making irregular cuts on wider pieces of stock. Never cut rod or wire on this machine, for they can nick the blade.

QUESTIONS AND ACTIVITIES

1. Define shearing.
2. Name three common styles of tin snips and tell what type of cut each is used to make.
3. What are the safety rules to follow when using the squaring shears?
4. Name the four most common shapes of chisels and tell what each type is used for.
5. Describe the safe and proper way to hold and use a chisel.
6. Name two machines used to shear heavy metals.

CHAPTER 37

Abrading Metal

Abrading is a form of removal cutting. Small mineral grains wear away the surface of the metal until the desired size or effect is achieved. Grinding, polishing, and buffing metal are types of abrading. *Grinding* is the process of removing large amounts of material or fine finishing a metal surface by precision methods. *Polishing* is the process of removing burrs, deep scratches, and nicks from metal. *Buffing* is the process of final finishing a metal piece to improve its appearance. In this chapter you will learn about metal grinding and heavy polishing and the kinds of abrasives used. Light polishing and buffing are a part of metal finishing and will be covered in Chapter 55.

KINDS OF ABRASIVES

Abrasives are hard substances used to wear away softer materials. Whether you are grinding metal or sanding wood or plastic, you are abrading, and the cutting action is the same. The tiny particles of abrasive act as chisels to cut material. The kinds of abrasives are many. Here are some important ones.

Emery

Although many people call all abrasive papers emery papers, actually very little emery is used today except for polishing. Emery is a natural abrasive; it is used as it is mined from the earth. It is dull black and rocklike.

Flint

Flint is another natural abrasive. It is whitish and is not as good as garnet or emery. It is used primarily in woodworking.

Garnet

This is a red, very sharp and hard natural abrasive used in woodworking. It is more durable than flint. All the natural abrasives are used mainly for abrasive paper (often called "sandpaper").

Aluminum Oxide

This gray-brown abrasive is artificial. It is made from bauxite, a mineral from which aluminum is also made. A very tough, durable abrasive, it is made into grinding wheels and coated abrasives for working steel and other hard materials.

Silicon Carbide

Another artificial abrasive, this shiny black material is used on aluminum, copper, and other soft materials, including wood. It is made from coke, sand, and salt.

GRADING ABRASIVES

To grade abrasives means to separate them according to the size of the abrasive grains. This size is called the "grit number" and is determined by the number of holes in a sifting screen. For example, abrasive grains that will sift through a screen having 60 openings per square inch are said to have a grit size of 60.

Abrasive Grit Numbers

	Very Coarse	Coarse	Medium	Fine	Very Fine
ALUMINUM OXIDE & SILICON CARBIDE	12 16 20 24 30	36 40	50 60 80 100	120 150 180	220 240 280 320 360 400 500 600
GARNET	3 2½	2 1½	1 ½ 0 2/0	3/0 4/0 5/0	6/0 7/0 8/0

37-1. *Comparative abrasive grit numbers. Emery and flint papers are not generally sold according to grit size. Instead, the terms* **very coarse, coarse,** *etc., are used.*

37-2. *Coated abrasives are made by gluing abrasive grains to backing materials such as paper, cloth, or fiber. They come in many forms such as disks, belts, sleeves, and sheets.*

37-3. *Belt sanders (left) and disk sanders (right) are used for metal polishing. The abrasive is aluminum oxide.*

Abrasive grit numbers range from very fine (No. 600) to very coarse (No. 12). Fig. 37-1.

USING ABRASIVES

Abrasives are made into grinding wheels, coated papers of many shapes, sharpening stones, and polishing grains. The uses of these abrasives in metalworking are described here.

FOR YOUR SAFETY . . .

● Always wear eye protection when abrading metal.

● Avoid loose clothing. It could get caught in the machinery.

● When you are abrading small pieces, the wheel may touch your fingers. Wear gloves to protect yourself.

Polishing

To remove scratches, nicks, and burrs from metal, coated abrasives are used. These are made by coating sheet materials with abrasive grains. Fig. 37-2. Some coated abrasives are waterproof. To use them in hand methods, wrap a small piece around a block or a file, add a few drops of light machine oil, and rub back and forth gently.

Machine polishing is done on machines such as those shown in Fig. 37-3. Aluminum oxide belts, disks, and drums are used. Do not use machines reserved for sanding wood or plastic. The metal workpiece will ruin the belts or disks. Be careful that the machine does not tear the work-

37-4. *Hold the workpiece securely when machine polishing. Wear safety glasses.*

piece from your hands. Hold it securely. The workpiece should be held near the bottom of the moving belt. Fig. 37-4. Use gloves if necessary, and be sure to wear safety glasses.

Grinding

Grinding is a quick way to rough-shape metal and to sharpen tools. The floor grinder, Fig. 37-5, is used in the shop. Grinders have abrasive wheels. These wheels are made of abrasive grains bonded (stuck) together with resin, rubber, shellac, or ceramics. Make sure the wheels are free of cracks or nicks before using. The wheels are available in many shapes and sizes. For hard and brittle metals, use grit numbers 60 or 80. For softer metals, use numbers 36 or 46.

"Off-hand" grinding is so called because the workpiece is held by hand. To do off-hand grinding, the tool rest should be locked tightly in position 1⁄16″ or 1⁄8″ away from the wheel. Hold the work tightly, wear safety glasses,

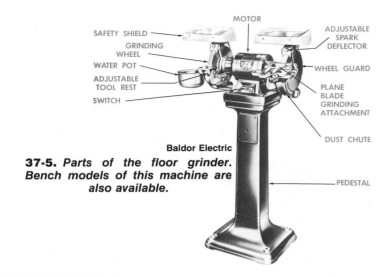

SAFETY SHIELD
GRINDING WHEEL
WATER POT
ADJUSTABLE TOOL REST
SWITCH
MOTOR
ADJUSTABLE SPARK DEFLECTOR
WHEEL GUARD
PLANE BLADE GRINDING ATTACHMENT
DUST CHUTE
PEDESTAL

Baldor Electric

37-5. *Parts of the floor grinder. Bench models of this machine are also available.*

Rockwell-Delta Company

37-6. *Hold the workpiece firmly when grinding. Press gently against the face of the wheel. Pressing against the sides of the wheel may cause the wheel to crack.*

and be certain to tuck in loose clothing. Fig. 37-6. A water pot on the grinder is used to cool the metal while grinding. Dip the metal into the water often to avoid overheating it.

A clean wheel cuts faster. Wheel dressers are used to free wheels which have become loaded with metal chips. Dressing also removes gouges from the wheel surface. Mechanical disk

dressers are used, as are diamond tip dressers. Fig. 37-7.

Sharpening Tools

Edge-cutting tools such as cold chisels, punches, drills, tool bits, and scribers should be regularly sharpened on a grinder. Use a fine wheel and be certain that you do not overheat the tools because this will soften them. Use the water pot frequently for cooling. There are special clamps available for sharpening chisels and drills. Fig. 37-8. These hold the tools at the proper angle for grinding. If such a clamp is not available, hold the tools carefully at the proper angle and press lightly against the wheel. Lathe tool bits are also sharpened on grinders.

Rockwell-Delta Company

37-7. *A diamond tip wheel dresser. This one is locked in a special clamp. Move the clamp back and forth slowly with the machine running. Move the dresser forward into the wheel as needed. Wear safety glasses.*

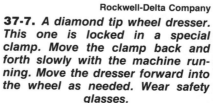

Rockwell-Delta Company

37-8. *Sharpening a cold chisel on a grinder. This special clamping device holds the chisel at the correct angle.*

ULTRASONIC MACHINING

Ultrasonic machining, Fig. 37-9, is a special way to abrade hard materials quickly, economically, and accurately. The machine used has three parts: a frame with an adjustable table; a power and frequency generator; and a pump for circulating a liquid (called a *slurry*) which contains abrasive particles.

The pump circulates the slurry

37-9. *Ultrasonic machining.*

PUMP NOZZLES

TOOL DRIVES ABRASIVE GRAINS AGAINST WORK-PIECE

ABRASIVE SLURRY

WORKPIECE   CAVITATION IS PRODUCED BY ABRASIVE CHIPPING ACTION

between the face of the tool and the workpiece. The power unit causes the tool to move up and down about 20 000 times per second. Under these conditions the tool drives against the workpiece with great force. The impact of the abrasives on the workpiece does the actual cutting (cavitation) by chipping away small pieces of the material. Typical ultrasonic machining operations include drilling, shaving, slicing, and cutting unusual punch and die shapes. Other materials, such as ceramics and plastics, can also be machined ultrasonically.

QUESTIONS AND ACTIVITIES

1. What are abrasives?
2. What two types of abrasives are most commonly used in metalworking?
3. What determines the grit number of an abrasive, and what does this number tell you about the abrasive?
4. What are coated abrasives?
5. What purposes do the water pot and wheel dresser serve on the grinder?
6. What are three safety rules to follow when doing "off-hand" grinding?
7. How does the ultrasonic cutting machine work?

CHAPTER 38

Shaping Metal

Shaping is a removal cutting process used to smooth rough surfaces and to cut grooves in metal. For example, some rough foundry castings are smoothed by shaping. The metal shaper, Fig. 38-1, has a tool which moves back and forth through a stationary (nonmoving) workpiece.

Metal planing is similar to shaping, except that in planing the workpiece moves through a stationary tool.

Shaping and planing are ways of producing smooth and accurate surfaces on metal parts. Industry now does most shaping operations on a milling machine.

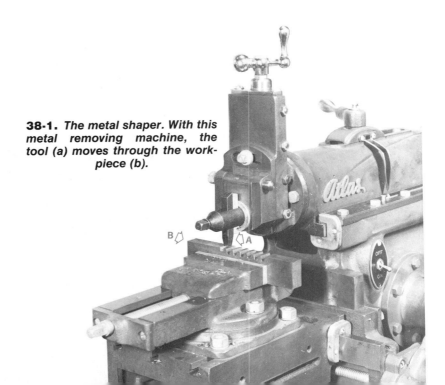

38-1. *The metal shaper. With this metal removing machine, the tool (a) moves through the workpiece (b).*

QUESTIONS AND ACTIVITIES

1. What are shapers used for?
2. How does the metal shaper work?
3. How does planing differ from shaping?

4. What machine is used in industry to do shaping operations?

CHAPTER 39

Drilling Metal

Drilling is the cutting of round holes in material. It is an important first step in preparing pieces to be bolted or riveted together, in tapping (threading holes), and in reaming (making holes smooth and accurate). Industry uses many kinds of hand and automatic drilling machinery, including robots.

DRILLING TOOLS

The common drilling tools which are found in the shop are shown in Fig. 39-1. *Hand drills,* both manual and electric models, will generally take straight shank twist drills up to ¼" in diameter. Larger models of hand drills are available for heavier work.

Twist drills (the cutting tools that fit into the drill) are generally made of high-speed steel. (That is the meaning of the marking HSS or HS found on many of these drills. High-speed steel is so called because cutting tools made from it can be run at much higher speeds than tools made of ordinary high-carbon steel.) The drill size is also marked on the shank of the drill. The size is shown as whole num-

bers, letters, or fractions. Sizes shown by whole numbers range from the largest, No. 1, to the smallest, No. 80. In lettered drills, Z is the largest, A the smallest. Fractional drills are most commonly used; they range from ¹⁄₆₄″ upwards by 64ths.

It is necessary to understand all three size systems because the twist drill you need may be available in only one of them. For example, if you need a drill with 0.368″ diameter, you will not find it in the fractional system, which skips from ²³⁄₆₄″ (0.359) to ⅜″

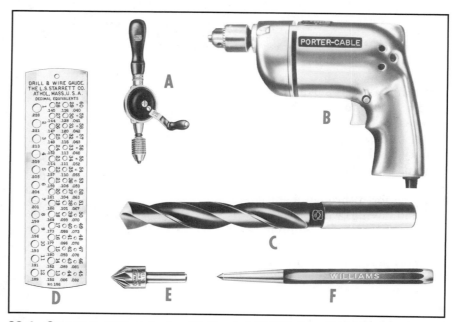

39-1. *Common drilling tools: (A) Hand drill. (B) Portable electric drill. (C) Straight-shank twist drill. (D) Drill gauge for numbered drills. (E) Countersink. (F) Center punch.*

39-2. *Mark the hole to be drilled with a center punch and ball-peen hammer.*

breakage. Hold the tools firmly and straight. Remember to locate holes with the center punch before drilling them. Otherwise the drill bit will tend to skip around, and it will be difficult to drill in the right place.

To use the hand drill, fasten the workpiece firmly in a vise. Hold the tool straight and place the drill point in the center punch hole. Turn the handle slowly, being careful not to force the drill. A drop or two of cutting oil will make drilling easier. Fig. 39-3.

The drill press is used for heavy drilling. Fig. 39-4. To operate the drill press, fasten the

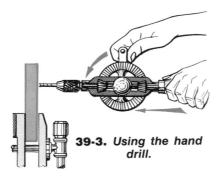

39-3. *Using the hand drill.*

(0.375). However, letter size "U" is exactly 0.368″ in diameter. A drill chart, including metric sizes, is found in Table 4 of the Appendix. Attention to such tiny differences in size is important, especially in tapping and reaming.

Drill gauges are used to check the sizes of number, letter, or fractional twist drills.

Use the *center punch* to mark holes to be drilled. Fig. 39-2.

A *countersink* is used to enlarge one end of a drilled hole to a cone shape. This is done to accommodate a flathead rivet or machine screw. The top of the rivet or screw will then be even with the surface of the workpiece.

FOR YOUR SAFETY . . .

● Always wear eye protection when drilling metal.
● Avoid loose clothing. Tuck in loose shirts; roll up long sleeves.
● Wear gloves when drilling small pieces.
● Inspect the twist drill before you start working. It should not be dull, worn, or bent.
● Apply steady, even pressure. Otherwise, you may break a drill and cause an accident.

DRILLING HOLES

Both the hand drill and the electric hand drill should be operated carefully to reduce drill

39-4. *Parts of the drill press. This machine drills accurate holes quickly and easily.*

Clausing Corporation

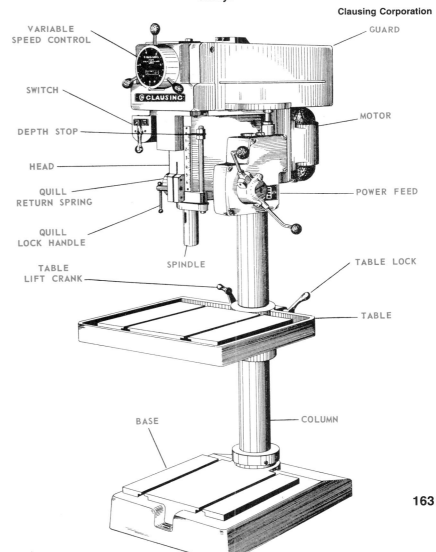

VARIABLE SPEED CONTROL
GUARD
SWITCH
MOTOR
DEPTH STOP
HEAD
POWER FEED
QUILL RETURN SPRING
QUILL LOCK HANDLE
SPINDLE
TABLE LIFT CRANK
TABLE LOCK
TABLE
BASE
COLUMN

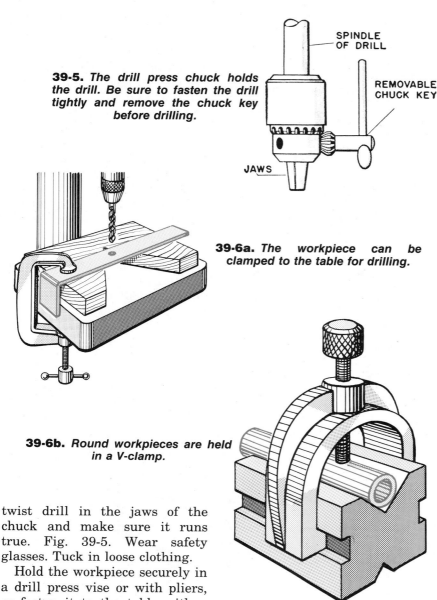

39-5. *The drill press chuck holds the drill. Be sure to fasten the drill tightly and remove the chuck key before drilling.*

SPINDLE OF DRILL

REMOVABLE CHUCK KEY

JAWS

39·6a. *The workpiece can be clamped to the table for drilling.*

39·6b. *Round workpieces are held in a V-clamp.*

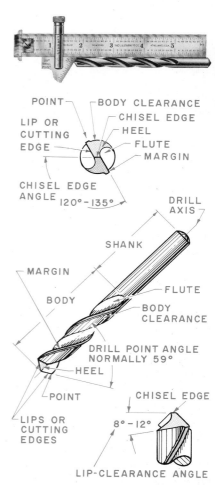

POINT — BODY CLEARANCE

LIP OR CUTTING EDGE — CHISEL EDGE — HEEL — FLUTE — MARGIN

CHISEL EDGE ANGLE 120°-135°

DRILL AXIS

SHANK

MARGIN

BODY

FLUTE

BODY CLEARANCE

DRILL POINT ANGLE NORMALLY 59°

HEEL

POINT

LIPS OR CUTTING EDGES

CHISEL EDGE

8° - 12°

LIP-CLEARANCE ANGLE

The L. S. Starrett Company

39-7. *Study the parts of the drill to make sure you are sharpening it correctly. Use the drill gauge to check the angle of the drill point.*

twist drill in the jaws of the chuck and make sure it runs true. Fig. 39-5. Wear safety glasses. Tuck in loose clothing.

Hold the workpiece securely in a drill press vise or with pliers, or fasten it to the table with a clamp. Fig. 39-6. The drilling speed varies with the diameter of the twist drill and the kind of material to be drilled. For example, a ¼″ hole in aluminum or brass should be drilled at about 4500 RPM (revolutions per minute); in mild steel, 1500 RPM. For a ½″ hole in aluminum, operate at 2200 RPM; in mild steel, 700 RPM. In other words, use fast speeds for small diameter drills and soft materials and slow speeds for large drills and hard materials.

With the power on, apply enough pressure to produce chips. A cutting oil will improve the cutting action of the drill.

SHARPENING TWIST DRILLS

A dull, worn, or bent drill is dangerous to use. Inspect it before using. If the drill needs sharpening, dress it carefully with a grinder. Be careful not to overheat the drill. Use the water-

pot often. Check the drill with a gauge frequently to be sure it is being ground at the proper angle. Fig. 39-7.

REAMING

Reamers are used to enlarge holes and make them true. A drilled hole is usually slightly oversize, which is all right for rivets and bolts. When greater accuracy is required, the hole is first drilled undersize and then reamed to the proper size. Reaming is done with a reamer held in a tap wrench. Fig. 39-8.

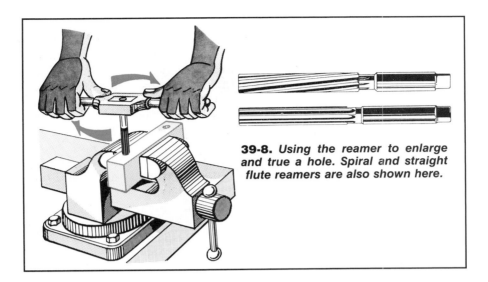

39-8. *Using the reamer to enlarge and true a hole. Spiral and straight flute reamers are also shown here.*

QUESTIONS AND ACTIVITIES

1. Define drilling.
2. What are the three systems for identifying drill sizes? What is the smallest size drill in each system?
3. What size metric drill comes closest to the ⅜″ customary size?
4. What is a countersink used for and why is it important?

5. How do you mark holes to be drilled? Why is it important to do this?
6. What safety rules should you keep in mind when operating the drill press?
7. What rule should you keep in mind when determining the speed to use on a drill press?
8. What are reamers used for?

CHAPTER 40

Milling Metal

MACHINE MILLING

Milling metal with a machine is a removal cutting process in which a wide cutter revolves against a workpiece. Fig. 40-1. The machine in Fig. 40-2 is called a horizontal mill because of the position of the tool. The vertical mill works like a drill press and can mill surfaces or true holes at many angles. Both are accurate machines for producing smooth surfaces and for cutting grooves and slots.

FILING

Filing is a common hand milling operation which is widely used in the shop. A file works much like the mill. If you took a milling cutter and stretched the teeth out on a flat surface, you would have a file. Fig. 40-3.

Files are used to remove scratches and nicks from metal, to remove burrs, to shape material, and to sharpen tools.

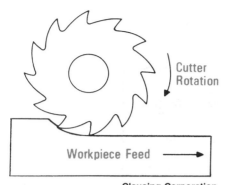

Clausing Corporation

40-1. *The cutting action of the milling machine.*

Clausing Corporation

40-2. *The horizontal milling machine removes metal quickly and accurately. The metal workpiece (A) moves against the revolving cutter (B).*

Types of Files

Files are classified according to length, shape, type of tooth (cut), and coarseness (number of teeth in relationship to the file length). Files range in length from 3″ to 20″. An 8″ or 10″ file is best for general use. Some common shapes are shown in Fig. 40-4. Files are made with various types of teeth for different kinds of cutting. Fig. 40-5. A double-cut file has two rows of teeth which cut in opposite directions. There are also single-cut files. Rasp-cut files are for woods and very soft metals, as are curved-tooth files. File coarseness grades range from rough to dead smooth.

Use mill files for general-purpose work. Mill files have a rectangular shape and are slightly tapered in width and thickness. They have single-cut teeth. Fig. 40-4, A. Needle, or jeweler's, files have finely cut teeth. They are used for filing on jewelry, or in special die work. Figs. 40-6 & 40-7.

Using Files

Before using a file, make certain it is clean and free of oil and grease. When the teeth of a file become clogged with chips and filings, the file must be cleaned. This is best done with a file card and brush. The card loosens the metal in the teeth, and the brush clears the metal from them. Fig. 40-8.

Always place the proper-size handle over the tang of the file. (The *tang* is the sharp end of the file.) Check to see that the handle is tight. Never use a file without a handle.

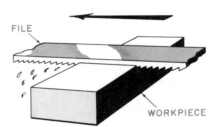

40-3. *The cutting action of the file. This tool is like a stretched-out milling cutter. Note that the cutting is done on the forward stroke of the file.*

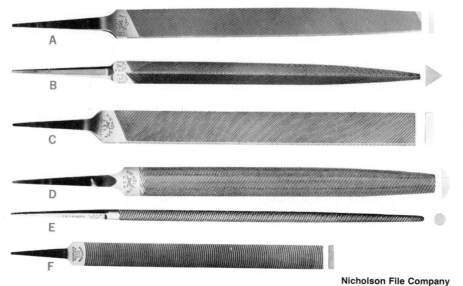

Nicholson File Company

40-4. *Common shapes of files: (A) Mill. (B) Triangular. (C) Hand. (D) Half-round. (E) Round. (F) Curved-tooth.*

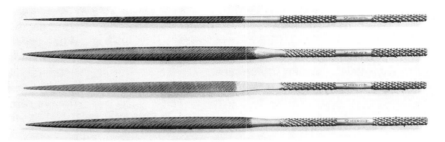

Simonds File Company

40-6. *Some examples of jeweler's files*

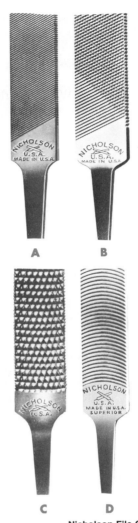

Nicholson File Company

40-5. *The cut of the file shows how the teeth are cut on the file face: (A) Single cut. (B) Double cut. (C) Rasp cut. (D) Curved cut.*

40-7. *Using the jeweler's file. The file cuts on the forward stroke.*

40-8. *To get best results from your files, keep them clean.*

40-9. *Draw filing produces a smooth, true surface on a workpiece.*

The workpiece should be fastened securely in a vise. Grasp the file by the handle and the tip and press gently against the workpiece. The file cuts on the forward stroke.

Draw filing is done to get a flatter (more even) and more finely finished surface than ordinary filing (cross-filing) provides. Edges can be finished by this method. A mill file is generally recommended for draw filing. To draw file, hold the file with one hand on each end of the file, and your thumbs about ½″ to ¾″ from the work. Hold the file flat against the work and move back and forth along the work, pressing down only on the forward stroke. Fig. 40-9.

FOR YOUR SAFETY . . .

● Always put a handle over the tang of the file before using it.

● Use the right file for the job.

● Never use a file for prying or hammering. The file could break, sending sharp chips in all directions.

● Keep file teeth clean. A clogged file could slip off the workpiece and injure you.

QUESTIONS AND ACTIVITIES

1. What are two kinds of milling machines? What tool is used to mill by hand?

2. How are files classified?

3. List four common file shapes.

4. Describe a mill file. What stroke does it cut on?

5. What is another name for a needle file, and what is it used for? What stroke does it cut on? What type teeth does it have?

6. When must a file be cleaned? How is it cleaned?

7. What is the tang of a file? What should you always remember to do with the tang?

CHAPTER 41

Turning Metal

The metal-cutting lathe is a machine which removes metal by a process called *turning*. In this process the cutting is done by revolving the workpiece against a fixed, single-edge tool. The lathe is used for many kinds of cutting. For example, it can be used for accurate turning to precise dimensions, for cutting threads on shafts, for cutting tapers, and for drilling holes. Figure 41-1 shows the parts of the lathe. The size of the lathe is determined by the swing—that is, the largest diameter which can be turned on it—and by the length of the bed.

The lathe is so basic in metalworking that it is sometimes called the father of all machine tools. Industry uses several kinds of complicated lathes for production work, including the turret lathe, the automatic screw machine, and the tracer lathe. The operation of the type of lathe generally used in the school shop is described in this chapter.

LATHE ACCESSORIES

Lathe dogs are used to hold the drive (headstock) end of the workpiece when turning between centers. The three types of lathe dogs are shown in Fig. 41-2.

The *faceplate* fastens to the headstock spindle and holds the lathe dog in place. Fig. 41-3.

Chucks are used to hold workpieces for chuck-turning, drilling, reaming, and tapping. The two types of chucks are shown in Fig. 41-4.

Tool bits are small rectangular cutting tools that are made of hard, high-speed steel. They are held against the workpiece as it revolves on the lathe. Tool bits come in several shapes for different types of cutting. Fig. 41-5.

Toolholders hold tool bits and are in turn held in the *tool post*. There are several types of toolholders, as shown in Fig. 41-6.

The *knurling tool*, Fig. 41-7, is used to produce a cross-hatched design in metal. Knurling is done on tool handles to give the tool a better grip.

MEASURING TOOLS

There are a number of different measuring tools used in machine shop work. A steel rule (see Chapter 33) has many uses, as do the inside and outside calipers. Typical uses of these tools are shown in Figs. 41-8 & 41-9. One very important precision tool is the micrometer. Fig. 41-10.

USING THE LATHE

Common lathe operations include turning between centers and chuck turning. The workpiece should be about 1″ longer and ⅛″ larger in diameter than the finished size. Be careful when using the lathe. Follow the safety rules given here.

FOR YOUR SAFETY . . .

• Always wear eye protection when using the lathe.

• Tuck in loose clothing. Roll up long sleeves.

• Keep your hands away from the rotating stock. Be sure to keep your left elbow high so it won't be hit by the rotating lathe dog.

• Use the correct speeds and feeds. The cutting action should produce short chips.

• The metal chips are very sharp. Don't touch them.

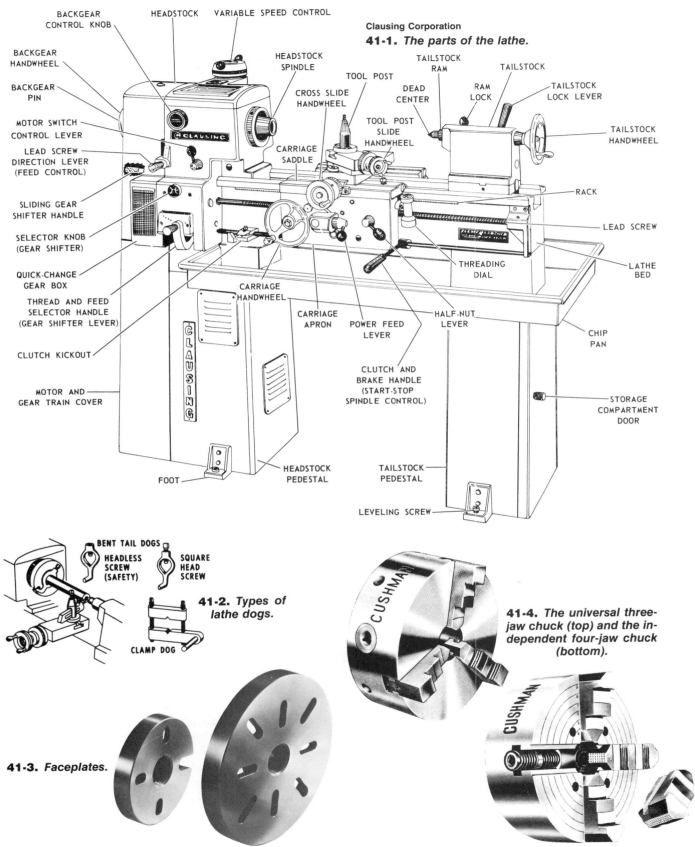

BACKGEAR CONTROL KNOB

HEADSTOCK

VARIABLE SPEED CONTROL

Clausing Corporation
41-1. *The parts of the lathe.*

BACKGEAR HANDWHEEL

BACKGEAR PIN

MOTOR SWITCH CONTROL LEVER

LEAD SCREW DIRECTION LEVER (FEED CONTROL)

SLIDING GEAR SHIFTER HANDLE

SELECTOR KNOB (GEAR SHIFTER)

QUICK-CHANGE GEAR BOX

THREAD AND FEED SELECTOR HANDLE (GEAR SHIFTER LEVER)

CLUTCH KICKOUT

MOTOR AND GEAR TRAIN COVER

FOOT

HEADSTOCK PEDESTAL

HEADSTOCK SPINDLE

CROSS SLIDE HANDWHEEL

TOOL POST

CARRIAGE SADDLE

TOOL POST SLIDE HANDWHEEL

CARRIAGE HANDWHEEL

CARRIAGE APRON

POWER FEED LEVER

CLUTCH AND BRAKE HANDLE (START-STOP SPINDLE CONTROL)

TAILSTOCK PEDESTAL

TAILSTOCK RAM

DEAD CENTER

RAM LOCK

TAILSTOCK

TAILSTOCK LOCK LEVER

TAILSTOCK HANDWHEEL

RACK

LEAD SCREW

THREADING DIAL

HALF-NUT LEVER

LATHE BED

CHIP PAN

STORAGE COMPARTMENT DOOR

LEVELING SCREW

BENT TAIL DOGS

HEADLESS SCREW (SAFETY)

SQUARE HEAD SCREW

CLAMP DOG

41-2. *Types of lathe dogs.*

41-3. *Faceplates.*

41-4. *The universal three-jaw chuck (top) and the independent four-jaw chuck (bottom).*

169

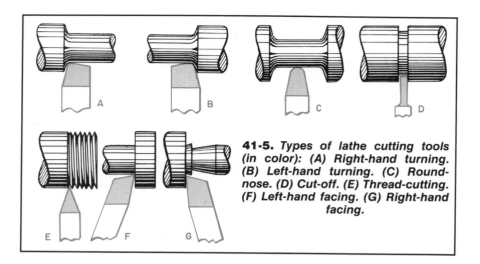

41-5. *Types of lathe cutting tools (in color): (A) Right-hand turning. (B) Left-hand turning. (C) Round-nose. (D) Cut-off. (E) Thread-cutting. (F) Left-hand facing. (G) Right-hand facing.*

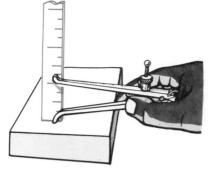

41-8. *Using the inside calipers.*

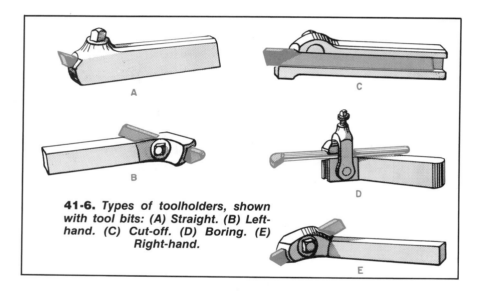

41-6. *Types of toolholders, shown with tool bits: (A) Straight. (B) Left-hand. (C) Cut-off. (D) Boring. (E) Right-hand.*

Clausing Corporation

41-7. *Using the knurling tool.*

Work Held between Centers

1. The main way of holding work in a lathe is to mount it between centers. Fig. 41-11. To do this, drill countersunk holes in the ends of the work. Locate these as shown in Fig. 41-12. Drilling methods are illustrated in Fig. 41-13.

2. Clamp a lathe dog on the drive (headstock) end of the workpiece so that the tail of the dog will fit in a slot in the faceplate. Make sure the tail of the lathe dog does not bind in the faceplate slot. Align the hole in

the other end of the workpiece with the dead center (tailstock center). Turn the tailstock hand-wheel until the dead center meets the hole in this end of the workpiece. The dead center should fit the countersunk hole just loosely enough so the workpiece will revolve, but not so loosely that it will wobble. Place a little lubricant on the dead center to reduce friction.

3. Place a sharpened tool bit in the toolholder. Lock this in the tool post. Set the cutting edge of the tool so that it is slightly above the centerline of the workpiece. The tool should be turned slightly away from the headstock.

4. Adjust the speed so that it is correct for the kind of metal and the diameter of the workpiece. Generally, fast speeds are needed

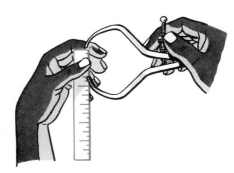

41-9. *Using the outside calipers.*

41-10. *The micrometer must be used carefully. Tighten the barrel against the workpiece lightly. Do not force it, or you will damage the tool. See Chapter 8 for information on reading a micrometer.*

41-11. *Work held between centers on the metal lathe. What type of lathe dog is being used?*

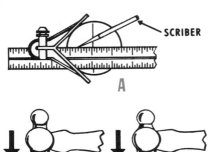

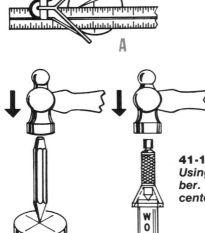

SCRIBER

A

PUNCHING THE CENTER

B

BELL CENTER PUNCH METHOD

C

41-12. *Ways to locate centers: (A) Using the centering head and scriber. (B) Marking the center with a center punch. (C) Using the bell punch.*

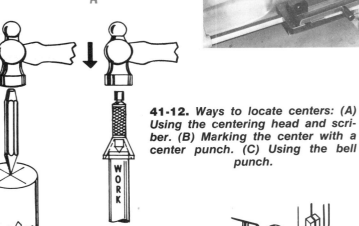

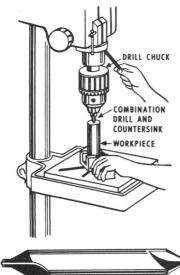

DRILL CHUCK

COMBINATION DRILL AND COUNTERSINK

WORKPIECE

COMBINATION DRILL AND COUNTERSINK

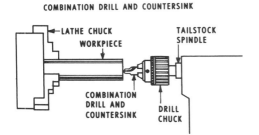

LATHE CHUCK
WORKPIECE

TAILSTOCK SPINDLE

COMBINATION DRILL AND COUNTERSINK

DRILL CHUCK

41-13. *Two methods of drilling center holes.*

for soft metals and smaller diameters, and slow speeds for large diameters and hard metals.

5. Move the carriage back and forth by turning the carriage handwheel. The point of the tool should clear the right end of the workpiece, and it should not touch the lathe dog. The carriage should move from the tailstock to the headstock when the power is on.

6. The outside calipers should be set about 1/16″ larger than the finished diameter.

7. For rough cutting (turning the workpiece to "roughly" the finished size), move the carriage over until the tool point clears the right end of the workpiece. Start the lathe. Turn the cross slide handwheel in and the carriage handwheel toward the headstock until the tool begins to remove a small chip. Continue to feed the carriage handwheel by hand for a short distance. Stop the lathe and check the trial cut with the calipers to make sure you have left enough stock for the finishing cut—usually about 1/16″. If the diameter is satisfactory, restart the lathe and engage the power (longitudinal or lengthwise) feed lever. Check the cutting action. The chips should roll off in short sections—long

chips can be dangerous. Make sure there is no chattering (vibration caused by slight jumping of the workpiece away from the cutting tool). Continue the cut until over half the length of the stock has been cut. At the end of the cut, turn the cross slide out and release the power feed lever at the same time.

8. Return the carriage to the right-hand starting position. Repeat the rough cutting operation if more material must be removed.

9. If the full length of the workpiece is to be turned, remove the workpiece and turn it end for end. Remember to lubricate the dead center. Rough cut the remaining length. Resharpen the tool bit, and set the lathe for a higher speed and finer feed. Check the dead center to see if it needs more lubricant. Finish turning the workpiece, checking carefully with the caliper now set for the finished diameter. After the finish cut, again turn the workpiece end for end and finish cut the second half.

Work Held in a Chuck

Holding the workpiece in a chuck makes it possible to turn short pieces without using the tailstock for support. This method also permits drilling, boring, reaming, tapping, and cutoff operations.

1. Mount a three- or four-jaw chuck on the lathe. Mount the workpiece in the chuck. Make sure it is seated properly; then tighten with the chuck wrench. Fig. 41-14.

2. To face the end of the workpiece, place a left-hand facing tool in the tool post. Set the point of the tool at the center of the workpiece. Move the carriage so that the tool clears the workpiece. Turn the lathe on. The rough cutting should start at the outside of the workpiece and move into the center. For the finish cut, start at the center and move outwards.

Remember to lock the carriage by turning the carriage lock screw. The tool is fed by hand for most cutting. For large diameters, use the power cross feed.

3. Turning the chuck-held workpiece to diameter is done as you would when working between centers. Be careful not to run the tool into the chuck jaws.

4. Drilling is done as shown in Fig. 41-15. Set the lathe for the proper drilling speed and feed the drill slowly. Clear the chips from the drill frequently and use cutting oil.

5. The collet chuck is shown in Fig. 41-16. This is a special chuck available in a number of diameters. It speeds the chuck turning operation.

FILING AND POLISHING

Both chuck-held and lathe dog-held workpieces can be filed and polished on a lathe. Choose a smooth file, clean and free of nicks. Set the lathe for a faster speed and slowly press the file along the surface of the workpiece. Be very careful. Make certain your arm clears the machine. Roll up your sleeves and wear goggles. Fig. 41-17. Use a file card to clean the file frequently.

Polishing is done in the same way, using strips of abrasive cloth and machine oil.

Clausing Corporation

41-14. Here the workpiece is being held in a universal three-jaw chuck.

Clausing Corporation

41-15. Drilling on a lathe with the workpiece held in a universal three-jaw chuck. Reaming and tapping are done in a manner similar to this.

Clausing Corporation

41-16. *Using the collet chuck.*

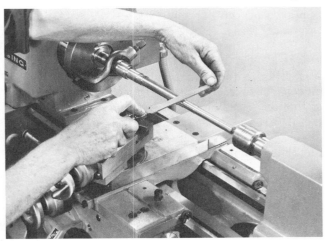

Clausing Corporation

41-17. *Filing on the lathe. This is a dangerous operation. Be sure your sleeves are rolled up. Hold your arm out of the way of the revolving lathe dog.*

QUESTIONS AND ACTIVITIES

1. Define the metal removal process called turning.

2. Name two operations besides turning that can be performed on the lathe.

3. What are lathe dogs used for? Name the types shown in this chapter.

4. What are tool bits? Name three types.

5. What is knurling and what is it used for?

6. When is a chuck used? Name the two types of chucks.

CHAPTER 42

Etching and Thermal Cutting of Metal

Not all metal cutting involves the use of sharpened tools. Chemicals and heat are also used in both separating and removing metal. In *etching,* corrosive chemicals (such as acid) are used to eat away unprotected areas on the surface of the metal. Etching is used to decorate the surface of

metal. *Chemical milling,* an industrial process similar to etching, is a method of cutting away metal from complex shapes or very large surfaces.

In *thermal cutting,* a section of a metal workpiece is melted to separate it into two parts. Thermal cutting is closely related to

welding. The main difference is that, in welding, the molten metal remains to join two workpieces together. Industry uses flame, carbon arc, powder, and laser thermal cutting processes. Flame cutting is commonly used in school shopwork. In flame cutting, metal is heated and oxygen

173

is played on the hot spot to melt the metal away.

ETCHING

Etching in the school shop is a process for decorating the surface of metal. In this process, the metal article to be etched is placed in an etching solution. Metal is removed or eaten away by the action of the solution. Those parts of the surface which are not to be etched are covered by a *resist*. This material prevents the etching solution from eating away the metal. Fig. 42-1.

The following steps describe how etching is done:

1. The article to be etched should be polished, buffed, and cleaned. Clean it by wiping with a soft cloth dipped in lacquer thinner. Once the article has been etched, it should not be buffed again, as this will destroy the effect of the etching.

2. Carefully transfer the design you want to the metal, using carbon paper and a firm pencil.

3. Apply asphaltum varnish resist with a fine brush. Flow the material on so that everything is covered except the areas to be etched. Allow this to dry for 24 hours. Plastic self-adhering wallpaper, beeswax, and masking tape may also be used as resist materials.

4. Mix the etching solution. For copper and brass, mix one

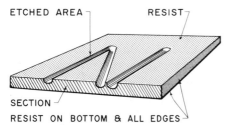

42-1. The process of etching. The etching solution will not attack those areas covered by the resist.

ETCHED AREA — RESIST —
SECTION —
RESIST ON BOTTOM & ALL EDGES —

42-2. A design etched on a metal flower vase. A deep etch like this requires that the piece be left in the solution longer.

part nitric acid with one part water. For aluminum, use one part muriatic (hydrochloric) acid and one part water. **CAUTION:** *Always pour the acid into the water. Pouring water into acid is dangerous.*

There are safer etching solutions called mordants. Mordants are available for etching most metals and should be used whenever possible in the shop. Wear goggles, gloves, and a rubber apron when working with acids or mordants.

5. Dip the article into the etching solution. The longer the article is left in the solution, the deeper the etch. When the etching is deep enough, remove the article from the solution with plastic or wooden tongs. Rinse with water and remove the resist. Asphaltum varnish can be removed with paint thinner, lacquer thinner, or turpentine. Gently rub the surface dry with a soft cloth. Apply a suitable finish.

(See Chapter 54.) A completed workpiece is shown in Fig. 42-2.

FOR YOUR SAFETY . . .

Follow these safety rules when etching metal.

• Always wear protective clothing when working with acids—gloves, face guard, sleeve covers, apron. This clothing should all be heat and acid resistant. It should be heavy canvas, not ordinary cloth.

• Always pour acid into water, never the reverse. Pouring water into acid will cause the solution to bubble and splash.

• Use the "buddy system." Never work alone.

• Always work in a well-ventilated area.

• Use plastic or wooden tongs, not your hands, to dip articles into the etching solution.

THERMAL CUTTING

Flame Cutting

Flame cutting is used in manufacturing, construction, and in scrapping and salvage work. Police and fire departments use flame cutting in rescue work, such as freeing people from wrecked vehicles.

Oxyacetylene cutting is one type of flame cutting. In this process metal is first heated, using an oxyacetylene torch. Fig. 42-3. When the metal is red hot, the oxygen lever on the torch is pressed. This causes pure oxygen to be fed to the tip of the torch, and the cutting begins. Other fuels besides acetylene can be used. Some of these are propane, natural gas, and hydrogen.

Arc Cutting

In arc cutting, metal is melted by the heat of an electric arc that passes between the metal and an

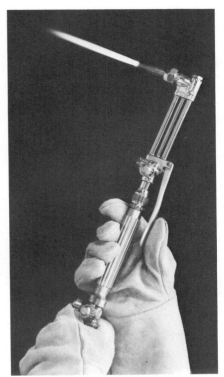

42-3. *The oxyacteylene cutting torch. The flame is a mixture of oxygen and acetylene gas.*

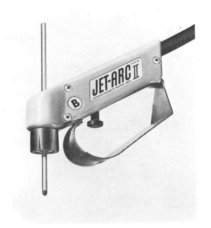

42-4. *A carbon arc cutting torch has a carbon electrode. The torch melts metal, and a blast of air blows it away from the cutting path.*

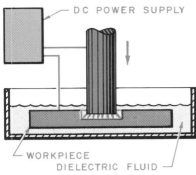

DC POWER SUPPLY

WORKPIECE

DIELECTRIC FLUID

42-6. *Electrical discharge machining removes metal by melting it.*

electrode. Oxygen, compressed air, or inert gas may be used in addition to the arc. Fig. 42-4.

Arc cutting usually does not produce work of the same quality and smoothness as flame cutting. However, it can be used on all types of metals, whereas flame cutting is suitable only for ferrous metals.

Laser Cutting

In laser cutting, the heat of a concentrated beam of light is used to separate or remove metal. One example is the use of lasers to manufacture blades for circular saws. In this process, a laser cutting machine is computer-programmed to produce the size and shape of the teeth. The laser vaporizes the metal rapidly, without warping or cracking it. Fig. 42-5.

Electrical Discharge Machining

Electrical discharge machining (EDM) is the removal of metal by the energy of an electric spark that arcs between a tool and a workpiece. The tool and the workpiece are immersed in a dielectric fluid such as oil. *Dielectric* means that the fluid will not conduct direct electric current. The tool is placed at a slight distance from the workpiece to be machined. Rapid pulses of electricity are delivered to the tool, causing sparks to jump from the tool to the workpiece. The heat from each spark melts away a small amount of metal. As the metal is removed, it is cooled and flushed away by the fluid being circulated through the spark gap. The spark also removes material from the tool, so the tool is slowly consumed as machining progresses. Because the tool is destroyed, complex shapes are not often produced by EDM. The greatest use of the process is to produce holes. Fig. 42-6.

42-5. *A laser cutting machine in operation.*

QUESTIONS AND ACTIVITIES

1. Define etching. What is this process used for in the school shop?

2. Define thermal cutting.

3. What is a resist used for? Name three resist materials.

4. List the safety rules to follow when etching metal.

5. What is arc cutting? Name one advantage and one disadvantage of arc cutting compared to flame cutting.

6. What is laser cutting? What is an advantage of this form of cutting?

7. Define electrical discharge machining. What is the most common use of this process? What is a disadvantage of this process?

CHAPTER 43

Metal Forming Principles

Very simply, *metal forming* is giving a shape to a piece of metal without adding to or removing any of the metal. You can also give shape to metal by turning it on a lathe, but you would have to remove some metal. For example, a bar of metal 2″ square and 1′ long could be turned and drilled on the lathe to make a piece of pipe 1′ long. However, a better way of producing the pipe is to *form* it by extrusion. The difference between these two ways of producing pipe illustrates the effectiveness of the forming process. The pipe is formed quickly without any metal removal or waste.

The first four basic processes shown in the chart in Fig. 43-1 are commonly done in the shop. They are also used in product manufacture. Drawing, extruding, and rolling are mainly metal fabrication processes used to produce raw materials for industry.

Forging in the shop is mainly done by hand. For instance, you might flatten the end of a piece of rod to form a screwdriver. In this case the anvil becomes a kind of

43-1.

Forming

Forming is the process of shaping a material without adding to or removing any of the material.

KINDS OF FORMING	DEFINITION	EXAMPLES
Bending	Forming by uniformly straining metal around a straight axis.	Operations on the bar folder, sheet metal brake, box and pan brake, bender, forming rolls.
Casting	Forming by pouring molten metal into a hollow cavity and allowing it to harden.	Operations such as die casting, sand casting, investment casting, powdered metal sintering (special form of casting).

(Continued)

Forming (Continued)

KINDS OF FORMING	DEFINITION	EXAMPLES
Forging HAMMER WORKPIECE ANVIL OR DIE	Forming by applying blows or steady pressure to a heavy workpiece, forcing it to take the shape of a die.	Operations such as hand forging, drop forging, automatic closed die forging.
Pressing PUNCH (MALE PART) — WORKPIECE DIE (FEMALE PART)	Forming by forcing sheet material between two dies.	Operations such as stamping, embossing, hand raising and sinking, spinning, metal tooling.
Drawing ROD (WORKPIECE) DIE GRIPPERS WIRE	Forming by pulling a metal rod or ribbon through a die to reduce it to a wire or form a tube.	Tube and wire drawing operations.
Extruding RAM DIE EXTRUSION BILLET (WORKPIECE)	Forming by forcing metal through an opening (or die) which controls its cross-sectional area.	Molding and channel extrusion operations.
Rolling ROLLER WORKPIECE ROLLER	Forming by passing metal between rollers which change its cross-sectional area.	Metal sheet and bar rolling operations.

die which controls the shape of the metal. In industry, forging is done by machines.

Industrial pressing is done with huge machinery to form such items as pots and pans and automobile fenders. In the shop, typical pressing operations might involve sinking a copper ashtray into a wooden form by hammer blows, or tooling a piece of aluminum foil with a modeling tool.

As you read the following chapters on metal forming, try to learn something about how industry might do these same kinds of forming, but under mass production conditions.

QUESTIONS AND ACTIVITIES

1. Define metal forming.
2. What is the advantage of metal forming over metal cutting?
3. List the seven basic kinds of metal forming; give an example of each.

4. Which forming processes are commonly done in the shop?

CHAPTER 44

Bending Metal

As you read in the chart in Chapter 43, *bending* is the process of forming by uniformly straining metal around a straight axis. There are many ways of bending metal. Machines used to bend light sheet metals are different from those used for heavy rods or bars. There are also many hand tools which are used in bending. But the purpose of all bending is the same—to change the form of the metal so that it can be made into usable products. Metal formed by this process is bent in one direction only. This chapter shows some

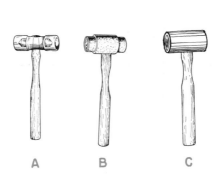

Niagara Tools

44-1. Metal bending stakes are used for light- and heavy-gauge metals: (A) Conductor. (B) Double seaming. (C) Beak-horn. (D) Candle mold. (E) Creasing. (F) Blowhorn. (G) Needlecase. (H) Stake holder.

44-2. Hammers for bending light metals: (A) Plastic-tipped hammer. (B) Fiber mallet. (C) Wooden mallet. The mallets shown here will not mar or nick the surface of the metal.

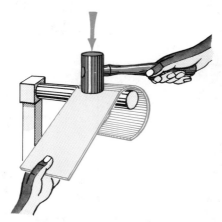

44-3. Making a curved bend on a conductor stake. Do not use metal hammers on stakes.

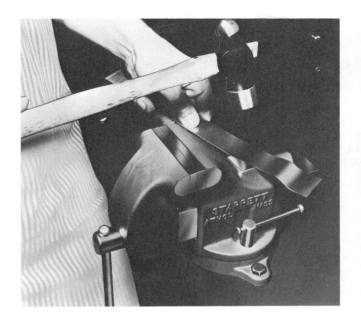 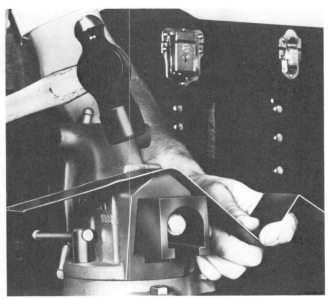

The L. S. Starrett Company

44-4. *Bending sheet metal on the vise. The metal can also be clamped in the vise jaws to make sharp bends.*

ways of changing the form of light and heavy metals.

BENDING LIGHT METALS

Simple hand bends can be made by forcing sheet metal around stakes made for this purpose. Fig. 44-1. Mallets and hammers can be used to work the metal on these stakes. Fig. 44-2. Do not use metal hammers for sheet metal stake bending. They will mar and dent the stake and

the metal. Fig. 44-3. The metal can also be clamped on a bench or between blocks of wood in a vise and folded over by beating.

The *metal vise* can be used for many kinds of bends. Curved bends are made on the special vise surfaces. Fig. 44-4. Sharp bends are made by clamping the workpiece between the jaws of the vise and striking with a hammer.

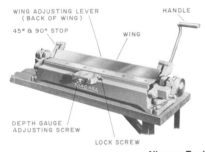

Niagara Tools

44-6. *Parts of the bar folder.*

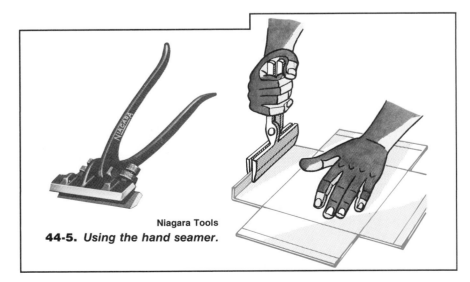

Niagara Tools

44-5. *Using the hand seamer.*

The *hand seamer* is used to make straight, shallow bends in light-gauge sheet metals. Fig. 44-5.

Bar folders work much like the hand seamer except that they can make longer bends in heavier sheet metals. Fig. 44-6. Shop models of this machine usually are 30″ wide. They can make bends from ⅛″ to 1″ wide in metals up to 22 gauge. Bends, hems, and open folds for grooved seams or wired edges can be made on this machine. Fig. 44-7.

To use the bar folder, set the depth adjusting gauge to the size

179

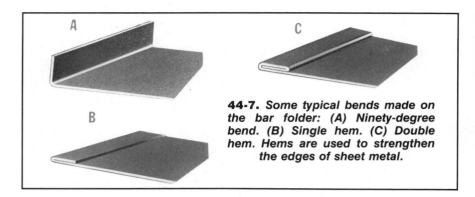

44-7. Some typical bends made on the bar folder: (A) Ninety-degree bend. (B) Single hem. (C) Double hem. Hems are used to strengthen the edges of sheet metal.

44-10. Using the box and pan brake. It is a good idea to bend a little past the desired angle to allow for springback in the metal.

Peck, Stow, and Wilcox Company
44-8. Parts of the box and pan brake.

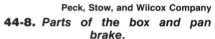

GAGE ADJUSTMENT KNOBS
BENDING WING
CLAMPING BAR WITH REMOVABLE FINGERS
CLAMPING BAR HANDLE
BENDING WING LEVERS

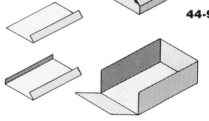

44-9. Typical bends made on a box and pan brake.

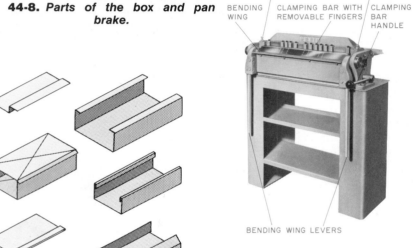

of bend desired. Set the wing adjusting lever for a sharp or rounded bend, as needed. Set the 45° or 90° stop, if necessary. Carefully place the workpiece in the opening, hold it firmly, and pull the handle to make the bend. Caution: Handle the metal with care to avoid painful cuts.

The *box and pan brake,* Fig. 44-8, is used for deeper bends in sheet metals. It is used much like the bar folder and comes in lengths up to eight feet. The machine can make many kinds of bends for a variety of product applications. It has removable fingers, which permit the bending of boxes and trays. Fig. 44-9. Larger machines of this type are called cornice brakes and do not have removable fingers.

To use the box and pan brake, lift the clamping bar handle and insert the metal in the brake. Tighten the clamping bar handle with the layout line marking the

bend directly under the clamping bar. Lift the bending wing until the bend is made. Fig. 44-10.

Cylinders are formed on the *slip roll forming machine.* Fig. 44-11. To use the machine, adjust the bottom roll to the thickness of the metal workpiece. The metal should just slip between the top and bottom rolls. The back, or idler, roll is adjusted to the diameter of the cylinder to be formed. Insert the metal and crank it through to form the cylinder. Several passes may have to be made. Readjust the back roll as needed. Trip the release handle to remove the formed workpiece. Fig. 44-12.

BENDING HEAVY METALS

Sharp bends can be produced in heavy metals by placing the material in a sturdy vise and striking it with a ball-peen hammer. To prevent the vise jaws from marring the metal, use vise jaw protectors. Metal can also be twisted in the vise by turning the workpiece with an adjustable wrench. Fig. 44-13. (To prevent

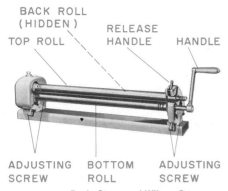

BACK ROLL (HIDDEN)
TOP ROLL
RELEASE HANDLE
HANDLE
ADJUSTING SCREW
BOTTOM ROLL
ADJUSTING SCREW

Peck, Stow, and Wilcox Company

44-11. The slip roll forming machine is used for rolling metal into cylinders or cones.

the twist from becoming crooked, slip a piece of pipe over the workpiece.) Round bends can be made by hammering the metal over a piece of heavy pipe or by using the bending fork. The fork is also used for forming scrolls. Fig. 44-14.

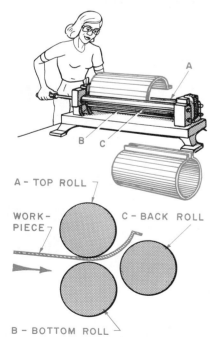

A - TOP ROLL
WORK-PIECE
C - BACK ROLL
B - BOTTOM ROLL

44-12. Using the slip roll forming machine. The rolls are adjusted for thickness of metal and diameter of bend: (A) Top roll. (B) Bottom roll. (C) Back roll.

BENDING BARS, RODS, AND TUBES

Metal bars, rods, and tubes are easily and accurately bent with special *bending machines*. All bending machines merely provide a way to apply power, either manually or mechanically, for the bending operation. They also supply mountings for the bending tools. Fig. 44-15. These tools are a form or radius collar having the same shape as the desired bend, a clamping block or locking pin that securely grips the material during the bending operation, and a forming roller or follow block which moves around the bending form.

When bending materials such

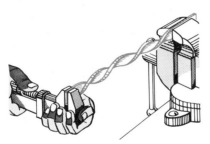

44-13. Metal can be twisted as shown here.

as tubing, channel, or angle, the bending form should exactly fit the contour of the metal to provide support during the forming operation. This is also true of the clamping block and forming roller. Only by completely confining the metal can a perfect bend be made. As with any metal operation, the results obtained will be in direct proportion to the care taken in properly tooling the machine for the job to be done.

44-14. Use the bending fork for forming scrolls and rounded bends.

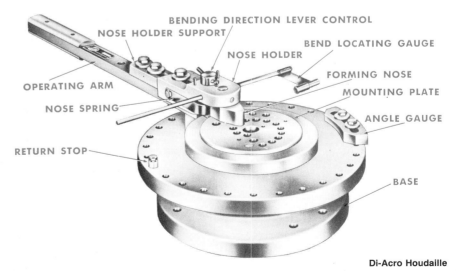

BENDING DIRECTION LEVER CONTROL
NOSE HOLDER SUPPORT
BEND LOCATING GAUGE
NOSE HOLDER
FORMING NOSE
OPERATING ARM
MOUNTING PLATE
NOSE SPRING
ANGLE GAUGE
RETURN STOP
BASE

Di-Acro Houdaille

44-15. The parts of the bending machine. Several styles and models of these machines are available, both power and manually operated.

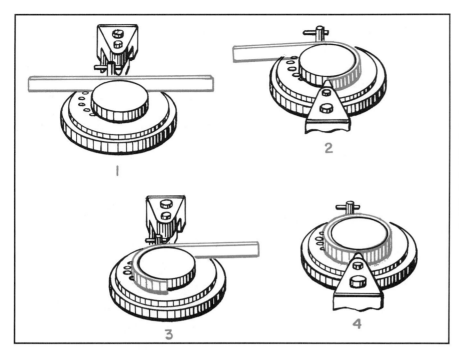

Di-Acro Houdaille

44-16. *Forming a circle on a bending machine: (1) Set forming nose against material and clamp material against radius collar with locking pin. (2) Advance operating arm until forming nose reaches extreme end of material. (3) Relocate material and clamp with locking pin at a point where radius is already formed. (4) Advance operating arm until forming nose again reaches extreme end of material.*

A complete circle can also be formed in one operation by clamping the material at one end and revolving the operating arm 360°. However, this method produces longer unformed lengths of material where the ends meet than when using the two-operation method shown in Fig. 44-16. Study the manual included with each machine to find out what other shapes can be formed.

Making a circle is a typical bending operation. Fig. 44-16. Although a circle can be easily formed with this bending machine, you must remember that most materials "spring back" after they have been formed. To allow for this, you must use a radius collar having a smaller diameter than that of the circle to be made. Actual size can best be determined by experiment, as the "springback" varies in different materials. Material should be precut to exact length before forming.

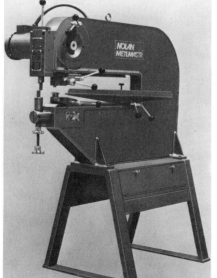

The Nolan Company

44-17a. *This industrial machine is used to cut and bend metal. It has a number of interchangeable jaws which move up and down rapidly to form or cut sheet metal. It can be used to cut and dish metal disks, make many different bends, cut slots, and punch holes.*

The Nolan Company

44-17b. *The machine in Fig. 44-17a can be fitted with jaws to produce an offset bend on a sheet metal part, as shown here.*

INDUSTRIAL MACHINES FOR BENDING

This chapter has discussed many of the machines that can be used in the school shop to form metal by bending. Industry uses similar metal-forming machines. These machines are power operated and can bend metals as much as one inch thick. Industry also uses many specialized machines to bend metal. Fig. 44-17.

FOR YOUR SAFETY . . .

● Be careful when handling sheet metal. Sharp edges can cause painful cuts. Use gloves when necessary.

● Bending machines are powerful. Be careful not to get your fingers caught in the jaws.

QUESTIONS AND ACTIVITIES

1. Define metal bending.
2. Name four kinds of metal bending stakes.
3. What are three types of hammers that can be used for bending light metals?
4. What are bar folders used to make? How are they different from the hand seamer?
5. What is the purpose of making hems on sheet metal?
6. What machine can be used to form a piece of sheet metal into a tray or a box? A cylinder?
7. What is the bending fork used for?
8. Describe how to form a circle on a bending machine.

CHAPTER 45

Casting Metal

If you dig a hole in the ground, fill it with cement, and then remove the cement piece after it has hardened, you have made a casting. This is basically all there is to the process. Many kinds of castings are used in industry: investment casting, shell molding, die casting, and sand casting.

Investment casting is used to make jewelry, gold fillings for dental work, and other products requiring precision. A wax or plastic pattern is imbedded in an "investment" consisting of wet plaster or a sand mixture. This is baked to melt the wax or plastic, which runs out, and to dry the mold. Later the mold is filled with molten metal. After cooling, the investment is broken away and the casting is cleaned.

In *shell molding*, molten metal is poured into a thin shell of sand and resin binders. The shell may be backed by other materials, such as foundry sand.

In *die casting*, molten metal is forced into a mold (die) under pressure. As the metal solidifies, it takes the shape of the die. Fig. 45-1.

In *sand casting*, a pattern is packed into damp sand. When the pattern is carefully removed, an impression is left in the sand. This impression is the mold into which the molten metal is poured

45-1. *Three stages in the production of a die-cast model truck. Only nonferrous metals and alloys are suitable for die casting. Aluminum, zinc alloys, copper, and magnesium are commonly used.*

to make the casting. The engine block of the automobile is one product made by this inexpensive method of casting.

SAND CASTING

Sand casting is commonly done in the school shop to produce interesting projects. Fig. 45-2.

First, you will need a pattern, which will be imbedded in the sand to make the desired imprint. This imprint will be the mold. Wooden or plastic patterns of the object to be cast can be made. You may also use plastic, ceramic, or metal objects purchased in stores as patterns. Sim-

ple one-piece patterns can be made of pine, then waxed so that the sand doesn't stick to them. The sides should be tapered slightly to make the pattern easier to remove. (The taper is called *draft*.) More complex objects are made of two-piece, or split patterns. Patterns can also be made of Styrofoam plastic. These can be left in the sand, as the molten metal will melt them away.

Some of the common casting tools are shown in Fig. 45-3.

Instructions and illustrations showing how to make a simple mold are shown in Figs. 45-4 through 45-12. Study these carefully to learn proper procedures.

The metal to be poured is melted in a foundry furnace. Fig. 45-13. When the metal is molten, it must be carefully poured into the cavity. Wear safety goggles or a mask, heat-resistant leggings, apron, and gloves. Fig. 45-14. Melting temperatures of some

45-2. *These attractive projects were made by sand casting.*

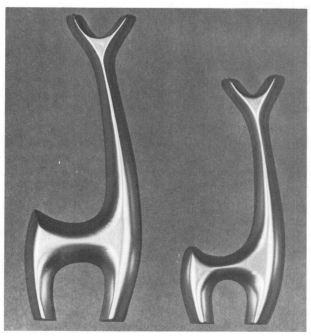

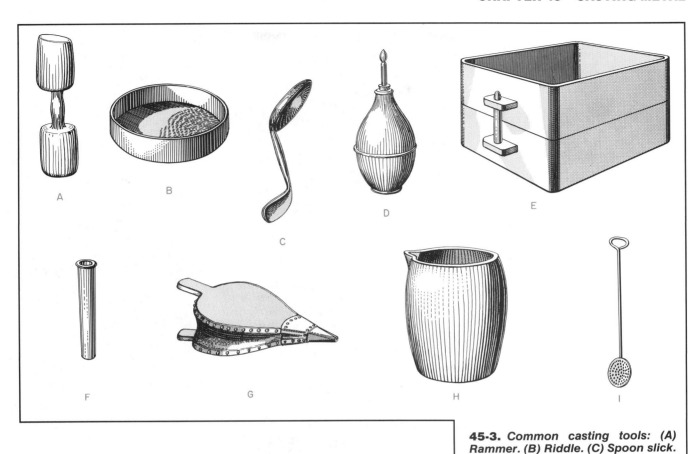

45-3. Common casting tools: (A) Rammer. (B) Riddle. (C) Spoon slick. (D) Bulb. (E) Flask. (F) Sprue pin. (G) Bellows. (H) Crucible. (I) Skimmer.

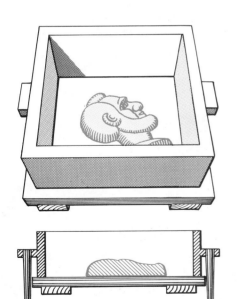

45-4. Place drag (bottom half of flask) on moldboard, pins down. Center pattern on moldboard. Dust pattern lightly with parting dust and blow off excess with bellows.

45-5. Temper (mix) sand with water until a handful squeezed will break cleanly. Shovel sand into riddle and shake until pattern is covered. Fill drag with sand and ram firmly.

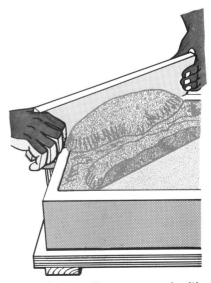

45-6. Strike off excess sand with a steel bar.

185

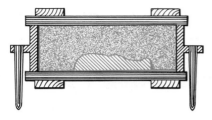

DRAG READY TO BE TURNED OVER

45-7. *Place a second moldboard over the drag. Turn top and bottom moldboards and the drag over.*

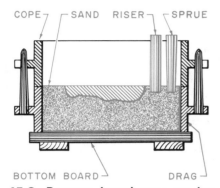

COPE — SAND RISER — SPRUE

BOTTOM BOARD — DRAG

45-8. *Remove board now on top and place cope (top half of flask) in position. Dust with parting dust. Place sprue and riser in position. Riddle sand over pattern, fill cope with sand, ram, and strike off excess. Vent pattern with piece of wire. Vent hole should not touch pattern. Remove sprue and riser pins.*

45-9. *Carefully remove cope and set it aside on edge. Use a wire lifter to remove pattern.*

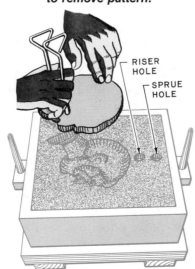

RISER HOLE

SPRUE HOLE

45-10. *Cut a gate with piece of sheet metal bent to a V shape. The molten metal will run down the sprue and through the gate into the cavity left by removing the pattern.*

45-11. *Make any repairs to mold with bulb sponge and spoon slick. Clean mold by blowing gently with bellows.*

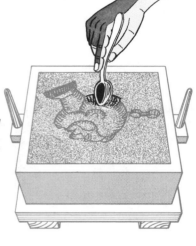

SAND — VENTS GATE
COPE — RISER — SPRUE

BOTTOM BOARD — DRAG

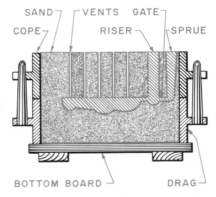

45-12. *Drawing shows flask assembled. The mold is ready for the molten metal.*

Johnson Gas Appliance Company

45-13. *The crucible of molten metal must be carefully removed from the furnace with tongs. Before pouring, sprinkle some powdered casting flux into the molten metal. This will cause the impurities in the metal to collect on the surface. Remove these with the metal skimmer.*

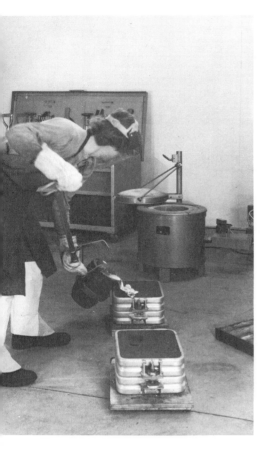

McEnglevan Heat Treating and Manufacturing Company

45-14. *Pouring molten metal into a mold to make a casting. Stand to one side, and pour the metal slowly to avoid spills. The metal will take the exact shape of the mold into which it is poured.*

45-15.
Melting Temperatures of Common Metals

Metal	°F	°C
Lead	621	328
Aluminum	1218	659
Bronze	1675	914
Brass	1700	927
Copper	1981	1082
Cast Iron	2200	1204

common metals are shown in Fig. 45-15.

After the mold has cooled, remove the casting and use a hacksaw to cut off the metal that filled the sprue, riser, and gate. Grind and clean the cast piece as needed.

FOR YOUR SAFETY . . .

Making metal castings is fun, but you need to follow these safety rules. Otherwise, you could get badly burned.

• Wear protective clothing —gloves, arm protectors, goggles or face shield, leggings, and an apron.

• Practice the "buddy system." Do not work alone.

• Work in a well-ventilated area.

• Your instructor should be present whenever you are melting or pouring metal.

• When pouring metal, be sure to stand to one side of the mold so that you don't lean over it.

• Keep moisture away from molten metal. It can cause the metal to explode.

QUESTIONS AND ACTIVITIES

1. List and describe four kinds of metal casting.
2. What is draft and why is it needed?
3. List at least five of the tools used in sand casting.
4. Name and describe the two parts of the flask.

5. What is the flux used for in metal casting?
6. What safety clothing should be worn when casting?
7. What is the Fahrenheit melting temperature of brass? Aluminum? Bronze?

In *forging,* steady pressure or repeated blows are applied to metal, forcing it to take the shape of a die. Forging increases the strength of the metal by changing its grain structure. No metal is removed, and the pieces gain strength in the process. Fig. 46-1.

In industrial forging, huge drop, steam, or trip hammers are dropped automatically on hot (or sometimes cold) workpieces to form tools, automobile crankshafts, airplane propellers, and many other metal parts. Fig. 46-

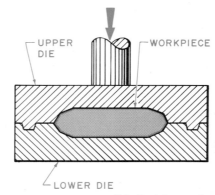

U.S. Steel Corporation

46-2. Forging aluminum airplane propeller blades. The huge hammer (A) beats the blade (B) into a die shape. The forging diagram illustrates the process.

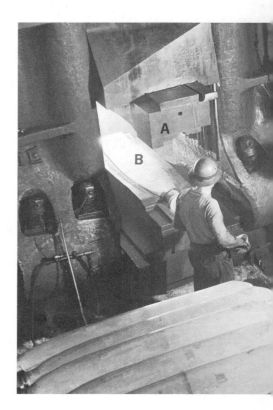

2. Both open and closed dies are used in forging.

Forging in the shop is limited to the hot forming of screwdrivers, chisels, and other small items and to the hot bending of metal rings and other shapes. Drawing may also be done. In *drawing,* metal is made longer and thinner by placing it over the round edge of an anvil and striking the metal with a heavy hammer. For example, you may do drawing to change the shape of a metal rod or decrease its diameter. In addition to drawing and bending, metal may also be "upset." This involves striking the ends of metal rods to increase their thickness (the opposite of drawing).

Although heat treating is not a forging process, the two are closely related because many

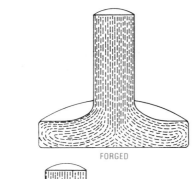

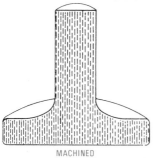

46-1. Flow lines in forged and machined parts. The flow lines show the direction of the metal grains. The forged part is stronger because the flow lines follow the contour of the part.

tools must be strengthened by heat treatment after forging. For this reason both processes are included in this chapter.

FORGING

The tools needed for shop forging are shown in Fig. 46-3. Forging should be done on the anvil and not on sheet metal stakes. The anvil acts as a kind of a die which controls the shape of the metal. The anvil hardy (A, Fig. 46-3) fits into the hardy hole on the anvil. It is used for cutting metal bars and rods. Tongs are used to hold the metal while forg-

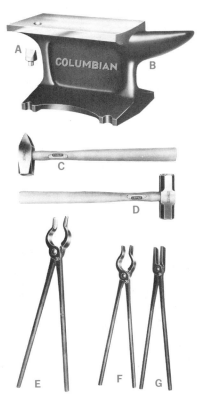

46-3. *Forging tools: (A) Anvil hardy. (B) Anvil. (C) Cross-peen hammer. (D) Engineer's or sledge hammer. (E) Curved-lip tongs. (F) Pick-up tongs. (G) Flat-lip tongs. A heavy ball-peen hammer could also be used for forging.*

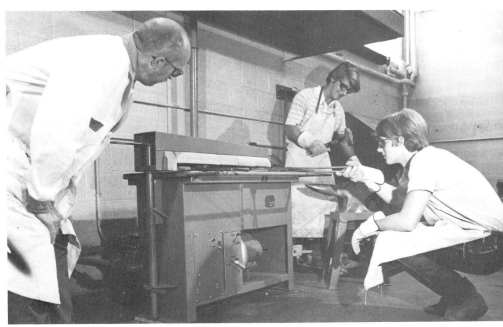

Johnson Gas Appliance Company

46-4. *Using the forge. Caution should be used in lighting the gas forge. Ask your instructor to show you how to do this. To avoid burning someone, heated metal should be carefully moved to the anvil for working.*

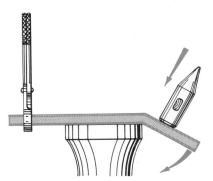

46-5. *Bending hot metal on an anvil. Use tongs to hold the metal workpiece.*

Lindberg Company

46-6. *The heat-treating furnace. This model is used for small workpieces. Larger furnaces are also available.*

ing. Hammers are used to strike repeated blows against the hot metal to force it to take the desired shape. Heavy hammers work the metal more easily than light ones.

Remember to wear eye protection, gloves, and an apron when forging metal.

When forging mild steel, heat it in the forge until it is cherry red. Fig. 46-4. Work it before it begins to cool. Fig. 46-5. Tool steel requires slightly less heat.

HEAT-TREATING

In the forging process, the metal becomes soft when you heat it and hard but brittle (easily broken) when you work it. In order to make chisels and other tools usable, they must be heated and cooled through a specific temperature sequence. This process of heating and cooling metal to change its properties is called *heat treatment,* and is usually done in a heat-treating furnace. Fig. 46-6. (It can also be done in a forge, but less conveniently.) Here, for example, are the steps

Steel Temperature Chart

Color	Temperature	
	°F	°C
White	2200	1205
Lemon	1825	997
Orange	1725	940
Bright red	1650	899
Cherry red	1325	718
Dark red	1175	635
Faint red	900	482
Pale blue	590	310

46-7. *The temperature of steel as it is heated can be accurately estimated by observing its color.*

Tempering Chart

Tools	Color	Temperature	
		°F	°C
Scribers	Pale straw yellow	450	232
Center punches	Spotted red brown	510	266
Cold chisels	Spotted red brown	510	266
Screwdrivers	Light purple	530	277

46-8. *Temper colors and temperatures for common shop tools.*

in heat-treating a screwdriver made of tool steel.

First, *anneal* the workpiece by heating until cherry red, then cooling slowly in air. Fig. 46-7. This process softens the metal and relieves the strains of forging. File smooth and polish with abrasive cloth and oil.

Next, *harden* by reheating about ¾″ of the screwdriver tip until it is cherry red. This is called the critical temperature. Plunge the metal quickly into water, moving it with a circular motion. Test its hardness with a file. If the file nicks the screwdriver, anneal and harden again.

Finally, *temper* the metal by polishing the tip with abrasive cloth and applying heat about 1″ above the tip. When the temper color (light purple in this case) reaches the tip, plunge the tool into water. Fig. 46-8. Tempering reduces the brittleness of the piece.

Mild steel is heat-treated by a process called *case hardening*. This is done by adding a special carbon hardening powder to the surface of a metal workpiece, and then hardening this outer case.

To case harden metal, heat the workpiece to a bright red. Remove any scale with a wire brush. Dip, roll, or sprinkle the powder on the workpiece. The powder will melt and adhere to the surface, forming a shell around the work. Reheat to bright red, hold at this temperature for a few minutes, and then quench in clean, cold water. This will give the workpiece a completely hard outer surface, or case.

FOR YOUR SAFETY . . .
- Wear protective clothing —safety glasses or goggles, gloves, and an apron.
- Work in a well-ventilated area.
- Use the "buddy system." Don't work alone.
- Make sure that the hammer handles are not loose.

QUESTIONS AND ACTIVITIES

1. Define forging.
2. What is the process called drawing?
3. What is the process called upsetting?
4. Describe the job of the anvil, the hardy, the tongs, and the hammer in forging.
5. What safety clothing should be worn while forging?
6. What color should mild steel be heated to before forging?
7. What is heat treatment?
8. What is annealing? Why is it done?
9. List the steps in the heat-treating of a screwdriver.
10. Describe tempering. Why is it done?
11. What is case hardening?

CHAPTER 47

Pressing, or press forming, is a process whereby sheet metal is forced to take the shape of dies (forms). Fig. 47-1. Pressing can be done by hand or machine. Huge machines form automobile hoods by pressing the metal between dies. Smaller pieces such as kitchen utensils are also formed this way. Fig. 47-2.

Many kinds of pressing operations are done in the school shop. The most common are sinking, raising, tooling, chasing, stamping, and spinning. These forming methods are a part of art metalwork.

FOR YOUR SAFETY . . .
● Always be careful when handling sharp metal pieces.
● Make sure hammer handles are not loose.

SINKING
Sinking, or low raising, is a method of forming shallow dishes or trays by beating down sheet metal into a form. Figure 47-3 shows typical projects made by sinking. Small bowls are made from 24- or 22-gauge metal. Use 20- or 18-gauge metal for large trays, pans, etc.

Sinking is done by nailing a piece of sheet copper, brass, or aluminum to a wooden or metal form which has been hollowed out to the desired shape of the dish. Fig. 47-4. The metal is beaten down by striking it with a wooden forming hammer. Fig. 47-5. Work the metal down near the edges of the form first. Then move to the center and work back toward the edges. Fig. 47-6.

After sinking, the piece is removed and trimmed with tin snips. Any dents or imperfections can be removed by planishing. (Planishing is a way of finishing metal by hammering it lightly. See Chapter 56 for details.)

When metal is pounded or bent, it *work hardens*. That is, it

47-3. These shallow dishes for candy or nuts were made by sinking.

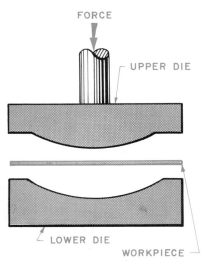

47-1. The metal pressing process.

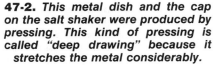

47-2. This metal dish and the cap on the salt shaker were produced by pressing. This kind of pressing is called "deep drawing" because it stretches the metal considerably.

47-4. Wooden or metal forms are used for sinking metal. You could make a form by turning wood on a lathe or by cutting a hole in a piece of ¾" plywood and nailing it to a wood base.

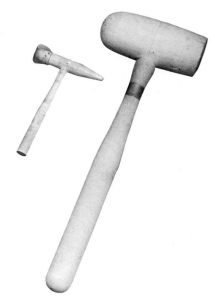

47-8. Metal raising hammers come in many shapes and sizes.

47-5. *Wooden forming hammers are generally used for sinking, although metal hammers may also be used. The hammers should be smooth and free of dents. Sand them if necessary. They may also be capped with leather to prevent marring the metal. The blunt end is used for truing the bottoms of the trays or flattening their edges.*

cup of hydrochloric acid into four cups of water. The solution must be mixed in a ceramic or plastic container. CAUTION: Always pour the *acid* into the *water.* Pouring water into acid causes dangerous splattering.

Copper can be annealed and pickled at the same time. To do this, heat the copper to a faint red color, remove from heat, and plunge it into the pickling solution. Rinse in cold water and wipe dry.

RAISING

Raising is a method for forming deeper dishes. Fig. 47-7. A piece of 20- or 18-gauge brass or copper is cut to the desired size disk. This disk is held at a slight angle over a shallow cavity or depression in a wooden forming block. It is then beaten with metal raising hammers (Fig. 47-8) and slowly rotated over the depression after each blow of the hammer. Fig. 47-9. A strong can-

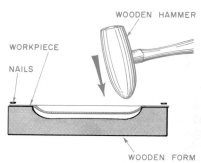

47-6. *Sinking is done by beating metal down into a form.*

becomes stiff and brittle. It must be softened before you can continue to work it. To soften the metal you must anneal it. To anneal copper, brass, or nickel-silver, heat it to a faint red color in a soldering furnace or use a torch. Avoid overheating. Allow the metal to cool in air. To anneal aluminum, rub the surface with cutting oil and heat until smoke appears. Cool the metal in air.

When copper is heated, it becomes tarnished and scaly. It can be cleaned by pickling in a mild acid solution. A pickling solution for copper is made by pouring one

47-7. *Deeper bowls such as these are made by raising.*

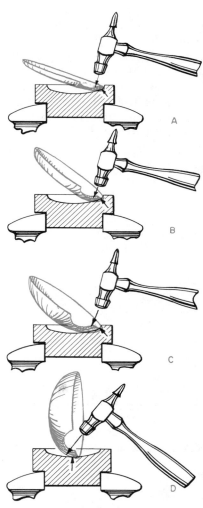

47-9. *Metal raising differs from sinking in that the metal is moved by hand while it is being hammered. (The broken-out section is to help you see the inside of the bowl.)*

47-14. *Raising the design. Using the flat end of the smoothing tool within the raised design lines, work the design out from the back. Raising the design in this manner is often called "embossing."*

47-10. *A tooled-metal wall plaque and the tools used to make it: (A) Hardwood smoothing tool. (B) Modeler. (C) Hardwood tracing tool.*

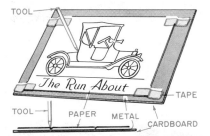

47-11. *Transferring the design. Fasten the paper pattern to the metal with masking tape. Place the metal on a piece of cardboard. With the tracing tool, press the lines of the design firmly, thus transferring the design to the metal.*

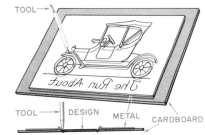

47-13. *Interlining the design. Turn the metal over and place it on the cardboard. Using the modeler, mark lines just inside the traced design. Be careful to line just inside the original design lines, not on them.*

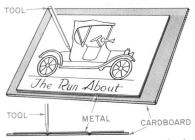

47-12. *Deepening the design. Check the back of the metal before removing the design to be sure all lines have been traced. Then remove the paper and deepen the traced lines by going over them directly on the metal. Use the tracing tool or modeler.*

47-15. *Flattening the background. Turn the metal so that the front side is up and place it on a smooth hard surface such as glass or Masonite fiberboard. Using the smoothing tool, carefully smooth down the background. Use even pressure and be careful that you do not scratch the surface.*

vas sandbag can be used in place of the forming block. It must be only partially filled with fine sand so that it will fit the contour (shape) of the workpiece. As in sinking, the metal should be annealed and pickled as it work hardens. Planishing should also be done after raising.

TOOLING

In metal tooling (or embossing), a design is raised on a piece of thin sheet metal (about 36 gauge). Figure 47-10 shows a tooled project and the simple tools used in making it. The steps involved in metal tooling are shown in Figs. 47-11 to 47-15.

The finished piece can be colored and highlighted by dipping in liver of sulfur (see Chapter 57) and rubbing with steel wool. Brush or spray on a flat lacquer finish.

The finished tooled workpiece can be mounted on a piece of wood or heavy cardboard and framed. The back of the workpiece should be filled with clay, plaster, or some other solid material to prevent denting the formed piece.

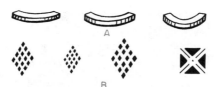

47-16. Chasing tools and hammer. Typical working ends of chasing (A) and matting (B) tools. Matting tools can be used to give the background a decorative design.

47-17. Metal being chased in a pitch pan. The pitch is a mixture of plaster of paris, rosin, and beef tallow.

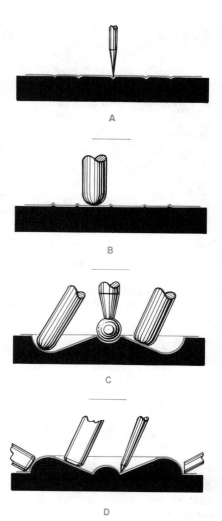

47-18. Chasing metal on pitch: (A) The design is traced on the face of the metal. (B & C) The metal is turned face down. Using various chasing tools, the design is worked up within the raised outline. (D) The metal is again placed face up on the pitch to finish the design surface.

CHASING

Chasing is similar to tooling except the heavier gauge metal used in chasing is so thick that metal tools and a metal hammer are needed to form the design. Fig. 47-16. Chasing tools look much like small punches and chisels with specially shaped blunt, rounded, and polished ends. The chasing hammer is lightweight and has a broad, flat face which allows the craftsperson to strike it continually without looking away from the workpiece. The work is held in a pan of chaser's pitch or on a block of lead. This supports the work while allowing it to give under the hammer blows. Fig. 47-17. A good choice of metal for a shop project is 24-gauge copper.

First, the design must be transferred onto the workpiece with a scriber or a scratch awl. Then a pan at least 1½" deep is filled with chaser's pitch. Oil the back of the metal, and gently heat both the metal and the pitch. Place the metal into the pitch, oily side down.

Allow the metal and pitch to cool. Place the thin, blunt-edged tracing tool on the outline so that the top of the tool is slanting slightly away from you. Gently, but steadily, tap the tool with the chasing hammer, forcing the tool

gradually toward you. This will result in a smooth recessed line that protrudes on the other side. Fig. 47-18, A.

When the outline has been completely traced into the metal, heat the pitch and remove the workpiece. Clean any pitch from the metal with turpentine and steel wool. Oil the face of the metal and place this side down on the pitch. To model the design, choose a heavier rounded chasing tool that resembles the finished curve as nearly as possible. Hold the tool at right angles to the curve being modeled. With overlapping blows work up the design between the outline lines until the desired depth is reached. Fig. 47-18, B & C. The work can again be reversed for retracing and smoothing, if necessary. Fig. 47-18, D. Whenever

SUN SYMBOL HAPPINESS

SUNRAYS CONSTANCY

HORSE JOURNEY

THUNDERBIRD SACRED BEARER OF HAPPINESS

RAIN CLOUDS GOOD PROSPECTS

CACTUS FLOWER COURTSHIP

47-19. Typical stamped designs.

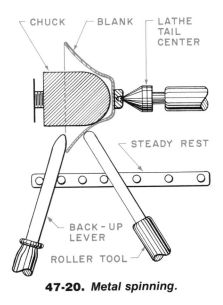

47-20. *Metal spinning.*

necessary, the metal can be annealed for easier working.

STAMPING

In a general sense, *stamping* can mean any press forming process. *Stamping* is also the name of a process for making decorative patterns in metal by striking it with a punch. The end of the punch has a design on it. The Indians of the American Southwest stamp designs in metal with steel punches, repeating or alternating their markings in rows. Fig. 47-19. The designs depend upon the shapes of the punches and the skill of the craftsperson.

Workpieces are usually of 16-gauge sheet copper. Anneal the copper and flatten it with a rawhide mallet on a hardwood block before stamping.

Place the copper on a metal plate. Hold the tool perpendicular to the work and give it a sharp blow with a ball-peen hammer. The impression should be about $1/32''$ deep.

When the stamping is complete, bend to shape (as for a bracelet) over a forming stake. With a rawhide mallet, begin from the ends and work toward the middle. If the bracelet is the closed type, fit the joint, bend it, and solder. Smooth the edges with a file and emery cloth and finish with spray lacquer.

Metal tapping is similar to stamping except that nails, ice picks, pricks, or center punches are used as tapping tools. Lighter, 30-gauge metals are more commonly used.

SPINNING

Spinning, one of the oldest of the metalworking arts, is still widely used. In this process a circular metal blank or shell is pressed against a chuck rotating at high speed on a lathe. Pressure applied manually or mechanically with forming tools forms the blank to the shape of the chuck. Fig. 47-20.

Typical articles produced by spinning include cooking utensils, water pitchers, bowls, reflectors, kettle shells, ring molds, tapered pans, perforated cones, milk cans, and parts for aircraft and street lights.

INDUSTRIAL METHODS

Explosive Forming

Explosive forming, Fig. 47-21, is much like exploding a firecracker under a tin can, causing the sides of the can to expand outward. Needed for explosive forming are a tank of water, a die, a blank of metal to be formed, and some explosives. The metal is clamped over the die cavity and the air is removed from the cavity through a vacuum tube. The metal and the die are lowered into the water. The charge of explosive sets off a powerful wave traveling from 250 to 500 feet per second with a force of five to six million pounds per square inch. This deforms the metal so fast it doesn't have a chance to break. Huge pieces of metal can be formed this way. Typical explosive-formed items are jet engine covers, railroad tank cars, and radar reflectors.

Hydrospinning

Hydrospinning is a forming process in which a workpiece is forced to take the shape of a hardened, rotating piece called a mandrel. The workpiece turns with the mandrel and is shaped

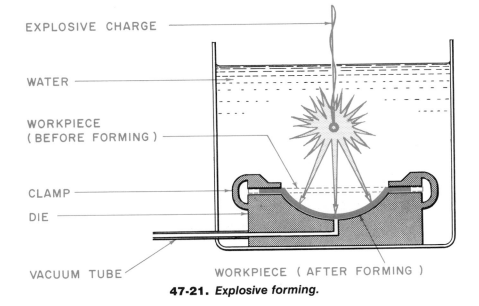

47-21. *Explosive forming.*

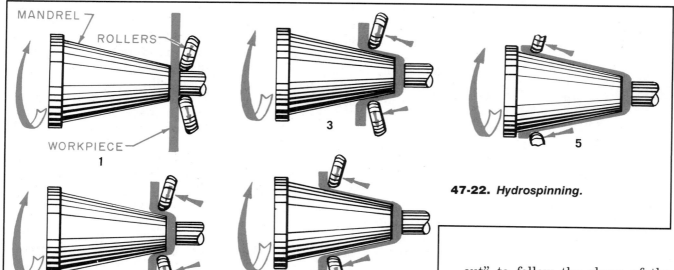

MANDREL
ROLLERS
WORKPIECE
1

2

3

4

5

47-22. *Hydrospinning.*

by one or two hardened, polished rollers which stretch and form the workpiece around the rotating mandrel as the rollers move along the mandrel's length. Fig. 47-22.

Hydrospun parts are stronger than forgings. During machining or forging, the grain structure of the part is changed very little. During hydrospinning, however, the grain structure is "stretched out" to follow the shape of the part, making the part even stronger. Hydrospun parts have a finish comparable to a good commercial grind finish. The parts are made from a simple workpiece using about ¼ as much material as for ordinary machining. And, the hydrospun part is produced more quickly and at less cost.

QUESTIONS AND ACTIVITIES

1. Define press forming.
2. What kinds of objects are formed by sinking? What type of metal is used? What type of hammer is used?
3. Define the sinking process.
4. What happens to metal when it work hardens? What must be done to the work hardened metal, and why?
5. Describe pickling and tell why it is done.
6. Briefly describe the raising process.
7. What type of metal is used in raising? What type of hammer is used? How is raising different from sinking?

8. Make a chart showing the steps in metal tooling.
9. What is a good metal to use in chasing? How is chasing different from tooling?
10. Describe the decorative metal forming process of stamping.
11. What is spinning?
12. Briefly describe explosive forming.
13. What are some advantages of hydrospinning?

CHAPTER 48

Drawing, Extruding, and Rolling Metal

DRAWING METAL

Metal products other than sheet and strip are produced by a process called drawing. This is done by pulling a rod, tube, or bar through a hole in a die. The hole is smaller than the starting size of the workpiece. As it is pulled through the die, the workpiece is stretched out and reduced in cross section. Fig. 48-1.

One end of the workpiece is usually pointed by grinding or some other method so that it can be inserted in the die to begin drawing. Grippers grasp the pointed end extending from the die and pull the rest of the piece through. A variety of shapes can be drawn by using different dies. For example, drawn wire of exact diameters, which is used in the making of nails, wire mesh, fencing, and so forth, can be made by this process.

Deep drawing is done on mechanical or hydraulic presses. The bottom die on the press contains a deep depression (hole). The bottom die is usually called a draw ring. The upper die is in the form of a punch. The flat workpiece, called a blank, is placed over the hole in the draw ring. The punch is lowered, pressing the blank through the draw ring.

The simplest deep-drawing operation is that used to make cylindrical cups. In that case, the hole in the draw ring is round, and the punch is a cylinder. The blank is a circle of flat stock. During forming, the blank is stretched through the draw ring and wrapped around the punch. The drawn article then has the shape of the punch. Some machines not only do deep drawing but also other forming operations. Fig. 48-2.

Drawing is an industrial process not usually done in the shop, except in some art metalwork and forging projects.

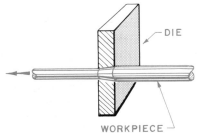

48-1. Metal drawing.

EXTRUDING METAL

In the basic metal extrusion process, a ram (punch) forces a heated billet (block) of metal through a die to produce a rod or bar of a given shape. Fig. 48-3. This is much like squeezing toothpaste out of a tube. Industry uses this process to form hundreds of different metal shapes to be used in manufacturing such items as storm and screen windows, frames of many kinds, and decorative moldings.

The process is called forward extrusion if the metal flows in the same direction that the punch travels. It is called back-

Waterbury Farrel

48-2. All these parts were produced automatically, using one machine and various accessories.

197

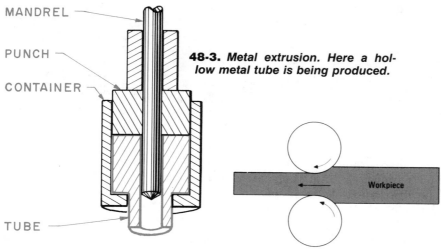

48-3. *Metal extrusion. Here a hollow metal tube is being produced.*

48-4. *Metal rolling.*

Aluminum Company of America

48-5. *Hot-rolling aluminum. Often weighing over 10 tons, the huge blocks are rolled by both hot and high-speed cold finishing mills into coils of sheet miles long and as thin as 0.016 inch.*

ward extrusion if metal flow is opposite to the motion of the punch. Other directions of metal flow are possible, depending on the location of the hole in the container. To extrude parts that require metal flow in directions other than straight forward or backward, the container must be made in two pieces so that the workpiece can be removed. Many shapes can be made by these extruding processes.

Very little metal extruding is done in the shop because of the heavy machinery required. Plastics and ceramics can also be formed by extrusion.

ROLLING METAL

Rolling is the process of passing slabs of metal, up to 26″ thick, between huge cylinders which reduce them to heavy metal plates or to thin sheets and foils. Fig. 48-4.

Industry uses both hot and cold metal rolling operations. Hot rolling is mainly a roughing operation. Its chief use is to shape and refine the large cast ingots that come from the refining processes. Fig. 48-5. The rolling operation puts the metal in a form that can be further changed by other methods. With grooves in the rolls, rolling mills can produce a variety of shapes. Some of the more common ones are sheet, round rods, flat bars, and angles. Other shapes produced by hot rolling are half-rounds, channels, T-bars, and I-bars.

Cold rolling is usually used to produce thin, flat products such as sheet and strip. Cold-rolling mills usually have smaller rolls than hot-rolling mills. Cold rolling requires much more force, and smaller rolls are required to concentrate the force over a smaller area. These rolls also enable thinner material to be rolled. Cold rolling is seldom used to produce finished products. Cold-rolled sheet and strip are made into finished articles by other methods.

Rolling is seldom done in the school shop. However, it is an important process in industry.

QUESTIONS AND ACTIVITIES

1. How is metal drawing done? List some products made from drawn wire.

2. Describe the basic process of extrusion. List some objects formed by extrusion.

3. What is metal rolling? What kinds of metal shapes are formed by rolling?

CHAPTER 49

Metal Fastening Principles

Fastening is the process of joining materials together. There are many ways of fastening things together, as shown in Fig. 49-1. Study these and become familiar with them, so that the following chapters on fastening methods will mean more to you. You will use many of them in your shopwork. Fig. 49-2.

Fastening methods most commonly used by the metalworker include riveting, soldering, and welding. The decision to use mechanical, adhesive, or cohesive fastening methods depends upon what is needed for the product. For example, the wheels on an automobile are fastened to the drums by screwing nuts onto threaded studs. If they were riveted on or welded, changing a tire would be almost impossible. What is needed for the wheels is a safe, efficient, semipermanent method of fastening them to the drums. The nuts are safe because they can be drawn tight to hold

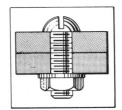

49-1.

Fastening

Fastening is the process of joining materials together permanently or semipermanently.

KIND OF FASTENING	DEFINITION	EXAMPLES
Mechanical WORKPIECE RIVET WORKPIECE	Permanent or semipermanent fastening with special locking devices.	Threaded fasteners, rivets, sheet metal joints.
Adhesive WORKPIECE SOLDER WORKPIECE	Permanent fastening by bonding like or unlike materials together with molten metal or cements.	Soldering, brazing, contact cementing, epoxy cementing.
Cohesive WORKPIECE WELD WELD WORKPIECE	Permanent fastening by fusing like materials together with molten metal or pressure.	Oxyacetylene welding, MIG and TIG welding, resistance welding, laser welding, submerged arc welding, plasma arc welding, explosive bonding.

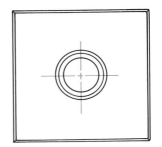

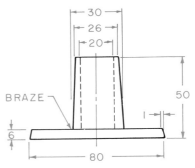

CANDLE HOLDER
C R S
ALL DIMENSIONS IN mm

49-2. The parts for this simple candleholder made of cold-rolled steel are joined by a fastening process called brazing. You will learn about brazing in Chapter 51.

49-3. *Arc welding is an important industrial fastening process. This worker is welding sections of pipe for a fuel pipeline.*

the wheel in place. They are efficient because they make wheel changing an easy task; and they are semipermanent because they can be screwed on and removed, over and over again. By contrast, heavy steel pipelines need to be permanently fastened. Therefore they are welded together. Fig. 49-3.

QUESTIONS AND ACTIVITIES

1. Define metal fastening.
2. What are the three kinds of fastening? Name two examples of each.

3. In your own words, what is the purpose of metal fastening?

CHAPTER 50

Mechanical Fastening of Metal

Mechanical fasteners join parts together, either permanently or semipermanently, with special locking devices. Threaded fasteners, such as screws, are semipermanent. The faceplate over a wall switch is fastened with screws. If the switch needs to be replaced, the faceplate can be easily removed and later reinstalled. Rivets and sheet metal joints are usually permanent. The fastened parts cannot be easily separated.

THREADED FASTENERS

Many kinds of threaded fasteners are used in metalworking. They allow machines and appliances to be taken apart and put back together again. Common threaded fasteners are shown in Fig. 50-1. As you can see, they have several head shapes. Hexagon, square, round, oval, and flat are the most common. Fig. 50-2.

Machine bolts and nuts have diameters up to 1″ and lengths up to 30″. They are often used to fasten machine parts together. They

50-1. *Threaded fasteners: (A) Hexagon-head machine bolt and nut. (B) Phillips-head machine screw and nut. (C) Hexagon-head cap screw. (D) Stud. (E) Lock washer. Washers are sometimes used with threaded fasteners to ensure tightness or relieve friction. (F) Plain flat washer. (G) Cone-point socket setscrew. (H) Flathead self-tapping screw.*

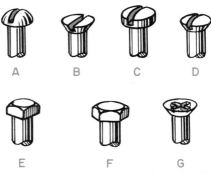

50-2. *Head shapes: (A) Round. (B) Flat. (C) Fillister. (D) Oval. (E) Square. (F) Hexagon. (G) Phillips. All slotted heads are either standard or Phillips.*

most common. These screws are generally used for such purposes as holding pulleys on shafts.

Self-tapping screws cut their own threads in sheet metal pieces to bind them together. Many head, point, and thread styles are available.

Washers are used to protect metal surfaces under the heads of fasteners or to lock the fasteners in place.

Some common tools used to hold and turn threaded fasteners are shown in Fig. 50-3. Adjustable, open-end, and box wrenches come in various sizes and are used to hold and turn nuts and bolts. Pipe wrenches are used on pipe and rod. Screwdrivers are designed for turning screw heads. DO NOT use them as pry bars or chisels. You will ruin the tip if you do. Pliers are also useful tools in the shop and around the home. There are many types and sizes of pliers. Some examples are shown in Fig. 50-4.

The press nut is a hollow fastener with internal threads. Fig.

hole in one piece and into a threaded hole in the second piece. The threaded hole acts as the nut. Cap screws are made in lengths up to 6″ and diameters up to 1″.

Studs are headless fasteners. Their use on automobile wheel drums, mentioned in the previous chapter, is typical of the ways they are used. They come in various lengths.

Setscrews are available with various head and point styles. Cone, oval, and flat points are

have either square or hexagon heads and are tightened with wrenches.

Machine screws and nuts are used much the same way as machine bolts, but generally for lighter work. They come in both fractional and wire gauge diameters up to about ½″ and in assorted lengths. They commonly have flat or round heads and are tightened with standard or Phillips screwdrivers. Machine screws are also used in tapped (threaded) holes for the assembly of metal parts.

Cap screws are similar to machine bolts except that they are not used with nuts. They hold parts of machinery together by passing through an unthreaded

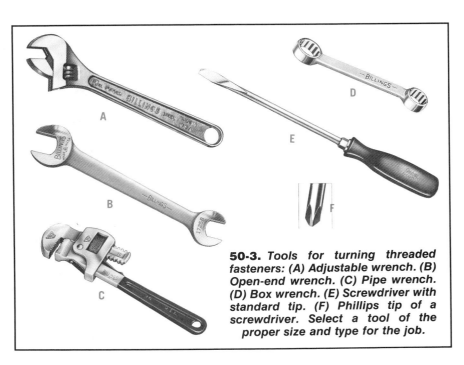

50-3. *Tools for turning threaded fasteners: (A) Adjustable wrench. (B) Open-end wrench. (C) Pipe wrench. (D) Box wrench. (E) Screwdriver with standard tip. (F) Phillips tip of a screwdriver. Select a tool of the proper size and type for the job.*

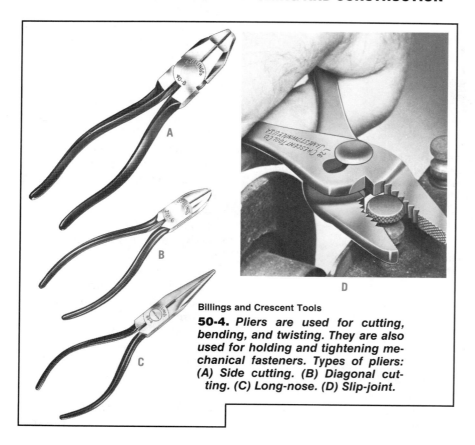

Billings and Crescent Tools

50-4. *Pliers are used for cutting, bending, and twisting. They are also used for holding and tightening mechanical fasteners. Types of pliers: (A) Side cutting. (B) Diagonal cutting. (C) Long-nose. (D) Slip-joint.*

Precision Metal Products

50-5. *Press nuts.*

50-5. It is used to join panels. First a hole is drilled or punched in one panel. Then the nut is pressed into the hole with a hydraulic or arbor press. This forces the metal of the panel around the nut, locking it firmly in place. Another hole is drilled in a second panel. A machine screw inserted through this hole and into the press nut holds the two panels together. Fig. 50-6.

Industry uses press nuts to join parts in electronic and aircraft assemblies. They are especially useful where panels must be fastened and unfastened many times.

Types of Threads

Industry makes threaded fasteners by cutting or rolling threads on them with automatic machines. In the shop, a die is

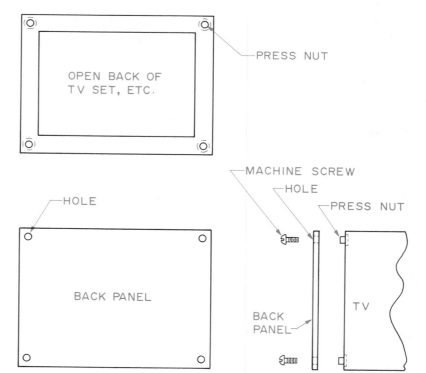

50-6. *Press nuts and machine screws hold these panels together. They can easily be separated by removing the machine screws.*

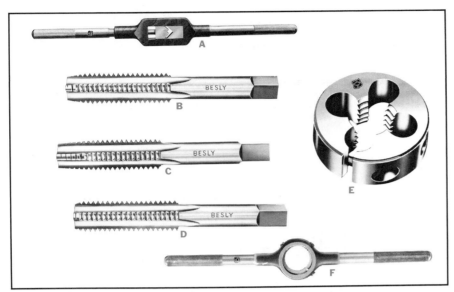

50-7. *Tools for cutting threads: (A) Tap wrench for turning taps. (B) Plug tap. (C) Taper tap. (D) Bottoming tap. (E) Die. (F) Die stock for turning dies. Remember to use a cutting oil when cutting threads. This makes threading easier and prevents tap and die breakage.*

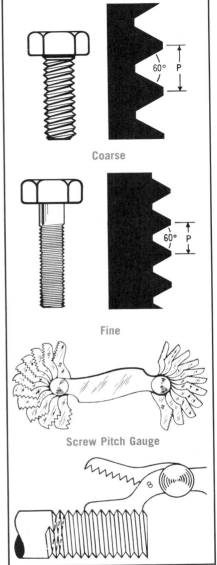

50-8. *The difference between coarse and fine threads is shown here. The screw pitch gauge is used to check thread sizes.*

used to cut threads on a rod (external threads) and a tap is used to thread a hole (internal thread). Fig. 50-7. A bolt has external threads; a nut has internal threads.

The most common kind of thread used in this country is the American National thread, in both the National Coarse (NC) and the National Fine (NF) series. The NF bolts have more threads per inch than NC bolts of the same diameter. Fig. 50-8. For example, a ¼″ diameter National Fine bolt has 28 threads per inch.

Tap Drill Chart

National Fine		National Coarse	
Size & Thread	Tap Drill	Size & Thread	Tap Drill
4-48	#42(0.0935)	4-40	#44(0.0860)
5-44	37(0.1040)	5-40	39(0.0995)
6-40	34(0.1110)	6-32	36(0.1065)
8-36	29(0.1360)	8-32	29(0.1360)
10-32	22(0.1570)	10-24	26(0.1470)
¼-28	3(0.2130)	¼-20	8(0.1990)
⁵⁄₁₆-24	"I"(0.2720)	⁵⁄₁₆-18	"F"(0.2570)
³⁄₈-24	"Q"(0.3320)	³⁄₈-16	⁵⁄₁₆(0.3125)
⁷⁄₁₆-20	"W"(0.3860)	⁷⁄₁₆-14	"U"(0.3680)
½-20	²⁹⁄₆₄(0.4531)	½-13	²⁷⁄₆₄(0.4219)

50-9. *This table tells which tap drill to use for various thread sizes.*

The National Coarse bolt has 20 threads to the inch. The designation for this bolt in the NC series would read: ¼-20 NC. Use a screw pitch gauge to check thread sizes.

Below ¼″ diameter, wire gauge sizes are generally used to designate the thread. For example, a 10-32 NF bolt is made of #10 wire (measured by the American Screw and Wire gauge) and has 32 threads per inch. Fig. 50-9.

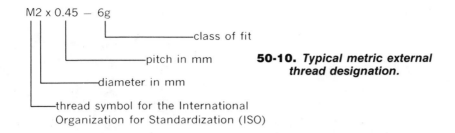

M2 x 0.45 — 6g

class of fit

pitch in mm

diameter in mm

thread symbol for the International
Organization for Standardization (ISO)

50-10. *Typical metric external thread designation.*

50-11.
ISO Metric Threads*

Diameter mm	Pitch mm	Tap Drill mm	Diameter mm	Pitch mm	Tap Drill mm
M 1.6	0.35	1.25	M 20	2.5	17.50
M 2	0.40	1.60	M 24	3.0	21.00
M 2.5	0.45	2.05	M 30	3.5	26.50
M 3	0.50	2.50	M 36	4.0	32.00
M 3.5	0.60	2.90	M 42	4.5	37.50
M 4	0.70	3.30	M 48	5.0	43.00
M 5	0.80	4.20	M 56	5.5	50.50
M 6	1.00	5.00	M 64	6.0	58.00
M 8	1.25	6.75	M 72	6.0	66.00
M 10	1.50	8.50	M 80	6.0	74.00
M 12	1.75	10.25	M 90	6.0	84.00
M 14	2.00	12.00	M 100	6.0	94.00
M 16	2.00	14.00			

*Not yet adopted in the United States.

Metric Threads

As more products are made to metric sizes, you will often be using metric as well as customary-inch threads. You can't tell these apart by looking at them because they do not look different. A typical thread designation is shown in Fig. 50-10. The thread dimensions are, of course, in millimetres. The metric pitch is the distance from the top of one thread point to the next. In customary threads, the pitch means the number of threads per inch. Metric thread designations also include the class of fit, or tightness. This is a number followed by a small *g* if it is an external thread, as on a bolt, or followed by a capital *H* if the thread is internal, as on a nut. In Fig. 50-10, the class of fit is 6g. This is a me-

dium or general-purpose fit for an external thread.

The metric threads which will probably be used in this country are shown in Fig. 50-11. There are about 56 different types of fine and coarse customary-inch threads used today. Note that there are only 25 metric threads (all coarse) which are used in the United States. Note too how easy it is to get the tap drill size: you just subtract the pitch from the

diameter. Metric drills are used in school shops and in industry. A customary-metric conversion chart for drill sizes is in the Appendix. A set of metric drills can also be made from customary drills. A list of these can be found in the Appendix.

Cutting Threads

To tap a hole, select the proper taper tap and secure it in a tap wrench. Look at a tap drill chart and select the proper tap drill. After the hole has been drilled, start the tap in straight and turn slowly until the thread catches the work. Apply a few drops of cutting oil and give the tap one more turn. Turn the tap gently, or it may break. Back the tap off slowly to clear the chips and then continue tapping. Back off again occasionally to remove chips. Fig. 50-12.

Threading a rod with a die is done in much the same way as tapping. Fig. 50-13. Be especially careful to turn both tap and die slowly, to lubricate well, and to remove chips frequently. This will produce sharp, clean threads

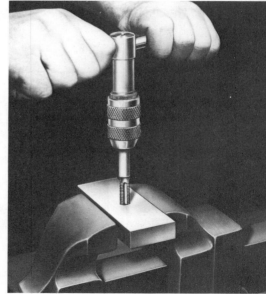

The L. S. Starrett Company
50-12. *Tapping a hole. The tap and the tap wrench are being used to cut internal threads.*

50-13. *Cutting external threads on a rod. A die and a die stock are being used. Hold the die stock square to the work and turn slowly.*

50-14. *A typical tap and die set.*

and will prevent tool breakage. The tap and die set generally found in the school shop is shown in Fig. 50-14.

RIVETS

Rivets provide a permanent mechanical fastening method for metal parts. Riveting is a simple process. A headed metal pin is placed in a hole through two or more metal pieces. It is then headed at the other end to clinch the pieces together. The "grip" of a rivet is the total thickness of metal held between the two heads.

Rivets are made of copper, aluminum, or steel. They come in many lengths and diameters. Select a rivet made of the same material as the workpiece to be joined. There are several different head styles. Fig. 50-15. Tinners rivets are very common in sheet metal work. The size of the tinners rivet is indicated by its weight per thousand. Common sizes are 10 ounce, 1 pound, and 2 pound. Others are sold by diameter and length.

Be sure to choose the proper length rivet to make the joint strong. The rivet should extend

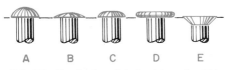

50-15. *Typical solid rivet heads: (A) Button, or round. (B) Truss. (C) Universal. (D) Tinners. (E) Flat.*

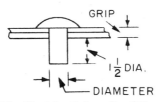

50-16. *The shank length of the rivet should equal the total thickness of the metal being joined (the grip) plus 1½ times the diameter of the shank.*

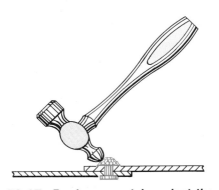

50-17. *For heavy metals, select the proper rivet and form the head with a ball-peen hammer.*

beyond the grip a distance equal to 1½ times the shank diameter of the rivet. This provides enough remaining material to form the head. Fig. 50-16.

The hole for the rivet is punched or drilled in the metal. Holes should be at least two rivet diameters (shank diameters) from the metal joint and three diameters from other rivets for strength and a tight hold.

On heavy metals, the rivet is inserted and the head formed by striking gently with a ball-peen hammer. Fig. 50-17. On sheet metals, the rivet set is used. Fig. 50-18. Place the open hole of the set over the rivet and gently strike with a riveting hammer to draw the pieces of metal to-

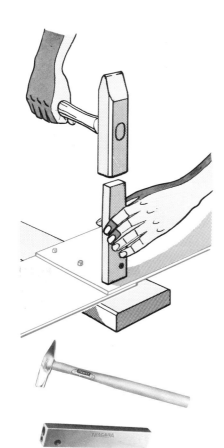

50-18. *Using the rivet set. Make sure the hole and the rivet diameter match.*

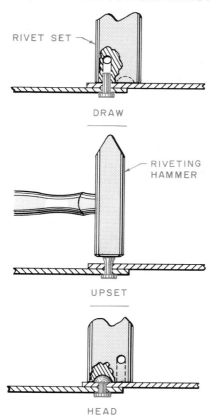

DRAW

RIVET SET

RIVETING HAMMER

UPSET

HEAD

50-19. *Steps in riveting.*

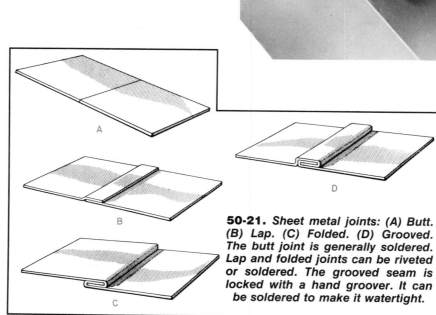

50-20. *Using the pop riveter.*

gether. Remove the rivet set. Strike the rivet end with the hammer a few times. This will make the shank expand to fill the hole. Then place the concave part of the rivet set on the end of the rivet and strike with the hammer to form the head. Fig. 50-19. When riveting a seam or long joint, install rivets at both ends. Complete the job by riveting from the center to both ends of the joint.

For blind holes and for general fast riveting, pop, or blind, rivets can be used. These rivets come in several sizes and materials. They are inserted with a tool called a pop riveter. To use the pop riveter, insert the rivet in the tool, put the tool in position, and squeeze firmly. Fig. 50-20. The result is a fast, strong joint.

50-21. *Sheet metal joints: (A) Butt. (B) Lap. (C) Folded. (D) Grooved. The butt joint is generally soldered. Lap and folded joints can be riveted or soldered. The grooved seam is locked with a hand groover. It can be soldered to make it watertight.*

SHEET METAL JOINTS

Common joints are shown in Fig. 50-21. They may be soldered, spot welded, riveted, or, for the grooved seam, clinched to form permanent joints in materials. Semipermanent joining is possible in the lap and folded seams by using threaded fasteners.

The grooved seam is formed by making two opposite folds in the metal. Select a hand groover of the proper size. Fig. 50-22. Place the hand groover over the seam and strike firmly to lock. Fig. 50-23. Remember to allow for the folding of the metal by adding three times the width of the grooved seam to the pattern layout.

Hand Groover Sizes

The small number stamped on the face of the groover indicates the size.	
Number	**Width of Groove**
6	1/8
5	5/32
4	7/32
3	9/32
2	5/16
1	11/32
0	3/8

50-22. The hand groover comes in various sizes.

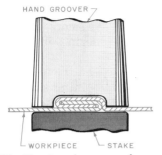

50-23. The hand groover is used to lock a grooved joint or seam together. Hold the groover square with the work to avoid unsightly dents in the work.

QUESTIONS AND ACTIVITIES

1. Define mechanical fastening.

2. List the three main mechanical fastening methods and give three examples of each.

3. Name at least five tools that are used in turning and holding threaded fasteners.

4. What tool is used in the school shop to cut external threads? Internal threads? Give an example of each.

5. What is the difference between a National Coarse and a National Fine bolt? What tool is used to measure thread size?

6. What does the pitch of a metric screw indicate? Of a customary screw?

7. Describe the basic riveting process.

8. How do you determine what size rivet to use?

9. Sketch the four kinds of sheet metal joints, and tell how each is made.

CHAPTER 51

Adhesive Fastening of Metal

A good example of fastening by adhesion is gluing two pieces of wood together. In the adhesion process, a layer of bonding material is put between the two workpieces to be joined. This layer penetrates the surfaces of the workpieces slightly and causes a tough bond or joint to be formed. (Cohesion forms a different type of joint. See Chapter 52.) In both industry and the shop, soldering, brazing, and cementing are common ways of joining materials by adhesion.

SOLDERING

In soldering, two pieces of metal are joined together with *solder,* a third metal with a lower melting point. The metal parts to be joined are called the *base metals.* They are not melted; only the solder is. The molten solder dissolves a small amount of the base metal; so a strong bond between the metals is formed upon cooling. Soft soldering is usually done at temperatures below 800°. Hard soldering is done at temperatures between 800° and 1400°. You will learn more about soft and hard soldering in the next few pages.

Soldering is an inexpensive, efficient way of joining sheet metal or electrical parts. It is widely used in the metalworking and electrical industries.

Equipment

The heat for soldering is provided by the equipment shown in Fig. 51-1. These devices are usually found in the school shop. They should be used carefully to avoid burns.

Common soldering materials are shown in Fig. 51-2. The solder used in soft soldering is an alloy (mixture) of tin and lead. The solder used in hard soldering is made of silver alloys. Solder materials for both soft and hard soldering are available in several alloys, forms, and sizes.

When metal is exposed to air, a film of oxide, or rust and tarnish, forms on it. This process of *oxidation* is greatly increased when the metal is heated. Oxidation must be prevented because it keeps the base metals from reaching the soldering temperature and from uniting with the solder. Acid and rosin fluxes are needed to prevent heated metals from oxidizing and to help the solder flow easily. Rosin fluxes are noncorrosive. That is, they won't continue to dissolve the metal and weaken the joint. Therefore, they are especially good in electrical work.

Procedures

Soldering is not difficult if it is done with care. There are four steps to be followed for successful soldering.

1. Clean the workpieces thoroughly.

2. Select the proper flux and solder for the job.

3. Select the correct soldering device and be sure it is in good condition.

4. Apply the proper amount of heat.

FOR YOUR SAFETY . . .

Follow these rules whenever you are doing soldering:

- Wear safety glasses.
- Handle soldering equipment carefully to avoid burns.
- Do not touch joints that have just been soldered. You could get burned.
- After soldering, wash your hands to remove all traces of flux.
- Let the soldering copper cool off before storing it. Otherwise, you could start a fire.

Soft Soldering

As mentioned earlier, soft soldering is done at temperatures below 800°F, and a tin-lead solder is used. The most common tin-lead alloy is called half-and-half, with 50 percent tin and 50 percent lead. Solder with more tin, such as 60/40 melts at lower temperatures and is more free-flowing.

Solder will not stick to a dirty, oily, or oxide-coated surface. To clean the workpieces, rub them with steel wool or a fine abrasive paper. Apply the flux with a swab or brush to these cleaned surfaces.

The soldering copper (or soldering iron, as it is often called) should be cleaned and tinned before using. To do this, carefully clamp the soldering copper in a vise and file it with a coarse file until it is smooth and free of any oxide. Remove the copper from the vise and heat it until it is yellow or light brown. *Never allow a soldering copper to become red hot.* Rub the copper back and

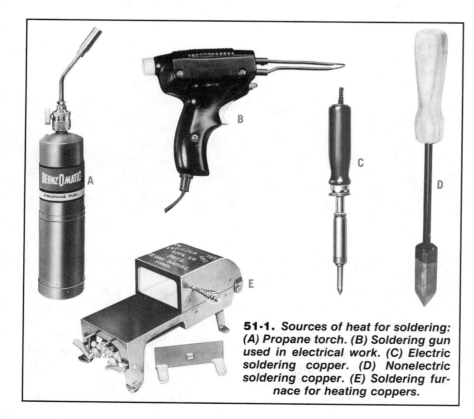

51-1. Sources of heat for soldering: (A) Propane torch. (B) Soldering gun used in electrical work. (C) Electric soldering copper. (D) Nonelectric soldering copper. (E) Soldering furnace for heating coppers.

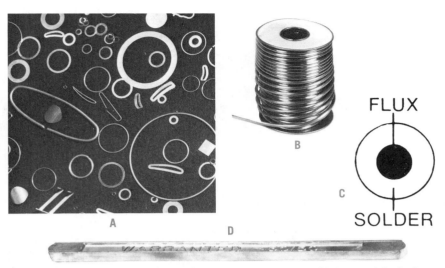

51-2. *Materials for soldering: (A) Solder preforms, widely used in industry. (B) Wire solder is available as solid wire or with a flux core. (C) Cross section of flux core solder. (D) Bar solder.*

51-5. *Move the soldering copper slowly along the joint. Hold the workpieces until the solder sets up. Reheat the soldering copper as necessary.*

forth on a sal ammoniac block (a corrosive flux). Apply a few drops of solder until the tip of the copper is covered with a light coating of solder (about ½″ up each face). Fig. 51-3. This coating process is called "tinning." After the copper is used for some time or it becomes overheated, the tip becomes covered with oxide and the copper will have to be tinned again.

SOLDERING A SEAM OR JOINT

Hold the workpieces with a file or scriber so they stay in place while soldering and until the solder hardens. Cylinders and similar structures should be wired together to hold them in place during the soldering process. Fig. 51-4.

Put the tip of the soldering copper at one end of the seam or joint to be soldered and hold it there until the flux begins to sizzle. Put a small amount of solder directly in front of the tip of the copper. Hold the copper with the tapered side flat against the workpieces until the solder begins to melt and flow into the seam. Move the copper along the seam slowly, and in one direction only (never back and forth). Fig. 51-5. Put more solder in front of the tip as it is needed. Be sure to keep the soldering copper hot at all times. Most poor soldering jobs are the result of insufficient heat.

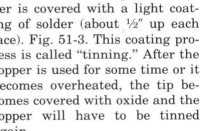

51-3. *A properly tinned soldering copper.*

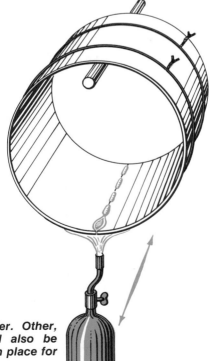

51-4. *Soldering a cylinder. Other, similar structures should also be wired to hold the pieces in place for soldering.*

SWEAT SOLDERING

Sweat soldering is soldering two or more pieces of metal onto each other, so that you cannot see any of the solder on the finished work. Sweat soldering is often used to fasten layers of metal on top of each other to make a heavier part, or to fasten a decorative piece of metal onto a plain piece of metal. Sweat soldering can also be used to fasten a joint when no solder is to be seen after assembly.

To sweat solder metal layers, flux one surface and apply a thin coat of solder. Then flux the opposite surface and clamp the pieces together. Fig. 51-6. Hold the clamped pieces over a propane torch or bunsen burner until the solder begins to ooze out at the edges. For a neat job, use small amounts of solder.

To sweat solder a joint, cover the surfaces to be joined with a flux and a thin layer of solder. Clip these pieces together and apply heat with a soldering copper until the solder melts and the pieces are joined.

Hard Soldering

Hard soldering (sometimes called silver soldering or silver brazing) is similar to brazing. (Brazing will be discussed in the next section.) Hard soldering is generally done at temperatures between 800° and 1400°F. The conventional soldering copper does not provide this much heat, so a propane torch, bunsen burner, or oxyacetylene torch must be used. Hard soldering produces a neater, more permanent joint than soft soldering.

Hard soldering is very useful in art metalwork. Follow the directions given here to hard solder jewelry and other small metal workpieces. Fig. 51-7.

1. Clean the metal parts thoroughly. File so that they fit together as closely as possible. Hold pieces together with clips.

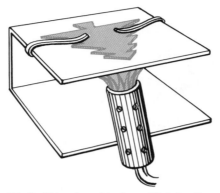

51-6. Sweat soldering. Metal clips hold the decorative piece of metal onto the plain piece of metal during the soldering process.

2. Apply a borax hard soldering flux along the joint. This can be bought ready made, or you can mix borax with water to make a creamy paste. Use a camel's-hair brush to apply the flux.

3. Select a hard solder to match the job; the solders melt at different temperatures. They are available in foil, wire, and powder forms.

4. Cut snippets of hard solder and lay them in place about ½ inch apart. Use tweezers or a camel's-hair brush to apply the bits along the joint.

5. Gently apply heat with a propane torch or bunsen burner, using the blue part of the flame to preheat the joint until the flux dries out. This will hold the solder in place. Play the inner point of the flame over the joint until the solder melts. Avoid putting the heat directly on the solder. Do not overheat, as the workpiece may melt and ruin the job.

6. Allow the workpiece to cool. Clean by dipping in acid (pickling) solution. Rinse with water.

BRAZING

Brazing is a way of joining metal workpieces by using a filler rod which has a lower melting point than either of the pieces to be joined. The most common filler rods are made of copper alloys such as bronze, or of silver alloys. In this process, the cleaned workpieces are heated cherry red. The filler rod is then touched to the heated area. It melts and begins to flow into the joint. Upon cooling, this leaves a strong bond. Brazing is used to repair metal and to make joints almost as strong as welded joints, without melting the workpieces. (In welding they are melted.)

The equipment for brazing is shown in Fig. 51-8. This equipment, if used wrongly, can be very dangerous. Follow instructions carefully. Report to your instructor any equipment which is damaged or not working prop-

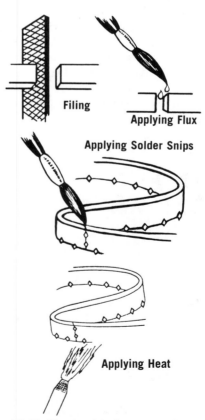

Filing

Applying Flux

Applying Solder Snips

Applying Heat

51-7. Hard soldering operation.

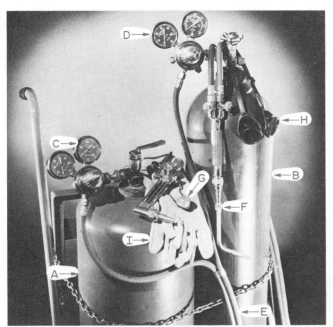

51-8. *Equipment for brazing: (A) Acetylene cylinder. (B) Oxygen cylinder. (C) Acetylene regulator. (D) Oxygen regulator. (E) Hoses. (F) Torch. (G) Sparklighter. (H) Goggles. (I) Gloves. (J) Brazing rods. (K) Brazing flux.*

erly. Never use faulty equipment. Wear gloves and welding goggles when brazing.

Refer to Fig. 51-8 and follow the directions given here for using the brazing equipment. These instructions are for general brazing of light materials.

1. Check to see that all hose, regulator, and torch connections are tight.

2. Slowly turn the oxygen cylinder valve wide open. Open the acetylene cylinder valve 1½ turns. Do not stand facing the regulators while you are turning the valves. Stand to one side.

3. Set the oxygen regulator to 20 lbs. pressure by slowly turning the regulator valve clockwise. In the same manner, set the acetylene regulator to read 5 lbs.

4. To light the torch, first open the acetylene valve on the torch slightly. Use a sparklighter to ignite the gas. (*Never use matches —you might burn your hands.*) Readjust the regulator pressure if necessary.

5. Open the oxygen valve on the torch until the flame is slightly oxidizing. Fig. 51-9. This flame is identified by a clear, well-defined white inner cone and a short heat envelope. When there is excess oxygen in the burning mixture, the flame is called an oxidizing flame. When there is excess acetylene, the flame is called a carburizing or reducing flame. The neutral flame has equal parts of oxygen and acetylene.

6. To turn off the torch, close the acetylene valve first, then the oxygen valve. Next, close the acetylene and oxygen cylinder valves. Open the acetylene valve on the torch until the regulator returns to zero, then close it again. Turn the acetylene regulator valve wide open to release the tension on the regulator. Do the same for the oxygen system.

Metal pieces to be brazed must be clean and must fit together closely for the best joint. Heat the metal until it is cherry red and hold the filler rod on the heated metal until it begins to flow. Never melt the filler rod directly with the torch. Keep the torch and the rod moving slowly along the joint. The filler rod must be dipped in flux before using. (Some filler rods come already coated with a flux.) This dissolves impurities (oxides) and permits the bronze to penetrate the metal pores. Hold workpieces

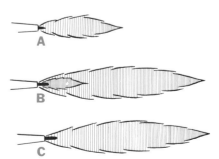

51-9. *Oxyacetylene flame patterns: (A) Oxidizing flame used for braze welding with a bronze rod. (B) Carburizing flame for hard-facing and welding white metal. (C) Neutral flame for welding steel and cast iron. The oxidizing flame is the correct flame for general brazing.*

in place with a clamping device to prevent movement while brazing.

FOR YOUR SAFETY . . .

Follow these safety rules when brazing:

- Cover the work area with heat-resistant material.
- Wear gloves and welding goggles.
- Have a fire extinguisher handy.
- Stand to one side when opening the valves on the oxygen and acetylene cylinders.
- Use a sparklighter, *not* a match, to ignite the acetylene gas.
- Use tongs or long-nose pliers for lifting or turning a hot workpiece.

CEMENTING

In modern metalworking processes, glues and cements are sometimes used for fastening metal pieces together. Three of the common adhesives used in industry and the shop are contact cement, epoxy glue, and hot melts.

Contact cements are used to join pieces of sheet metal. This is done by first cleaning the surfaces of the joint with lacquer thinner and then coating them with the cement. Coat both surfaces and allow them to dry. Then press the pieces together, place a block of wood over the joint and strike with a hammer. (No clamping is necessary.)

When using epoxy glues, be sure to work in a well-ventilated area. Mix the hardener and resin in equal proportions and apply the mixture to the cleaned surfaces. Clamp the surfaces together and allow to cure for about eight hours.

Nordson Corporation

51-10. *Assembling a fiberboard container with hot melt adhesive. Metal parts are joined in a similar way.*

Hot melts are applied with a special gun. Hot melt adhesive sticks are loaded into the gun and allowed to heat to a liquid state. A trigger squirts out a small amount of the hot adhesive onto the joint. The parts are held together for a few seconds until the hot melt cools. Most materials can be joined in this way, including wood, metal, plastic, and ceramics. Fig. 51-10.

QUESTIONS AND ACTIVITIES

1. Describe the basic adhesion process.
2. Name three sources of heat for soldering.
3. Define oxidation. Why must it be prevented in soldering?
4. What is the purpose of a flux in soldering?
5. What are the four steps to successful soldering?
6. Describe how a soldering copper is tinned.
7. Define sweat soldering.

8. What are the differences between soft and hard soldering?
9. What are most filler rods for brazing made of? Name and describe the type of flame used in brazing.
10. What are some safety rules for brazing?
11. Describe the three types of cementing processes discussed in this chapter.

In cohesive fastening of metal, the pieces are melted together to form one continuous mass. This process is called *welding*. Welding is done by heating the metal pieces to their melting temperature, causing them to flow and join together. Metal filler, usually in the form of rods, is often added to the molten pool to make the joint neater and stronger. Pressure is sometimes used to help join the workpieces. This is called *pressure welding*. Welding in which no pressure is required is called *fusion welding*. In both cases, the metal is melted to make the joint. This permanent fastening process is used in making small precision parts, toys, automobiles, steel bridges, oil pipelines, and many other products. The three most common kinds of welding are gas, arc, and resistance welding.

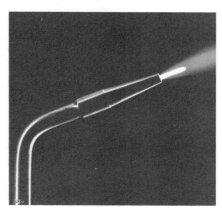

52-2. The neutral flame is used for most oxyacetylene welding operations. This flame has a rounded, well-defined blue-white inner cone which is surrounded by a light blue outer envelope.

GAS WELDING

A flammable gas can be burned with oxygen to produce the heat needed for welding. Acetylene is the gas most often used for this purpose, and the process is called *oxyacetylene welding*. Hydrogen, propane, and Mapp® gas (a liquified acetylene compound) can also be used. Gas welding is fusion welding and usually requires a filler rod to help form the joint. Fig. 52-1.

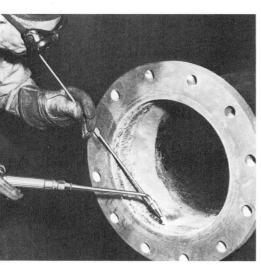

52-1. Gas welding is a form of fusion welding. Here oxygen and acetylene are being burned to supply the heat to melt the metal workpieces and the steel filler rod.

Welding equipment is similar to that used in brazing. (See Chapter 51.) Procedures for lighting and using this equipment are also similar. The filler rod, however, must be of the same metal as the material being welded. The neutral flame is used for most welding. This flame results from an equal mixture of oxygen and acetylene. Fig. 52-2. (See also Fig. 51-9.) Remember to wear welding goggles and gloves when welding.

There are five basic types of welded joints. Fig. 52-3. Most welding jobs will involve at least one of these. The edge weld is a good one to use for welding practice because it does not require the use of a filler rod. Begin by welding a tack at each end. (Tack welds are short welds spaced at intervals along a joint. They hold

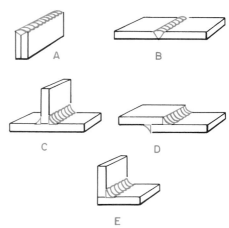

52-3. Basic welding joints: (A) Edge. (B) Butt. (C) Tee. (D) Lap. (E) Corner.

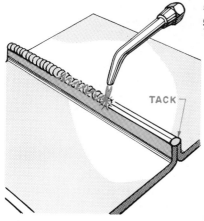

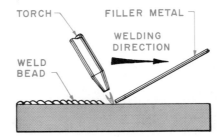

52-4. *Practice welding to develop a good zigzag pattern in your work. This will result in an even, strong welded joint.*

TORCH   FILLER METAL

WELDING DIRECTION

WELD BEAD

FOREHAND WELDING

52-5. *Welding can be done by either of these two methods. Practice them both to see which works better for you.*

FILLER METAL

WELDING DIRECTION   TORCH

WELD BEAD

BACKHAND WELDING

the workpieces in place for welding.) To weld the joint, play the flame back and forth in a zigzag pattern. Fig. 52-4. Draw the puddle (pool of molten metal) along, making certain that you leave no openings or voids.

You should next practice welding with a filler rod. The rod should match the material you are welding. Begin the weld by melting a puddle. Pull the puddle along, touching the filler rod to the puddle as needed to insure a full, firm joint.

Welding can be done by either the backhand or forehand method. In the forehand method, the torch points forward in the direction of travel. The torch is held between the weld and the filler rod. In the backhand method, the torch points back toward the weld. The filler rod is held between the weld and the flame. Fig. 52-5.

Oxyacetylene welding is not widely used today for manufacturing products. Arc welding methods are faster and easier. However, oxyacetylene welding

is still used for maintenance and repair work.

FOR YOUR SAFETY . . .

Fire is the most common hazard in oxyacetylene welding and cutting. Observe these safety rules:

● Remove all flammable and explosive materials from the area.

● Keep a fire extinguisher handy.

52-6. *In arc welding, the heat is supplied by an electric arc.*

● Be careful when handling and storing cylinders. Follow your teacher's instructions.

● Wear a fire-resistant apron, sleeves, and leggings to protect your clothes and yourself.

● Don't wear oily or greasy clothes. They can easily catch fire. Keep cuffs rolled down and pockets closed. Don't wear low cut or canvas shoes.

● Wear goggles with lenses of the proper shade. Your teacher will tell you which shade is right for the welding job you will do.

● Wear gauntlet gloves of heat-resistant material. Keep them away from oil and grease.

● Wear a respirator when welding or cutting material containing, or coated with, galvanized iron, brass, bronze, lead, zinc, aluminum, mercury, cadmium, or beryllium.

● Never try to blow dirt off your clothes with gas pressure. The oxygen or acetylene can saturate your clothes. One spark could make them catch fire instantly.

ARC WELDING

In arc welding, another fusion welding process, the heat for fus-

ing the metal pieces is supplied by an electric arc, or spark. The arc is produced by an electric current jumping an air gap between an electrode (welding rod) and the metal to be welded. Fig. 52-6.

There are many kinds of arc welding. A few of the more common ones will be discussed here.

Metal-Arc Welding (MAW)

In the metal-arc process, common in the school shop, the electrode itself melts away as a filler material. Fig. 52-7. The electrode may be bare or covered. Covered electrodes have a coating which turns into a gaseous shield. This shield helps to stabilize the arc and protect the weld area from atmospheric contamination. (When oxygen and nitrogen combine with molten metal, the weld is weakened.) When covered electrodes are used, the process is called shielded metal-arc welding (*SMAW*).

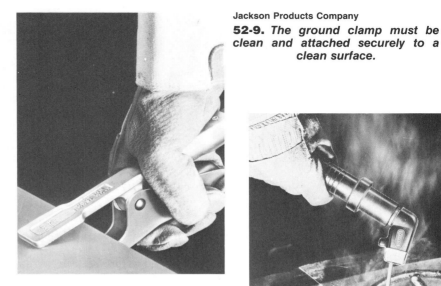

The equipment for metal-arc welding is shown in Fig. 52-8. Make certain that the ground clamp is securely attached. Fig. 52-9. If it is not, the welding cir-

52-9. *The ground clamp must be clean and attached securely to a clean surface.*

52-10. *These are two types of electrode holders.*

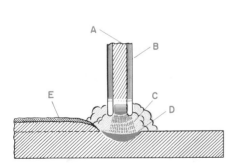

52-7. *Metal-arc welding diagram: (A) Electrode core. (B) Electrode coating. (C) Arc. (D) Molten metal puddle. (E) Slag. Slag is a nonmetallic layer that forms on top of the molten metal. It is removed after the weld has cooled.*

52-8. *Equipment for metal-arc welding. The heat of the electric arc between the electrode and the workpiece melts the metal.*

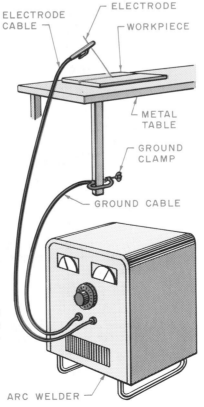

cuit will not be complete and arc welding cannot take place.

To arc weld, place an electrode in the holder (torch). Fig. 52-10. Remember to wear a helmet and protective clothing. Fig. 52-11. An arc is struck by lightly scratching the electrode tip against the workpiece and then withdrawing it. When the arc starts, move the electrode in a zigzag pattern similar to that of oxyacetylene welding. The electrode is fed to the molten puddle and consumed (becomes part of the weld bead). Be sure to maintain the correct arc gap (distance between electrode and work). A good arc gap is indicated by a sharp, crackling sound, an even transfer of the molten electrode

52-11. *Wear a protective helmet and gloves when arc welding.*

fed through the torch at preset speeds. This welding process is often called *MIG* (for metal inert gas) welding. Fig. 52-13.

Submerged-Arc Welding (SAW)

Submerged-arc welding is somewhat similar to MIG welding. A long rod of bare filler metal is used as the electrode and the filler metal is fed auto-

matically into the joint. The weld metal is shielded from the air by a loose, granular mixture of flux that surrounds the arc. Fig. 52-14.

Plasma-Arc Welding (PAW)

Plasma-arc welding is similar to TIG welding in that an electric arc is discharged between a workpiece and a nonconsumable tungsten electrode. The arc for plasma-arc welding, however, is conducted to the base metal by a controlled stream of plasma (electrically charged gas capable of conducting electricity). Fig. 52-15.

FOR YOUR SAFETY . . .

When doing arc welding, follow these safety rules:

• The electric arc produces ultraviolet and infrared rays which can be harmful to your eyes and skin. Whether you are welding or just observing, *wear an arc welding helmet* to protect your eyes, face, and neck.

• For the protection of others in the shop, welding should be done in a special room or a curtained welding booth.

metal across the gap, and a lack of spatter.

Gas-shielded Arc Welding

In this process, as in SMAW, a gas shields the weld area from contamination. However, in gas-shielded arc welding, the gas is emitted from the welding torch, not the electrode. The gas is usually argon or helium. The two most common gas-shielded arc welding processes are gas tungsten-arc welding (*GTAW*) and gas metal-arc welding (*GMAW*).

Gas tungsten-arc welding is done with a nonconsumable electrode made of tungsten. The process is often called *TIG* (for tungsten inert gas) welding. Tungsten is used for the electrode because of its resistance to high temperatures and for its electrical characteristics. No flux or coating is used, but filler metal may be added. Fig. 52-12.

In gas metal-arc welding the electrode melts and thus supplies the filler metal. The electrode, a continuous spool or coil of wire, is

52-12. *Gas tungsten-arc welding uses a nonconsumable tungsten electrode.*

52-13. *Gas metal-arc welding uses a consumable, continuous wire electrode.*

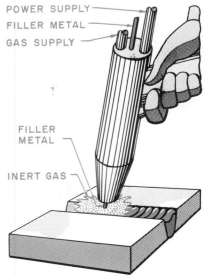

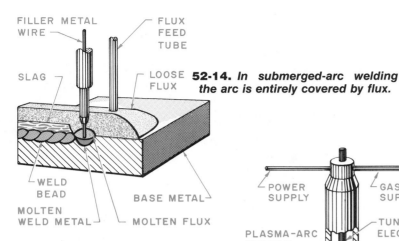

FILLER METAL WIRE

FLUX FEED TUBE

SLAG

LOOSE FLUX

WELD BEAD

BASE METAL

MOLTEN WELD METAL

MOLTEN FLUX

52-14. *In submerged-arc welding the arc is entirely covered by flux.*

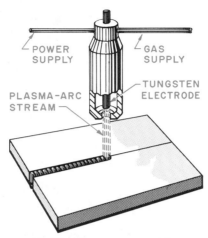

POWER SUPPLY

GAS SUPPLY

PLASMA-ARC STREAM

TUNGSTEN ELECTRODE

52-15. *Plasma-arc welding.*

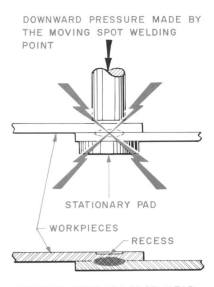

DOWNWARD PRESSURE MADE BY THE MOVING SPOT WELDING POINT

STATIONARY PAD

WORKPIECES

RECESS

SECTION THROUGH SPOT WELD

52-16. *Spot welding. The welded spot joins the two workpieces.*

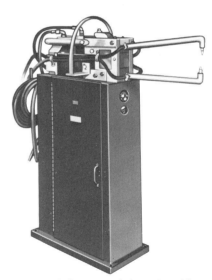

52-17. *A floor model spot welder.*

- Wear nonflammable protective clothing, such as an apron and gauntlet gloves.
- Wear leather shoes that are high enough to protect your ankles. Canvas shoes will not protect your feet from burns.
- Pant cuffs should not be turned up. Bits of molten metal could get caught in them.
- Remove flammable and explosive materials from the area. Welding sparks could ignite such material.
- Make sure the welding machine is grounded.
- Do not let the "live" metal parts of the electrode holder touch bare skin or wet clothing. You could get an electric shock.
- When changing electrodes, turn off the power first.
- Remember, water and electricity are a dangerous combination. Avoid standing on wet floors when using arc welding equipment. Never try to cool electrode holders by putting them in water.

RESISTANCE WELDING

Resistance welding methods use resistance to the flow of electric current to produce the heat for welding. This heat is localized in the weld area, producing a small fused section. Pressure is also used to insure a solid weld. Typical resistance methods are spot and seam welding.

Thin metal sheets or rods can be spot welded when they are held together under slight pressure between two copper electrodes. The electrodes pass a strong electric current through the metals. The resistance of the metals to the current causes both pieces of metal to melt in the spot where the current passes through. Fig. 52-16. Floor and hand-held spot welders are used. A common shop model is shown in Fig. 52-17. Spot welders are often water cooled to prevent warping of the workpieces. A timer usually controls the duration of the weld operation and the amount of heat.

Seam welding is a type of "continuous" spot welding. The welding wheels roll along the joint and produce a series of spot welds. Fig. 52-18. Cylinders and other containers are often welded this way.

Resistance welding methods are efficient and economical. They can be used in mass production techniques. Resistance weld-

Reaction Bonding

A new method of fastening metals to ceramics has been developed recently. In this method, called reaction bonding, the metal is neither melted nor glued. The pieces to be joined are fastened together and exposed to high temperatures in the presence of oxygen. The heat, oxygen, and the presence of the ceramic bring about a corrosionlike reaction in the metal. Tendrils of the metal become imbedded in the ceramic, creating a permanent bond.

The process can be used to fasten metal to most ceramics, glass, and some gemstones. Possible applications include bonding the Space Shuttle's heat-resistant tiles to its surface, making gold-coated ceramic wafers for semiconductors, and improving the bond between the ceramic cap and the gold base of dentures.

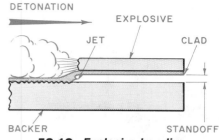

52-19. *Explosive bonding.*

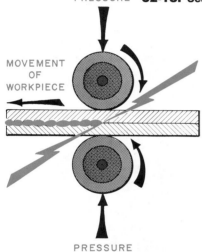

PRESSURE **52-18.** *Seam welding.*

MOVEMENT OF WORKPIECE

PRESSURE

ing produces relatively little heat and does not distort the metal.

EXPLOSIVE BONDING

This fastening method uses explosive energy to join two or more pieces of metal together permanently, without using fluxes, rods, or heat. It is a type of pressure welding. The technique was developed and patented by duPont; the trade name "Detaclad" is used for the products thus produced. The prime metal, or clad, is placed above the backer metal at a carefully controlled "standoff distance," or spacing. Fig. 52-19. The explosive (fastened to the clad) is then fired, causing the clad and the backer to collide and form a metallic jet (a stream of molten metal). This jet is trapped between the clad and backer and solidifies into a wavy bond joint.

Some applications for this bonding method are clad plates for the fabrication of chemical pressure tanks, heat exchangers for nuclear power plants, and heat-sealing jaws for plastic film packaging machines.

QUESTIONS AND ACTIVITIES

1. Define cohesive fastening.
2. What is the difference between the cohesive and the adhesive metal fastening processes?
3. Name and sketch the five basic types of welded joints.
4. What are tack welds and what purpose do they serve?
5. Define arc welding. Name four common types of arc welding. (Give both the full name and abbreviation of each.)
6. What is the difference between the GTAW and the GMAW processes?
7. Briefly explain how resistance welding works. What are two typical resistance welding methods? What are some advantages of resistance welding?
8. Name three things that are made by explosive bonding.

Most materials require some sort of protection if they are to last when exposed to weather, dampness, or similar conditions. Wood will rot and metal will rust if not covered by some protective coating. Such a coating is called a *finish*. But coatings, such as paint and lacquer, are only one kind of finish. The finishing chart, Fig. 53-1, shows three other kinds.

Remove finishing resembles both abrasive and chemical cutting. It is in fact a finishing method which uses a cutting action to produce a desired surface treatment. The buffer and polisher use abrasives, and etching tanks use chemical cutting as a finishing method.

Hammering and peening are examples of the displacement method of finishing. Metal knurling and decorative embossing are other common displacement finishes.

Remove and displacement finishes are often called *mechanical finishing*.

Coloring metal not only changes its appearance but also makes it more durable and serviceable. This is true of both the heat and chemical methods. Gun barrels are blued to prevent rust and to improve the looks of the weapon.

As you can see, many finishes both beautify *and* protect an object. Paint improves appearance and prevents rusting. A scratch finish or satin finish produced by the buffer adds interest to a sur-

53-1.

Finishing

Finishing is the process of treating the surface of a material for appearance and/or protection.

KIND OF FINISHING	DEFINITION	EXAMPLES
Coating*	Applying a layer of finishing substance to the surface of a material.	Lacquer, enamel, electroplate, flock, wrinkle finish, plastic dip, curtain-coating, electrostatic spraying.
Remove Finishing	Finishing by cutting the surface by abrasive or chemical action.	Grinder, wire brush, sand blaster, etching tanks, polisher, buffer, steel wool, coated abrasives.
Displacement Finishing	Finishing by the movement of surface material to a different position, to form a new appearance.	Hammer, peen, knurl, emboss, planish.
Coloring*	Applying penetrating chemicals or heat to a material to change its color.	Chemical coloring, bluing, tempering.

*Coating could be considered a coloring process, since coating usually does change a product's color. However, it is customary to consider coating and coloring to be separate processes, as defined in the chart above.

face and prevents minor nicks and fingerprints from showing. Peening a surface does the same thing as the satin finish. The example of gun bluing shows how chemicals are used to beautify and protect.

The tools, materials, and techniques for producing these four major kinds of finishes are discussed in the chapters which follow. The finishing chart will help you to see the differences and similarities among these surface treatments for metals.

QUESTIONS AND ACTIVITIES

1. Define finishing.
2. Name and define the four major kinds of finishes. Give one example of each kind of finishing.
3. Name two examples of finishes that both beautify and protect an object.

CHAPTER 54
Coat Finishing of Metal

Coating metal involves applying a layer of paint, lacquer, wax, metal, or glass to the surface to protect it, improve its appearance, or both. Coat finishes are widely used by industry. For example, coat finishes are sprayed on automobiles and metal furniture. This is ususally done automatically, often by programmed robots. See Chapter 5.

LACQUERS

Lacquers are quick-drying and provide a hard, durable finish that resists fading, scratching, and most mild chemicals. A lacquer finish is a tougher finish than paint. Lacquer can be clear (colorless) to allow the beauty of metal to show through, or it can be colored to add a decorative finish. All lacquers adhere better if the workpiece is first warmed *slightly* in the oven.

Clear lacquer is generally sprayed or brushed on art metal objects made from copper, brass, or aluminum. Projects must first be cleaned with lacquer thinner and dried with a soft cloth. Handle the workpiece with gloves or a clean cloth, since dampness from your hands can discolor many kinds of metals.

When applying lacquer with a brush, use a small, soft, camel's-hair brush. Use quick, even strokes to apply a thin coat over the entire surface. Fig. 54-1. Too thick a coat will produce a yellowish appearance and cause running. Apply a second thin coat, allowing about 20 minutes between coats. Clean the brush with lacquer thinner in a well-ventilated area. Be sure that the heel of the brush (where the bristles join the handle) is thoroughly clean.

Because lacquer dries so quickly, it is often better to apply it by spraying. Spray cans of lacquer are quick and easy to use. For good coverage, the nozzle should be held about 12″ from the workpiece throughout the spraying process. Fig. 54-2. If possible, the surface should be in a vertical position. Spray even strokes that run the length of the workpiece. The rule in coat finishing that two or more thin coats produce a better finish than one thick coat is especially true in spraying.

Colored lacquers are applied in the same way to projects made of ferrous metals (those containing iron). However, it is wise to apply a prime coat to the piece first, followed by two coats of lacquer, either sprayed or brushed on.

Always use caution when using lacquer. It gives off noxious

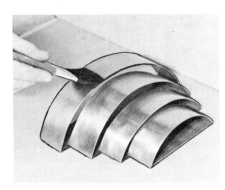

54-1. *Brush a thin coat of lacquer onto the entire surface of the workpiece with quick, even strokes.*

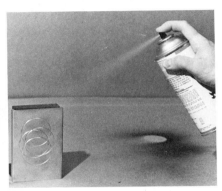

54-2. *Spraying a project. Hold the spray can about twelve inches from the workpiece.*

(harmful) and explosive fumes. It should always be applied in well-ventilated areas, and lacquer containers should be closed immediately after use. Lacquer is also extremely flammable and should be kept from open flames.

PAINTS

Paints are applied much the same as lacquers are, and for

similar purposes. The drying time for paints is generally longer, and brushes should be cleaned in turpentine, paint thinner, or water. Fig. 54-3.

Before a metal surface can be painted, it must be properly cleaned. Remove any old paint with a wire brush or scraper. Remove any grease or oil with a solvent. Sand smooth any shiny surfaces. Dust thoroughly. Apply a good primer paint as a first coat, to insure proper adhesion of the paint to the metal and to prevent rust. When the primer is dry, the desired finishing coats of paint can be applied.

54-3. *Brushes for oil-base paint are cleaned in turpentine or paint thinner. Lacquer brushes are cleaned in lacquer thinner. Brushes used with water-based paint are cleaned with water.*

Wrinkle finishes are also available for metal applications. They cover well, dry rapidly, come in a variety of colors, and can be brushed or sprayed on. These provide a hard, scratch resistant surface for tool boxes and tools. Some wrinkle finishes require oven baking; so be sure to follow the directions printed on the can.

Some paints that dry to a hard, high gloss or semigloss finish are often referred to as "enamel" paints. However, the actual process of enameling is a method of applying small grains of glass to metal, and fusing them to the metal surface with high heat. (This process will be discussed later in this chapter.) Do not confuse enamel paint with the "true enamel" used in enameling.

WAX

A semipermanent finish may be achieved by waxing. A coat of paste or liquid wax is applied to the warmed workpiece with a clean cloth, allowed to dry, and then buffed to a high gloss. This finish will wear off in time and must therefore be reapplied occasionally. Fig. 54-4.

54-4. *Wax can be applied to a project as a semipermanent finish.*

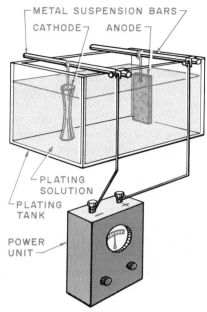

54-5. *The electroplating apparatus. The power unit includes a rheostat, ammeter, and voltmeter. For simple plating jobs, a dry cell may be used.*

ELECTROPLATING

Using electricity and chemicals to cover a metal object with a thin metal coating is called *electroplating*. This process is widely used in industry to apply chrome plating to automobile bumpers and other parts. It both beautifies and protects metal objects.

In electroplating, the object to be plated becomes the cathode (negative pole) in an electrolytic cell. This cell contains a solution of the salts of the metal to be deposited. The anode (positive pole) is the metal to be deposited. When an electric current passes through the cell, metal particles are separated from the anode and go into the electrolytic solution. They are then separated from the solution and deposited on the cathode.

A simple electroplating device is shown in Fig. 54-5. The tanks must be made of plastic, glass, or ceramic stoneware. The metal to be plated must be thoroughly

cleaned in a pickling solution or buffed and cleaned with lacquer thinner. The metal must be absolutely clean, or the plating will not hold properly. Copper, nickel, and cadmium are commonly used as plating (anode) materials.

Assemble the device as shown in Fig. 54-5. Remember that the plating solutions are different, depending on the type of metal anode used. The voltage required is from one to three volts. Adjust the machine properly and con-

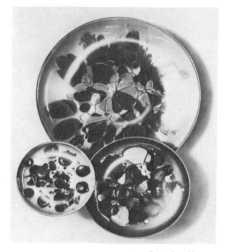

America House

54-6. *Copper enameled decorative dishes.*

tinue plating until the proper thickness of plate has been reached. Remove and rinse. **CAUTION:** *Wear safety glasses and protective clothing.*

ENAMELING

Enameling is the process of decorating metal by melting a layer of colored glass on the surface. The enamel is crushed and screened to size, then applied to the metal in powder or paste form. The enamels are fused to the metal by heating to about 1500°F (815°C). While steels and other metals are enameled industrially, copper is the metal generally used for enameling in the school shop. Fig. 54-6. The simple equipment used for enameling in the shop is shown in Fig. 54-7. Make sure the enameling powders are finely ground and free from dirt.

The procedures for enameling shown in Fig. 54-8 are few and simple:

1. Select a desired copper shape and clean thoroughly with any good scouring powder and water, or dip in pickling solution. Rinse and dry with a clean, soft cloth.

2. Lay the copper shape on a piece of clean, white paper. Then

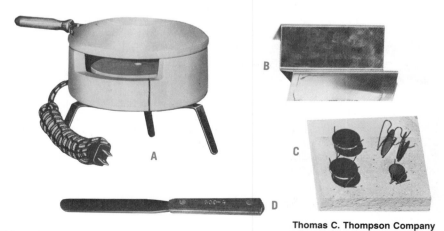

Thomas C. Thompson Company

54-7. *Equipment for enameling: (A) Kiln. (B) Trivet for large workpiece. (C) Trivets for small workpieces. (D) Spatula.*

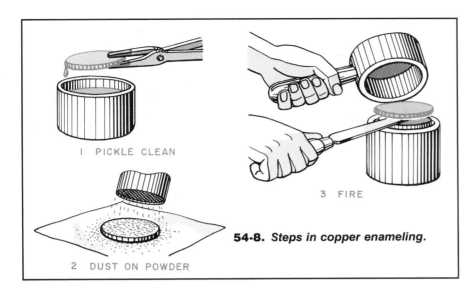

54-8. *Steps in copper enameling.*

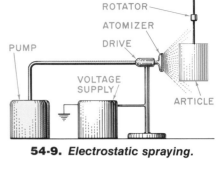

54-9. *Electrostatic spraying.*

opment of this process has made it possible for industries to reduce the amount of paint used by 30 to 80 percent, and to produce finishes of higher quality.

choose one of the enameling colors and cover the surface by sifting gently until the copper shape is completely covered.

3. With a spatula, very carefully place the copper shape on a trivet in a preheated kiln.

4. In a very few minutes the copper shape will take on a wavy glazed appearance. When this occurs, very carefully remove the copper shape with a spatula and allow to cool for about ten minutes. For additional color and design, repeat the process. Fig. 54-8.

The cooled workpiece should be cleaned of tarnish and scale with abrasive cloth or a file. Attach findings (tie tacks, stickpins, etc.) as needed.

ELECTROSTATIC SPRAYING

This process is a special method of coating products with paint or other finishes by giving the finish an electrical charge. The product to be coated is grounded, so that the electrically charged solution particles are attracted to the product. When the particles strike the metal product, they lose their charge and stick to its surfaces. The spraying unit consists of a grounded conveyor, an atomizer gun, a paint pump, and an electrostatic voltage supply. Fig. 54-9. The devel-

FOR YOUR SAFETY . . .

● Work in a well-ventilated area. Many of the chemicals used in metal finishing give off harmful fumes.

● Most finishing materials are flammable and/or explosive. Do not store them near heat or flames. Do not work near heat or flames.

● For electroplating, wear safety glasses and protective clothing to protect yourself from the chemicals used.

● Wear gloves, goggles, and protective clothing when painting or lacquering.

● Wear a respirator when using spray finishes.

QUESTIONS AND ACTIVITIES

1. Why should metal be heated before it is lacquered?

2. What are some safety rules to remember when using lacquer?

3. What liquid is used to clean lacquer brushes? Paint brushes?

4. What purpose does a primer paint serve?

5. Define electroplating. Give an example of an item which has been electroplated.

6. Define the process called *enameling*. Briefly describe the procedure for enameling in the school shop.

7. What is electrostatic spraying?

Remove finishing methods are used to produce smooth, bright surfaces on metals. Shiny or brushed finishes on brass hardware such as hinges, doorknobs, and drawer pulls are produced this way. Industry uses many automatic polishing and buffing devices. In the shop, this kind of finishing is done by hand or with buffing machines.

POLISHING

Polishing is done with hard, felt wheels coated with abrasive grains. Coated abrasive belts, disks, or sheets are also used. (See Chapter 37). A new polishing method uses wheels made from stiff plastic bristles. These are used for polishing industrial metal parts and for general shopwork.

55-2. *Materials for buffing. A stitched buffing wheel and compound bars are shown.*

Polishing removes nicks and scratches left after grinding or filing, and it is often used to remove burrs (thin ridges or rough spots produced in cutting or shaping metal). Fig. 55-1.

BUFFING

Buffing, which follows polishing, is done mainly to make the metal surface look better. Very little metal is removed in comparison to polishing. The buffing produces bright or satin finishes on metal.

Buffing wheels are made of loose pieces of flannel, cotton, felt, or leather stitched together. Buffing compounds are abrasive materials applied to the buffing wheel that enable the operator to finely polish and brighten the metal. They are usually bars of fine abrasive grains bonded with grease or wax. Fig. 55-2. Tripoli and rouge are common buffing compounds. Greaseless compounds made with glue binders are also available. These do not clog or dirty the wheels as much as other compounds. Industry also uses liquid spray compounds.

Choose the buffing compound according to the kind of metal being buffed. Coat the wheel with the compound by turning off the power and gently pressing the bar to the wheel as the machine is coming to a stop. This prevents the compound from spattering onto your clothes. When the surface of the wheel is coated, it is ready for buffing.

Hold the workpiece firmly on the underside of the front of the wheel. Fig. 55-3. This precaution is taken so that if the work is pulled from your hands, it will fly away from you instead of toward you. It will also help to keep the dust away from you. Press the

3M Company

55-1. *Removing burrs from a metal part.*

55-3. *The safe way to buff is to hold the workpiece firmly to the underside of the wheel.*

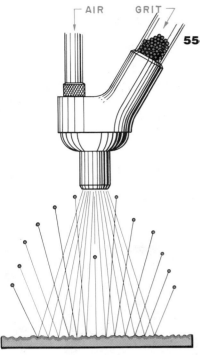

55-4. *Abrasive blasting.*

BASE METAL

Satin, or brushed, finishes may be produced by using special satin compounds or a fine wire wheel. When using wire wheels, select a fine wheel for a soft, smooth sheen. Use a coarse wheel for a frosted finish.

When buffing has been completed, clean the metal with lacquer thinner and dry with a clean, soft cloth. Apply a preserving finish such as lacquer or wax.

ABRASIVE BLASTING

One of the quickest ways to finish metal is with an abrasive blast. This type of finish uses a blast of air and abrasive particles. It is commonly called sand blasting, although abrasive grains other than sand are generally used. Fig. 55-4. Glass, steel shot, grit, and silica sand are the commonly used abrasives.

For this type of finishing, the workpiece is placed in a special cabinet, which is then shut tightly. The blasting gun is hand-controlled, and the operator must wear protective gloves. Fig. 55-5.

Inland Manufacturing Company

55-5. *Abrasive blasting is done in a tightly closed cabinet.*

Depending upon the abrasive used, the finish can range in smoothness from a satin-matte to a hammered or pebble-grain effect.

FOR YOUR SAFETY . . .

• Tuck in loose clothing and wear goggles when polishing or buffing.

• Wear a respirator if conditions are dusty.

• Whenever possible, use a safety shield on the machine.

workpiece lightly to the wheel. Pushing hard against the wheel creates friction and makes the workpiece hot. Turn and move the work as it is held against the wheel so every corner and curve is buffed. Continue until the piece is buffed to the desired finish. The wheel may have to be coated several times before buffing is completed.

QUESTIONS AND ACTIVITIES

1. What is the purpose of polishing? What materials are used in polishing?

2. What is the purpose of buffing?

3. What are buffing wheels made of?

4. What are buffing compounds? Name two.

5. What is a greaseless buffing compound, and what advantage does it have?

6. Explain how a buffing compound is applied to the wheel.

7. What is the safe way to hold the workpiece while buffing? Why should it be held this way?

8. Describe abrasive blasting. What type of abrasives are used?

CHAPTER 56

Displacement Finishing of Metal

Displacement finishing is a metal forming process. The appearance of a metal surface is changed without adding or removing any material. This is also called *mechanical finishing*. The hammered surfaces commonly found on wrought iron hinges or on metal ash trays are good examples of displacement finishes. Industry also embosses decorative patterns on sheet metals and metal tubing. The most common methods of displacement finishing in the shop are peening and planishing.

PEENING

Peening is simply producing a hammered effect on metal by striking it with the ball end of the ball-peen hammer. The metal is held on a flat surface and peened gently. By overlapping the peen marks slightly, a better-looking surface can be produced. Fig. 56-1.

PLANISHING

Planishing is using light hammer blows to stiffen and smooth metal which has been formed by sinking or raising. (See Chapter 47.) It is similar to peening except that the planish marks are finer and smoother. The planishing tools are shown in Fig. 56-2. Make certain that the hammers are free from nicks before using them.

Planishing is done by placing the metal object over a stake which has a similar shape. For

56-3. *Planishing a bowl. Work from the center toward the edges with gentle, even hammer blows.*

example, the metal bowl in Fig. 56-3 is placed over a round stake that is free of nicks. The planishing should be done with gentle, regular blows. Try to overlap the marks, as in peening, and work from the center of the bowl toward the edges. The edges of the bowl will become uneven due to the planishing. True them with a file after the planishing is complete. Never buff the piece after planishing, as this will cause the marks to run together and ruin the effect.

LACQUERING

After peening or planishing, clean the workpiece with lacquer thinner and a soft cloth. Apply lacquer as described in Chapter 54.

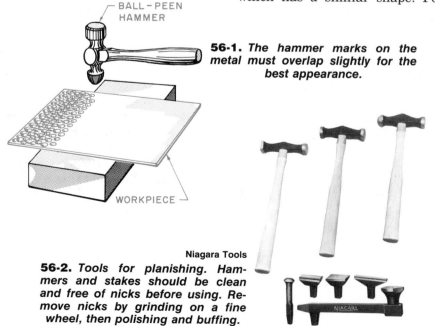

BALL-PEEN HAMMER

WORKPIECE

56-1. *The hammer marks on the metal must overlap slightly for the best appearance.*

Niagara Tools

56-2. *Tools for planishing. Hammers and stakes should be clean and free of nicks before using. Remove nicks by grinding on a fine wheel, then polishing and buffing.*

QUESTIONS AND ACTIVITIES

1. Define and describe peening.
2. Define planishing. How is planishing different from peening?

3. Describe how to planish a metal bowl.
4. Why should peen and planish marks overlap?

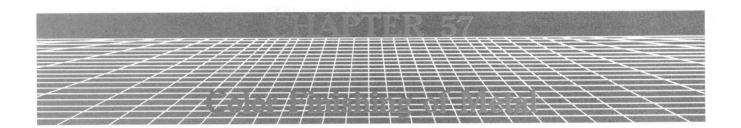

CHAPTER 57

Color Finishing of Metal

Metals can be given interesting color treatments by applying chemicals or heat to their surfaces. Shades of blue, brown, red, green, or black can be produced on metals to provide interesting accents and appearances. After coloring, metals should be coated with lacquer to preserve the finish.

CHEMICALS

The metals commonly colored in the shop are steel, copper, and brass. The chart in Fig. 57-1 shows chemical solutions for coloring copper and brass. All metals to be colored chemically must first be buffed, then cleaned in pickling solution and rinsed in clean water. Handle metal with wooden or plastic tongs to protect your hands and to prevent contamination of the chemical coloring solution. The solutions must be mixed in acid-proof containers. Slip the clean metal carefully into the coloring solution. When the desired color is obtained, remove the piece, rinse it with water, and allow it to dry. The metal may be shaded or highlighted by rubbing with steel wool.

Steel may be given a blue or a brown color, such as is found on the barrels of rifles and pistols. This is a very desirable finish for many projects you might make. Special equipment and solutions are available for these coloring operations.

HEAT

Cleaned metals can also be heated with a propane or welding torch to change their color. Watch the colors as they appear on the surface. Remove the heat as soon as the desired color is obtained. Most metals can be colored this way; however, the metal must be uniformly heated if a uniform color is desired. This can be done in a pot of molten lead, or in a special furnace. The workpiece should be inspected frequently and withdrawn from the heat as soon as the desired color is reached.

57-1.
Coloring Solutions for Copper and Brass

These formulas are given in amounts that are easy to handle. If you need more or less, you may change the amounts of ingredients, provided you keep the same proportions.		
BROWN 4 oz. iron nitrate 4 oz. sodium hyposulfite 1 qt. water	**ANTIQUE COPPER** 3 oz. potassium sulfide (liver of sulfur) ½ oz. ammonia 1 gal. water	**GREEN ANTIQUE** 3 qts. water 1 oz. ammonium chloride 2 oz. salt
RED 4 oz. copper sulfate 2 lbs. salt 1 gal. water	**PICKLING AND CLEANING** 1 pint sulfuric acid 1 pint nitric acid 4 pints water	**SATIN DIP** 1 pint hydrofluoric acid 3 pints water

FOR YOUR SAFETY . . .

The following safety rules should be observed when working with chemicals and heat:

- Wear protective clothing —safety glasses, gloves, and an apron.

- Always pour acid into water. Pouring water into acid causes spattering.
- Always work in a well-ventilated area.
- Use the "buddy" system so that someone will always be ready to help you if necessary.

- Be careful not to splash acid solutions. If a drop gets onto your skin, rinse with water at once and tell your instructor.

QUESTIONS AND ACTIVITIES

1. Is any protective coating needed to preserve color-finished metals? If your answer is yes, tell what kind of coating. If your answer is no, explain why no coating is needed.

2. What should you use to handle the metal in chemical coloring? Why?

3. What chemical solution is used to make copper red? Green antique?

4. Explain how heat coloring is done.

5. List three safety rules for working with chemicals.

PART 4

Production: Manufacturing and Construction

Section 2—Plastics

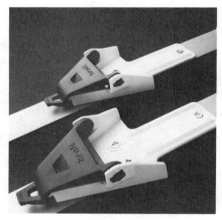

Pinso Sports, Ltd., Quebec

58-1. *These cross-country ski bindings are made of tough plastic.*

The plastics industry is one of the leading and fastest growing industries in the world. Today plastics are used for many products, especially those once made of metal or wood. Fig. 58-1.

Celluloid plastic, developed in 1869, was the first successful plastic. It was used as a substitute for ivory in billiard balls. Motion picture film and automobile windows were also made of Celluloid plastic.

The development of new plastics has grown rapidly since the 1940s. Today plastics are used in everything from ball-point pens to spaceships. For example, they are widely used for the housings (coverings) of tools and instruments. Figs. 58-2 & 58-3. Much furniture—including frames, foam cushions, and chair coverings—is made of plastic. Many easy-to-clean floor coverings are also made from plastics.

THE FAMILY OF PLASTICS

Unlike wood and metal, which are found in nature, plastics are man-made, or *synthetic*, materials. Plastics are usually made from some combination of carbon, oxygen, hydrogen, and nitrogen. These ingredients come mostly from coal or oil.

Plastics are actually a family of materials. Each branch of the family has certain characteristics. However, whatever the properties or form, all plastics fall into two groups: thermoplastic and thermosetting.

Thermoplastics soften when heated and harden when cooled. When heated, they may be shaped or molded. They can be reheated and changed to a different shape.

Thermosets, or thermosetting plastics, are cured (set) into permanent shape by heat. They get soft or burn if reheated, but they cannot be reshaped.

Figure 58-4a shows some common kinds of plastics and typical products made from them. Figure 58-4b describes some important properties of plastics and how these are tested.

COMPOSITES

A *composite* is a uniform material made up of two or more distinct components (ingredients). Composites are generally much stronger than their individual components. Concrete is an example of a composite. It is made of cement, sand, gravel, and water.

Plastic composites are usually made from a combination of a thermoset plastic and one or more types of high-strength fibers or other reinforcing additives. Some plastic composites can even be as much as 20 to 30

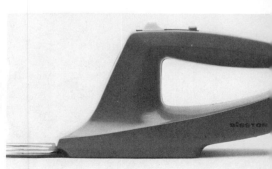

Disston Co.

58-2. *Many power tools are made with plastic housings. The plastic is safer than metal because it does not conduct electricity.*

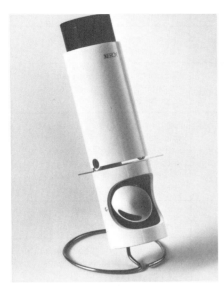

Robert P. Gersin Associates, Inc.

58-3. *This 100-power school microscope has a casing of plastic and metal.*

times stronger than the cured plastic alone would be.

One method of making composites is by lamination. In this process, mats of partially cured plastic embedded with strong fibers are stacked together in layers and bonded together, usually by means of heat and pressure. (You will learn more about lamination in Chapter 59.)

Plastic composites are light and tough, and they corrode less easily than metal. For these reasons they are often used in aircraft and space industries. Weight reduction is a goal of every aircraft design. Extra weight means that more fuel must be used in flight. Airplanes are usually made from light alloys of aluminum, magnesium, titanium, and steel. Composite materials are lighter and stronger than metals. Uses of composites in aircraft construction are shown in Fig. 58-5. Other composite parts, such as engine cowl doors, are made as a

NASA

Superplastics

Many new plastic materials are being developed to meet the needs of high technology. These plastics are lightweight, heat-resistant, and tougher than many metals. Here are some examples.

—An engine made of molded plastic is now being used in racing cars. It is about one-half the weight of a metal engine. Plastic engines cost less because they can be molded to shape, unlike metal which must be cast and machined.

—A process called coextrusion is being used to produce tough, flexible, puncture-proof bags for foods such as potato chips. The plastic is made as a five-layer sheet which blocks out light and prevents the loss of oxygen and flavor from the food. This plastic also is used to make aseptic (germ-free) packages for fruit juices and milk. These foods will keep for months without refrigeration.

—Heat-resistant plastic is available for use in products normally made of metal or glass. This plastic material can withstand temperatures of 300°F. It is used in medical instruments, milking machines, and hair dryers.

—Plastic composite materials, discussed in this chapter, are stronger and lighter than most metals. They are noncorrosive and can be easily molded into complex shapes. These composites are used in the aerospace industry for aircraft floors, ceilings, and doors. The Space Shuttle cargo doors are made from a graphite/epoxy resin composite.

—Thermal protection for the Space Shuttle is provided by a variety of plastic materials. One important example is the composite tile made from silica fibers. See Fig. 58-6. Over 31 000 of these tiles cover much of the outer skin of the Shuttle.

sandwich of aluminum honey-comb with graphite/epoxy facing sheets.

Composites are used in space-craft as well. The Space Shuttle is covered with a special kind of composite tile for protection against the very high temperatures (up to 2750°F) it meets when it returns from space to Earth's atmosphere. The tiles are made of silica fibers (made from sand) and other special fibers bonded together to create a kind of foam block. The tiles are then covered with either a black cured glass coating or a white silica-alumina oxide coating. Black tiles are used on the bottom of the vehicle where the temperature will be higher. White tiles cover the top of the vehicle, to better reflect the sun's rays and help keep the Shuttle cool when in orbit. The finished tiles are bonded to felt-like plastic pads, and then each is bonded to the spacecraft. Fig. 58-6.

CAREERS

The plastics industry has al-ways needed skilled people. One of the most important workers is the mold maker. Many plastic items are formed in molds. Fig. 58-7. Mold makers design and make these molds. Today CAD/CAM systems are commonly used to manufacture molds. See Chapters 3 and 4. Mold makers need new kinds of skills to use CAD/CAM equipment.

Another kind of worker is the

58-4b
Plastics Properties and Tests

Property	Description	Test
Tensile strength	Resists stretching.	Tensile tests. Specimen is gripped by its ends and pulled.
Toughness	Resists fracture under impact loads.	Charpy or Izod impact tests—a notched specimen is struck to fracture it.
Ductility	Can be deformed (changed in shape) without fracturing.	Bending tests (also called transverse flexure tests). Specimen is horizontally supported near its ends. A load is applied to the middle until specimen fractures. Tensile tests are also used.
Compressive strength	Resists squeezing.	Specimen is compressed (pushed together) until it fractures.
Fatigue strength	Resists fracturing under repeated loads.	Specimen is subjected to one or a combination of repeated loads, such as bending or twisting.
Elasticity	Returns to its original shape after removal of a deforming load.	Tensile tests.
Hardness	Resists penetration, scratching, cutting, and abrading. Hardness is actually a combination of basic properties: elasticity, ductility, brittleness, and toughness.	Brinell, Rockwell, Vickers, and microhardness tests. A metal ball or diamond-shaped point is pressed against a specimen.

58-4a
Plastics

Thermoplastics		Thermosets	
Kind	Common Uses	Kind	Common Uses
Cellulose acetate	Toys, electrical components	Melamine formaldehyde	Dinnerware
Cellulose acetate butyrate	Packaging, toothbrush handles	Urea formaldehyde	Adhesives, textile treatments
Cellulose nitrate	Varnishes, explosives	Polyesters	Reinforced plastics (used with glass or other fibers); alkyds for paint; dacron for texiles
Ethyl cellulose	Vacuum cleaner parts, tool handles	Polyurethane foams	Flexible: upholstery, bedding. Rigid: insulation
Vinyls	Floor tile, adhesives, inflatable toys	Phenolics	Electrical equipment, washing machine agitators
Acrylonitrile-Butadiene-Styrene (ABS)	Tool housings, luggage		
Polystyrene	Housewares, insulating foam, ice chests		
Polyvinyl chloride (PVC)	Plumbing pipes, printing plates, raincoats		
Acrylics	Aircraft windows, signs, contact lenses		
Fluorocarbons	Nonstick coatings, electrical insulation		
Polyethylene	Squeeze bottles, toys, packaging		
Polypropylene	Safety helmets, fishing nets		
Nylons	Fishing line, clothing		

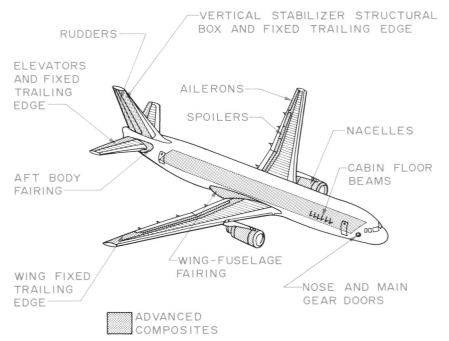

VERTICAL STABILIZER STRUCTURAL BOX AND FIXED TRAILING EDGE

RUDDERS

ELEVATORS AND FIXED TRAILING EDGE

AILERONS

SPOILERS

NACELLES

CABIN FLOOR BEAMS

AFT BODY FAIRING

WING-FUSELAGE FAIRING

WING FIXED TRAILING EDGE

NOSE AND MAIN GEAR DOORS

ADVANCED COMPOSITES

Boeing Commerical Airplane Company

58-5. *Shown on this drawing are some of the airplane parts made of composite materials.*

Lockheed Missiles and Space Company, Inc.
58-6b. *A technician examines one of the tiles. The code number indicates where the tile will be placed on the Shuttle.*

plastics materials engineer. Such a person usually has some experience in mechanical engineering technology, chemistry, and plastics. A plastics engineer must be experienced in all aspects of plastics production, from mold making to finishing the final product.

There are some schools where people can learn these skills, but many engineers go into the plastics factories to get experience on the job.

Other jobs in the plastics industry are like those in other industries. These include jobs like

machine operator, quality-control technician, supervisor, and inspector, to name but a few. The plastics industry will continue to grow and thus will need many more workers in all job skills.

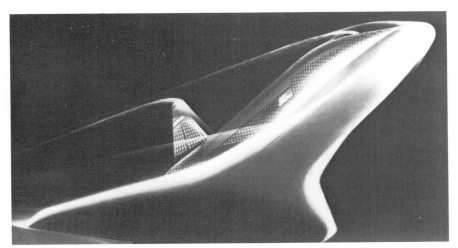

Lockheed Missiles and Space Company, Inc.
58-6a. *The Space Shuttle is covered with nearly 31 000 tiles. The underside of each tile is carefully milled to match the contour of the vehicle. No two tiles are alike.*

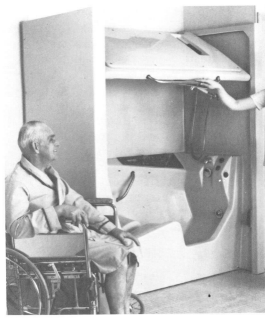

PPG Industries, Inc.
58-7. *This molded fiberglass bathtub has been designed for handicapped people.*

233

QUESTIONS AND ACTIVITIES

1. What are plastics usually made from?

2. What are the characteristics of thermoplastics? Give three examples of items made from thermoplastics.

3. Describe thermoset plastics. Give three examples of items made from thermoset plastics.

4. What are composites and what is the main advantage of making composites? What do plastic composites consist of?

5. Why are plastic composites useful to the aircraft and space industries? Name at least two parts of airplanes or spacecraft that are made from plastic composites.

6. Name at least two careers or jobs in the plastics industry.

CHAPTER 59

Processing Plastics in Industry

Plastic is processed, or worked, in several ways. Some manufacturers use only one technique, while others may use several in the same plant. The common industrial processes include:

- Extrusion.
- Calendering.
- Coating.
- Blow molding.
- Compression molding.
- Transfer molding.
- Thermoforming.
- Injection molding.
- Rotational molding.
- Solvent molding.
- Casting.
- High-pressure laminating.
- Reinforcing.
- Foaming.

EXTRUSION

Extrusion molding is used to form thermoplastic materials into continuous sheeting, film, tubes, rods, profile shapes, and filaments, and to coat wires and cables.

Dry plastic material is loaded into a hopper. It is then fed into a long heating chamber through which it is moved by the action of a continuously turning screw. At the end of the heating chamber the molten plastic is forced out through a *die*. This is a small opening with the shape desired in the finished product. As the plastic is forced through the die, it is fed onto a conveyor belt where it is cooled, most often by blowers or in water. Fig. 59-1.

In the case of wire and cable coating, the thermoplastic is extruded around a continuous length of wire or cable which, like the plastic, passes through the extruder die. The coated wire is wound on drums after cooling.

CALENDERING

Calendering is a process for making thermoplastics into film and sheeting. *Film* is plastic up

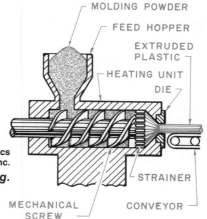

The Society of the Plastics Industry, Inc.

59-1. *Extrusion molding.*

MOLDING POWDER
FEED HOPPER
EXTRUDED PLASTIC
HEATING UNIT
DIE
STRAINER
CONVEYOR
MECHANICAL SCREW

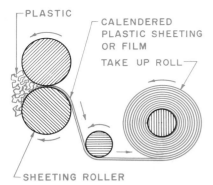

The Society of the Plastics Industry, Inc.
59-2. *Calendering.*

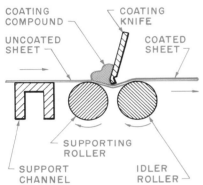

The Society of the Plastics Industry, Inc.
59-3. *Coating.*

to and including 10 mils in thickness. (A mil is $\frac{1}{1000}$ of an inch.) *Sheeting* refers to plastic more than 10 mils thick.

In calendering, heated plastic material is squeezed between heated rollers. The space between the rollers determines the thickness of the plastic. Fig. 59-2.

Calendering is used to make such items as electrical tape, window shades, and shower curtains.

COATING

Coating is the process of applying thermoplastic or thermoset plastic onto a support base such as metal, paper, wood, fabric, glass, ceramic, or plastic. Coating may be done with a knife, spray,

roller, or brush. Fig. 59-3. It can also be done by dipping.

BLOW MOLDING

Blow molding stretches a thermoplastic material against a mold and then hardens it. There are two types of blow molding: *direct* and *indirect*.

In the direct method a mass of molten (hot liquid) plastic is formed into a shape roughly similar to the finished product. While it is still warm and soft,

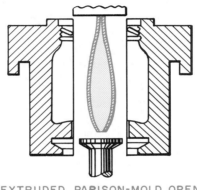

EXTRUDED PARISON-MOLD OPEN

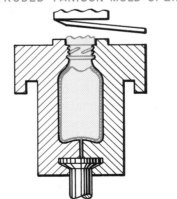

MOLD CLOSED & BOTTLE BLOWN

FINISHED BOTTLE

The Society of the Plastics Industry, Inc.
59-4. *Blow molding.*

this shape, called a *parison,* is placed into a mold. Air is blown into the plastic, like blowing up a balloon. The air forces the soft, warm plastic against the cold sides of the mold. The plastic is then cooled and hardened before being removed from the mold. Fig. 59-4.

The indirect method uses a plastic sheet which is first heated and then clamped between a die and a cover. Air pressure is used to force the plastic material against the die. The plastic is then cooled and hardened.

Blow molding is used to make hollow items with small openings, such as bottles.

COMPRESSION MOLDING

Compression molding is the most common method of forming thermosetting materials. Fig. 59-5. The plastic material is squeezed into the desired shape by using heat and pressure in a mold.

Plastic molding powder is mixed with fillers to add strength or other qualities to the product. This mixture is placed in a heated mold. The mold is closed, causing the plastic to flow to all parts. While the mold is closed,

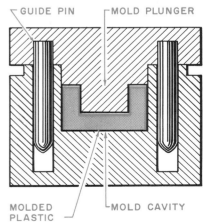

The Society of the Plastics Industry, Inc.
59-5. *Compression molding.*

235

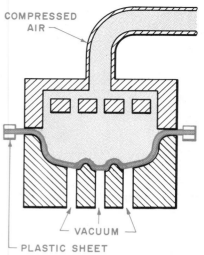

COMPRESSED AIR

VACUUM

PLASTIC SHEET

The Society of the Plastics Industry, Inc.
59-8. *A thermoforming mold.*

3M Company
59-6. *These molded plastic grill-work pieces are being cleaned with an abrasive cylinder brush.*

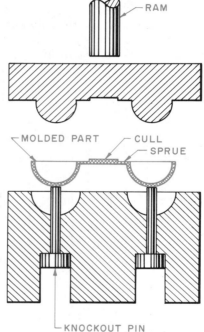

RAM

MOLDED PART — CULL — SPRUE

The Society of the Plastics Industry, Inc.
59-7. *Transfer molding.*

KNOCKOUT PIN

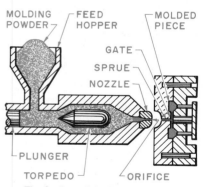

MOLDING POWDER — FEED HOPPER — MOLDED PIECE

GATE

SPRUE

NOZZLE

PLUNGER

TORPEDO — ORIFICE

The Society of the Plastics Industry, Inc.
59-9. *Injection molding.*

the plastic undergoes a chemical change that permanently sets its shape. After cooling, the mold is opened and the part removed and cleaned. Fig. 59-6.

Buttons, dinnerware, and cooking utensil handles are examples of compression molded products.

TRANSFER MOLDING

Transfer molding is usually used with thermosetting plastics. This method is like compression molding in that the plastic is cured in a mold under heat and pressure. It differs from compression molding in that the plastic is heated to a formable state before it reaches the mold, and it is forced into a closed mold by a plunger. Fig. 59-7.

Transfer molding was developed as a way of molding parts with small deep holes or numerous metal inserts. Automobile distributor caps are transfer molded.

THERMOFORMING

Thermoforming consists of heating a thermoplastic sheet or film to a formable state and then using air and/or mechanical assists to shape it.

There are many methods of thermoforming. One technique is simple vacuum forming. In this process, a plastic sheet is clamped over an airtight box. Heaters above the box soften the plastic. Then the air is drawn out of the box, pulling the soft plastic down on a "former" in the box. The plastic thus becomes shaped like the former. Another method uses compressed air to force the plastic into a former. Fig. 59-8.

Thermoforming is used to

make such products as luggage, briefcases, and utility trays.

INJECTION MOLDING

Injection is the most common method of forming thermoplastic materials. It can also be used on thermosetting plastics. In injection molding, plastic material is put into a hopper which feeds into a heating chamber. A plunger pushes the plastic through this long heating chamber where the material is softened to a fluid state. At the end of this chamber there is a nozzle which is placed firmly against the opening of a cool, closed mold. The liquid plastic is forced at high pressure through this nozzle into the cold mold. As the plastic cools and becomes solid, the mold opens and the finished plastic piece is released and removed from the mold. Fig. 59-9.

Injection molding is used to make a wide variety of products, including radio and television cabinets and control knobs. Disposable cups, knives, forks, and spoons are also made using this method.

ROTATIONAL MOLDING

This method is the only plastic molding process in which completely closed, seamless, hollow products (such as inflated balls) can be made. A measured amount of thermoplastic material is placed into a mold which is rotated in an oven. The movement spreads the material evenly throughout the inside of the mold. The mold is then cooled to harden the plastic into the shape of the mold.

Rotational molding can be used to make footballs, mannequins, and musical instrument cases.

SOLVENT MOLDING

In solvent molding a mold is dipped into a thermoplastic solution and withdrawn, or filled with a liquid plastic and then emptied. A layer of plastic film sticks to the sides of the mold.

Some items thus formed, like a bathing cap, are removed from the molds. Other solvent moldings remain permanently on the form. Examples are plastic coatings on pliers or wire cutters.

CASTING

In the casting process, no pressure is used. Plastic material is poured into a mold. After it *cures* (cools and becomes solid), it is removed from the mold. Casting may be done with either thermoplastic or thermosetting materials.

Furniture parts, boat propellers, gears, and even other molds are just a few of the things made by casting.

HIGH-PRESSURE LAMINATING

Thermosetting plastics are generally used in high-pressure laminating, which is done with high heat and pressure. The plastics are used to hold together the reinforcing materials that make up the body of the finished product. The reinforcing materials may be cloth, paper, wood, or fibers of glass. These materials are soaked in a plastic solution. After drying, layers of these materials are stacked between polished steel plates and subjected to heat and high pressure to bond them permanently together. Fig. 59-10.

The end product may be plain flat sheets; decorative sheets like those used for countertops; or rods, tubes, or formed shapes.

REINFORCING

Reinforced plastics differ from high-pressure laminates in that the thermoset plastics used require very little or no pressure in the processing. Fig. 59-11. In both methods, however, plastics are used to bind together the cloth, paper, or glass fiber reinforcing material used for the body of the product. The reinforcing material may be in sheet or mat form. The type of material used depends on the qualities desired in the end product. Reinforced plastics have very high strength, yet are lightweight.

Fiberglass is a very commonly used reinforced plastic, made from polyester resins and glass fibers. Fiberglass is used to make boats, aircraft sections, and safety helmets.

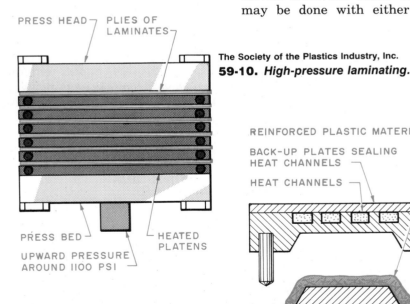

PRESS HEAD — PLIES OF LAMINATES

PRESS BED — HEATED PLATENS

UPWARD PRESSURE AROUND 1100 PSI

The Society of the Plastics Industry, Inc.
59-10. *High-pressure laminating.*

The Society of the Plastics Industry, Inc.
59-11. *Reinforcing.*

REINFORCED PLASTIC MATERIAL
BACK-UP PLATES SEALING HEAT CHANNELS
HEAT CHANNELS

HEAT CHANNELS
BACK-UP PLATES SEALING HEAT CHANNELS

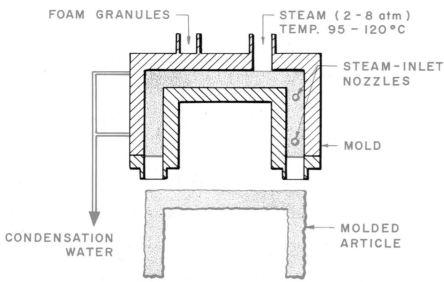

59-12. *One method of foaming plastics. The molded articles may be anything from dashboards to airplane parts.*

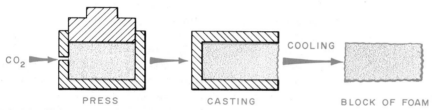

59-13. *Using carbon dioxide gas to make foamed plastics. The press shapes the material in which the expanding gas is dissolved. The casting is then heated. This is called thermal stabilization. The resulting product may be a simple form from which sheets or various shapes are cut, or it may be a molded item.*

FOAMED PLASTICS

Foamed or expanded plastics are materials made by forming gas bubbles in plastic material while it is in a liquid state. Several processes are used to make the bubbles. In most foamed plastics, the hollow bubbles (called cells) are separated from each other by "walls" of plastic. This makes the foams very good insulating material for heat or cold.

Foamed materials are made in two ways. One way uses beads of thermoplastic or thermosetting material that expand when heated. The heat causes a chemical reaction within the beads and produces a gas. It is this gas which causes the beads to expand and make the cell structure. Fig. 59-12. An example of a foamed plastic made by this process is Styrofoam®, a polystyrene plastic.

Another way to make foam is to blow a gas, such as carbon dioxide, into the plastic material while it is in a molten state and being formed. Fig. 59-13.

Plastic foams may be flexible or rigid. They are used in many forms and for a variety of purposes, from decorating to insulating. Some typical products are sponges, seat cushions, padding, insulation, automotive parts, and picnic coolers.

FORMING PLASTICS IN THE SCHOOL SHOP

Many of the forming methods discussed in this chapter require expensive, complex equipment. They can be done only in large factories. However, there are some types of plastics forming which can be done in the school shop. You will learn about these in the next chapter.

QUESTIONS AND ACTIVITIES

1. Name seven common industrial methods for processing plastics, and tell whether each uses thermoset plastics, thermoplastics, or either of the two. Whenever you can, name at least one type of product the method is used to make.

2. Briefly describe the direct method of blow molding.

3. What is the most common method of forming thermoset plastics? Of forming thermoplastics?

4. How are transfer molding and compression molding alike? How are they different?

5. How are reinforced plastics different from high-pressure laminates?

6. Briefly describe one method of making foamed plastics.

One of the outstanding characteristics of plastics is that they can be formed in many ways. They can be bent, cast, extruded, and molded into hundreds of different products. Fig. 60-1.

FOR YOUR SAFETY . . .

Whenever you are working with plastics, observe these safety rules:

● Many plastics produce toxic (poisonous) fumes when heated. Other plastics, such as polyurethane varnish, give off harmful fumes even without being heated. Always work in a well-ventilated area.

● Wear a face shield or safety goggles. This is especially important when using polyester resin catalysts, which can begin to destroy eye tissue within seconds after contact.

● When heating plastic or using liquid resins (plastics), wear gloves.

● Follow your teacher's instructions for safe use of tools and chemicals.

BENDING

One of the easiest ways to form plastic is by bending. When acrylic plastics such as Lucite® or Plexiglas® are heated, they soften. Remember, they are thermoplastics; when soft, they may be made into almost any shape.

Bending of plastic requires just a few tools. You need a device to heat the plastic, such as an electric hot plate or oven. Never use an open flame on plastic. The flame would cause the plastic to bubble and char. You also need a pair of cotton gloves to handle the hot plastic. If you are going to bend the plastic around a form, make the form before the plastic is heated. The form should be sanded smooth because the hot plastic will pick up any saw marks and they will appear on the finished piece.

Cut a piece of acrylic plastic the desired size. Smooth the edges. Be sure to remove all the masking paper from the plastic before it is heated.

Put the plastic on a piece of firebrick or other heat-resistant material. Place it on a hot plate, or in an oven at about 300°F. Heat until the plastic can be bent easily. Remove it from the heat and bend to the desired shape, either by hand or on a form. Be sure to wear gloves. Fig. 60-2.

When the plastic cools, it will retain this shape. If reheated, it will go back to its original form. This is useful if you did not get the desired shape on the first try. Just reheat the piece and bend it again.

In industry, strips of plastic are heated and bent to form letters for signs or to make decorations.

CASTING

Plastic casting is generally done by mixing liquid polyester resin with a catalyst and pouring it into a mold. (The catalyst helps the resin to harden.) Fig. 60-3.

After the mixture has hardened, the finished piece is removed from the mold. It can then be cut, drilled, sanded, or polished. Industry casts lenses and jewels by this method. In the school shop, you can use this method to make paperweights and other decorative items.

When casting plastic, be sure to wear gloves and safety goggles, and work only in a well-ventilated area.

1. The molds to be used must first be coated with a liquid release so that the finished piece can be removed easily. Allow the mold release to dry thoroughly before using the mold.

2. Mix the resin and catalyst. The amount to use will vary; read the directions on the containers.

3. Carefully pour the mixture into the mold.

4. Allow to harden overnight. Figure 60-4 shows the plastic molds (available from the resin supplier) and the finished pieces.

REINFORCED PLASTIC MOLDING

The most common type of reinforced plastic is *fiberglass*. Fiberglass is made from a number of layers of glass fibers saturated with a thermoset resin (usually polyester). The glass fibers give the material its strength and durability, so the more glass, the stronger the product (as long as there is enough resin to properly bond the fibers). You have probably seen many things made of fi-

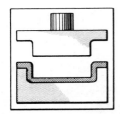

60-1.
Forming in the Shop

Forming is the process of shaping a material without adding to or removing any of the material.

60-2. *Bending plastic around a dowel form. The plastic must be heated before it can be bent.*

KINDS OF FORMING	DEFINITION	EXAMPLES
Bending	Forming by stretching heated plastic around a straight axis.	Letter-forming operations for signs and displays; forming heated plastic strips.
Casting	Forming by pouring liquid plastic into a hollow cavity and allowing it to harden.	Die casting and open-mold casting.
Molding	Forming by squeezing a heated plastic between two dies.	Compression molding, injection molding, blow forming and vacuum forming, foam molding.
Extruding	Forming by forcing the heated plastic through an opening (or die) which shapes the plastic.	Plastic molding and tubing extrusion operations.

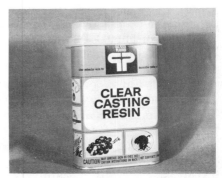

60-3. *Clear casting resin can be mixed with dyes to make many colors. The resin must also be mixed with a catalyst before using so it will harden properly.*

berglass. The body of the Corvette sports car is made entirely of fiberglass by a process called "matched molding." You can make a fiberglass fruit tray using a simple form of matched molding as a beginning shop project. Fig. 60-5.

CAUTION: *The catalyst mixed with the polyester resin in making fiberglass is very dangerous. If even one drop of the catalyst enters your eye, the eye tissue will begin to be destroyed unless the eye is immediately and thoroughly washed. Permanent damage and even blindness can result. Always wear eye protection when using a polyester resin catalyst.*

1. You will need two matching glass or metal trays to act as molds. The resin and glass fibers are applied to the bottom mold. The top mold nests securely on the bottom mold and presses the

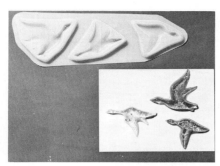

fiberglass material against the bottom mold to form the desired shape.

2. Coat the inside of the bottom mold and the outside of the top mold with a spray or paste mold release. Or tape a piece of cellophane, smooth and free of wrinkles, over each mold surface.

3. Cut a piece of glass cloth the same size as the mold.

4. Prepare a gel coat of polyester resin and catalyst according to the manufacturer's directions. Apply this to each half of the mold with a brush. Apply a heavy but even coat. If the coating is not smooth or even, let it dry and sand lightly. This coat keeps the glass fibers from coming to the surface.

5. Mix the resin and catalyst according to the manufacturer's directions. Apply a thick, even coat of this mixture to the bottom mold. Lay the glass cloth or fi-

60-4. *Plastic molds and finished castings. Molds may also be made of wood or metal.*

60-6. *Applying resin to glass fibers to make a fiberglass fruit tray. Notice the protective gloves.*

bers in position on top of this mixture. Use the brush to work it into the resin until the surface is even. Apply another coat of resin mixture to the glass fibers. Alternate layers of glass fibers and resin in this manner until the desired strength and thickness are reached. Fig. 60-5. The final coat or layer will be the resin mixture.

6. Set the top half of the mold in place on top of the fiberglass. Place a weight on the mold and allow the fiberglass to cure for two or three hours.

7. When the fiberglass is cured, remove the tray from the mold. Fiberglass has sharp edges; smooth these with a file or aluminum oxide cloth.

OTHER MOLDING PROCESSES

Industry *compresses* thermosetting plastic granules between dies to form distributor caps, ra-

60-5. *A student using a small injection molding machine for making a project in the school shop.*

dio cases, knobs, handles, and dishes. Thermoplastic materials are used in *injection molding* to form such things as appliance parts.

Compression and injection molding may also be done in the school shop. For each of these processes a mold or die is needed, along with the right machine. Figure 60-6 shows a student using a small injection molder. Figure 60-7 shows a student forming a part on a compression molding machine.

In industry, polystyrene or styrene sheets are *blow formed* or *vacuum formed* to make battery cases, trays, and assorted food containers. *Foam molding* is used to produce packaging cases, ice chests, and floating devices from expandable polystyrene beads. Foam molding, as well as blow and vacuum forming, is commonly done in the school shop.

60-7. *A student using a compression molding press.*

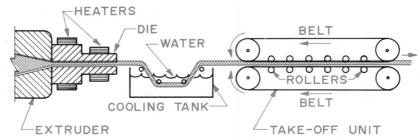

60-10. *This diagram shows the parts of a shop extruder. Attached to the die is a cooling tank. This could be constructed from a piece of gutter spout with hold-down fingers of soft sheet aluminum. At the end of the cooling tank come the take-off rolls.*

Dow Corning

60-8. *This model of a ship's hull is being removed from its silicone rubber mold. The model consists of a hand-laminated epoxy outer shell filled with polyurethane foam.*

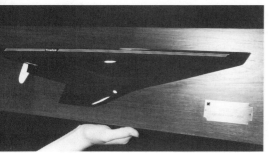

Dow Corning

60-9. *The model is hand finished and mounted on a plaque of satin-varnished teak veneer. This is a model of the hull of* **Freedom**, *1980 winner of the America's Cup.*

Silicone rubber molds can be used to make epoxy models of boats. Fig. 60-8. The molds are made by pouring the liquid rubber over a hand-carved wooden pattern. After being allowed to cure at room temperature for at least two days, the molds are ready to use. The boat model shown in Fig. 60-9 was made in a silicone rubber mold.

EXTRUDING

Extruding in the school shop is quite similar to the industrial extruding processes. Heated plastic is forced through a die in a continuous length. The shape of the die determines the shape of the extruded plastic. Dies can be purchased in many shapes, and they can also be machined in the shop for special shapes. The extruders used in the school shop are basically the same as the large industrial extruders, except the take-off systems often do not come with the smaller machines. They must be constructed in the shop for whatever the specific use of the extruder will be. Fig. 60-10. The following steps tell how to extrude low-density polyethylene and similar resins. In most cases, two or three people are needed for this operation.

1. Turn on the power for the heating devices. Set the dials for the temperature specified by the resin manufacturer.

2. When the heating units have reached the proper temperature, fill the hopper with the proper amount of extrusion grade plastic granules for the job. Allow the plastic to become soft.

3. Fill the cooling tank with water to cover the hold-down fingers.

4. Turn on the take-off rolls and adjust the speed.

5. Making sure the plastic is softened, turn on the extruder switch. With a wooden stick, guide the first extruded plastic under the hold-down fingers in the cooling tank.

6. As the extruded plastic comes to the end of the cooling tank, feed it between the take-off rolls. Use a wood stick to handle the plastic.

7. Adjust the speed of the take-off rolls so that there is slight tension on the extruded plastic.

8. When all the resin has been extruded, turn off the heating-devices, the extruder power, and the take-off rolls.

QUESTIONS AND ACTIVITIES

1. What are some safety rules to remember when working with plastics?

2. Define forming.

3. Name and define the four methods of forming plastics. Give an example of each method.

4. What is the purpose of a catalyst in casting?

5. Describe the four basic steps to follow when casting plastic.

6. What materials are needed to make fiberglass? What safety precaution is especially important to remember when working with fiberglass? Why?

CHAPTER 61

Cutting Plastics

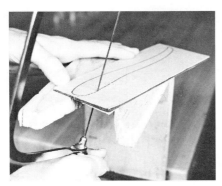

61-1. *Cutting plastic with a coping saw. Hold the plastic firmly on a V block. Use vertical (up-and-down) strokes. Note that the masking paper has been left on the plastic to protect it from nicks and scratches. The masking paper also provides an ideal surface on which to lay out the pattern you wish to cut.*

Plastics used in the shop may be cut in several ways. The tools and methods are similar to those used in woodcutting. Plastics dull tools quickly, so they will have to be resharpened often.

Shearing may be done on very thin sheets with scissors or tin snips. Thicker sheets, as well as rods and bars, are usually cut with a saw.

Sawing should be done very carefully and not too fast. If you saw too fast, you will heat the plastic, causing the saw to stick. Wear safety goggles when sawing plastics.

To saw plastic with hand tools, use a hacksaw, coping saw, or backsaw. A fine-tooth hacksaw is good for cutting rods, tubes, bars, and narrow sheets of plastic. A

coping saw or jeweler's saw may be used for cutting curved designs in sheet plastic. Fig. 61-1. When sawing plastic with a backsaw, it is best to clamp the piece securely in a vise or to a sawhorse.

Power tools can also be used to cut plastic. *Always ask the instructor before using any power machinery.* The jigsaw is good for cutting curved designs and irregular shapes from sheet plastic. Fig. 61-2. The band saw works well for cutting rods and tubing to length. It can also be used for making straight cuts on sheet plastic when a guide is used. The circular saw can be used for cutting sheets of plastic. It is used for squaring ends and cutting

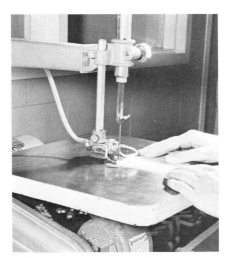

61-2. *Cutting an irregular shape on a jigsaw. You may need to use a clean oil can to apply a soap-and-water solution to the cut to prevent the plastic from overheating.*

61-3. *Plastic may be drilled with ordinary wood and metal drilling equipment. Do not force the tool, as this will cause the plastic to melt.*

grooves in sheet plastic. Cutting plastic on a circular saw should usually be done by the instructor.

Methods for *abrading,* or sanding, plastic are the same as for woodworking. Garnet and silicon carbide are good abrasive papers for finishing plastic edges. Sanding may be done by hand or with a disk or belt sander. These machines can also be used to shape irregular pieces. It is important to remember that the plastic will melt if it is pressed too hard on a sanding machine.

Twist drills used in woodwork or metalwork can also be used for *drilling* holes in plastic. Holes can be drilled with either a hand drill or a drill press. Fig. 61-3.

61-4. *After plastic has been cut, the edges should be finished smoothly. This can be done with a file, abrasive paper, or a buffing wheel as this student is doing. Note that the protective paper has been left on the plastic.*

Shaping and *smoothing* of plastic are done with files and other shaping tools as in woodworking. Fig. 61-4. When filing plastic edges, the cuttings will clog the file quickly; clean the file with a file card often. The plastic should be held firmly in a vise. Fig. 61-5. Be careful not to bend it, as this will cause crazing or cracking.

Plastic can also be *turned* on a lathe. The work must be held carefully in a chuck or between centers. Use a sharp tool and light cuts to reduce the heat generated by the friction of the turning tool against the plastic. Hard tough plastics like nylon are easier and safer to turn than brittle plastics like acrylic and polystyrene. Too much pressure on the cutting tool can cause plastic to break. This can be very dangerous; so check with your instructor before lathe-turning is attempted.

61-5. *Smoothing plastic sheets with a Surform shaping tool. The plastic is held between wooden blocks in a vise to keep it steady.*

The work must be held securely. Put a piece of wood on the underside to prevent chipping when the bit breaks through. With a center punch, indent the place where you want the hole. This will keep the drill from wandering and damaging the surface of the plastic. If you are drilling holes wider than ¼″ in thick plastic, drill a ⅛″ pilot hole first. When drilling holes in plastic, use a slow or medium speed. If you drill too fast, the drill will get hot and stick to the plastic. A soap-and-water solution applied to the area to be drilled will help prevent overheating and make drilling easier.

INDUSTRIAL CUTTING OF PLASTICS

Most plastic products are molded to the desired shape, so they require little machining or cutting, except for removal of small plastic sprues (projections) left by the mold. Laminates may be sawed or drilled and stock forms of plastic can also be machined. Screw threads can be cut onto extruded plastic pipe or rod.

Another industrial cutting process involves the use of the laser beam. As you learned in Chapter 6, laser beams can remove material by vaporizing it. Lasers can drill holes in plastic in less than a thousandth of a second. Industry uses lasers to bore holes in nylon aerosol nozzles, to perforate a moving plastic sheet for webbing, and to slit plastic film.

QUESTIONS AND ACTIVITIES

1. What tools are commonly used for cutting very thin sheets of plastic? For cutting rods, tubes, and bars?

2. Name two tools that might be used to cut curved designs in sheet plastic.

3. Why is it a good idea to leave the masking paper on plastic while cutting?

4. How do you start a ⅛″ hole in thin plastic? A ⅜″ hole in thick plastic?

5. What speed(s) should you use when drilling plastic? Why is this recommended?

6. What can you do to plastic to help keep it from overheating when you are cutting it with a jigsaw or a drill?

7. Name at least two tools that can be used to shape or smooth plastics.

Plastic materials can be joined together with glues and cements, and with mechanical fasteners. Permanent fastening is done with glues and cements, and semipermanent joints are made with fasteners such as screws. Fig. 62-1.

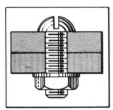

MECHANICAL FASTENERS

Plastic parts to be joined with screws or with nuts and bolts must be carefully drilled. Otherwise the plastic will fracture and split. The same techniques used for preparing holes for screws in wood and metal should be followed for plastic. Be sure to drill a hole for the shank (head of the screw) and a pilot hole for the

62-1.

Fastening

Fastening is the process of joining materials together permanently or semipermanently.

KIND OF FASTENING	DEFINITION	EXAMPLES
Mechanical Fastening MACHINE SCREWS IN TAPPED HOLES PLASTIC PARTS	Permanent or semipermanent fastening with special locking devices.	Fastening with machine screws, bolts and nuts, or self-tapping screws.
Adhesion GLUE PLASTIC PARTS	Permanent fastening by bonding like or unlike materials together with cements.	Epoxy cementing. Contact cementing.
Cohesion SOLVENT CEMENT PLASTIC PARTS	Permanent fastening by fusing plastic pieces together with softened or liquid plastic and pressure.	Solvent cementing. Thermal welding. Sonic welding.

thread. Self-tapping screws can be very effective for fastening plastic that is not too brittle. The drilled hole does not have to be threaded with a tap when using these screws. The drilled hole must be tapped when using machine screws or bolts.

To tap a hole in plastic, turn the tap once or twice, then reverse it (turn it in the opposite direction, the same number of times) to break the chip and to keep the tap from binding. Continue turning the tap and reversing it until the hole is fully tapped.

Plastic can be externally threaded with a die in the same manner—turning the die once or twice and reversing it, then repeating the procedure until the proper length is threaded. A soap-and-water solution makes a good lubricant for tapping and threading.

Flexible plastics can also be fastened with rivets, staples, hinges, and even nails in some plastic constructions.

ADHESION

Adhesives can be used to bond either thermosets or thermoplastics. They can be used to bond different types of plastics to each other, to bond plastics to other types of material (such as wood), or to bond plastics of the same type.

Adhesives used for wood usually do not work well for gluing plastics. The two general-purpose glues for plastics are epoxy glue

62-2. *Two pieces of plastic being held by spring clamps until the cement dries.*

stick so tight that they cannot be moved. No clamps are needed. However, when cementing a commercial laminate like Formica to another surface, the laminate should be pressed down firmly using a rolling pin or a block of wood and a hammer.

It is a good idea to experiment with adhesives on scrap materials before you use them on a shop project. Messy cement joints are very unattractive and can spoil the appearance of the entire workpiece.

COHESION

Thermoplastics can be fastened together by cohesion—that is, by fusing the two pieces of plastic

and contact cement. Epoxy glues come in two separate tubes. To use them, take equal amounts of material from each tube, mix them together, and apply the mixture to the surfaces to be joined. The pieces must be held or clamped until the glue sets. Fig. 62-2.

To fasten plastic with contact cement, coat both surfaces with the cement and allow to dry (about 20 minutes). Then bring the two surfaces together *carefully*. Once they touch, they will

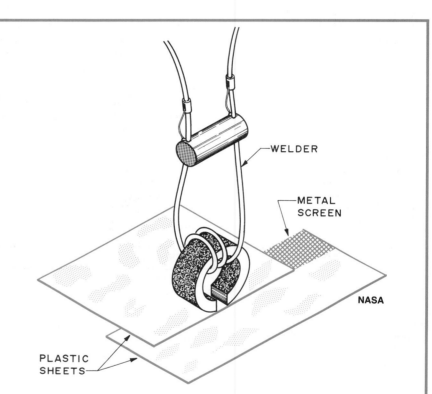

WELDER

METAL SCREEN

NASA

PLASTIC SHEETS

Branson

62-3. *This Branson sewing machine generates high frequency sound waves to bond nylon, polyester, polypropylene, modified acrylics, some vinyls, and most other synthetics. The synthetics can have up to 35 percent natural fiber content.*

Plastic Welding

For our aerospace construction program, NASA needed to find a practical way to fasten plastic parts in the airless environment of outer space. Adhesive bonding is not reliable in this type of vacuum, fusing and riveting often deform the plastic, and mechanical fasteners require both hole drilling and special fasteners. After much testing, they developed the "Induction Toroid Welder."

This welder uses low-voltage induction, or magnetic heating, to form a cohesive fusion of many types of plastic material, with little or no warping. An induction coil sends magnetic current through the

plastic to a metal screen which is sandwiched between the sheets of plastic to be joined. This alternating current heats the screen, causing the plastic materials to melt and flow into the mesh of the screen. This forms the welded seam or joint.

The low, 25- to 100-watt power needed for this system readily permits a solar-powered, self-contained unit. Various types of plastic welders may find use in areas other than the aerospace industry. Induction welders could also be used in the automobile, furniture, and construction industries.

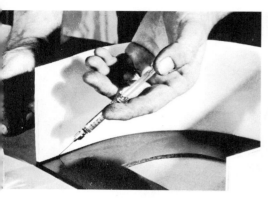

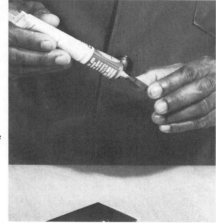

62-4. *Solvent cementing can be done by soaking or by injecting cement into the joint.*

62-5. *Applying glue to the edge of a piece of plastic.*

together. This may be done by solvent cementing or by thermal or sonic welding.

The most common example of thermal welding is the heat sealing of plastic food bags. Other thermal welding methods involve friction and hot gas. (Thermal welding is not generally done in the school shop.)

Another way of forming a cohesive bond in plastics is by sound waves. Very high frequency sound waves are used to "excite" the molecules of plastic. The movement of the molecules creates friction, which generates heat. The heat causes the materials to bond together. Fig. 62-3.

In school projects, solvent cement is used to fuse plastic. Since different plastics require different solvents for softening, both parts to be cemented must be the same type of plastic. Cementing is done by the following method.

First make sure the plastic is clean, using a file and sandpaper to smooth the edges to be joined. Always be sure that the joint fits well. (The solvent will not make up for a poor joint.) Then apply the cement to all surfaces to be joined. This is usually done by soaking the plastic in the solvent or applying the solvent with a hypodermic needle. Fig. 62-4. Solvent cement may also be applied to the edges of the plastic from a tube, as in Fig. 62-5. The solvent softens the plastic. When joining the pieces, use enough pressure to force out any air bubbles. Fasten the pieces with spring clamps until the solvent evaporates. If the pieces are moved before the solvent evaporates, the joint will be broken. As a substitute for clamps, you can make a special fixture for holding the parts. Fig. 62-6.

FOR YOUR SAFETY . . .

● Some solvents give off toxic fumes. When using solvent cements, work in a well-ventilated area.

● Care should always be taken to keep solvents from contacting your skin or eyes. Wear gloves and goggles when working with solvent cements.

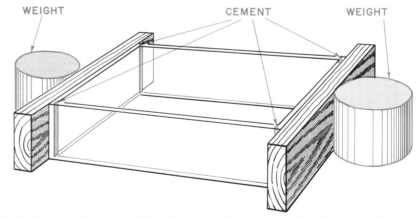

62-6. *One method of holding the cemented pieces while they dry. Make sure that the parts are held firmly in place.*

QUESTIONS AND ACTIVITIES

1. Define the three main kinds of fastening used with plastics and give an example of each.
2. Describe how to fasten plastic with epoxy.
3. Describe how to fasten plastic with contact cement.
4. Describe how to fasten plastic with solvent cement.
5. What are some safety rules to remember when fastening plastics?

Among the several unique properties of plastic materials is that they do not need to be protected with a coating of finishing material. Plastic is its own finish. It is waterproof and resists many stains. It is generally smooth and needs little attention after it has been formed into a product. However, if a decorative finish is desired, it may be applied by one of two basic methods—*coloring* and *coating*. Fig. 63-1. Plastic surfaces can also be decorated by scribing (cutting shallow lines in a decorative pattern) and by gluing on decals.

NOTE: Before plastic is finished, it should be polished and buffed. Remove deep scratches with abrasive paper; then buff to a high luster on a buffing wheel. Use a light pressure on the wheel, and keep the plastic moving to prevent burning. Fig. 63-2.

COLORING

Most plastic sheet materials (acrylics) can be bought in a variety of colors. Clear plastic can, however, be colored with special dyes. These dyes can be bought

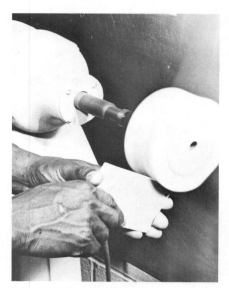

63-2. Buffing plastic on a cloth buffing wheel. Be sure to keep the plastic moving.

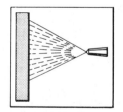

63-1.

Finishing

Finishing is the process of treating the surface of a material for appearance and/or protection.

KIND OF FINISHING	DEFINITION	EXAMPLES
Coloring* DYE PENETRATES WORKPIECE	Applying penetrating dyes or chemicals to a material to change its color.	Dyeing sheet materials and liquid resins.
Coating* WORKPIECE	Applying a layer of finishing substance to the surface of a material.	Brush and spray lacquering, glazing, and painting.

*Coating could be considered a coloring process, since coating usually does change a product's color. However, it is customary to consider coating and coloring to be separate processes, as defined in the chart above.

in liquid or powdered form. Some dyes must be heated, but others can be used directly from the container or mixed with tap water. Use a porcelain or earthenware container large enough to contain the entire surface to be colored, so the piece can be soaked and dyed evenly. The plastic should be buffed smooth and cleaned with detergent and water. It is a good idea to test the tint by placing a cleaned piece of scrap plastic in the dye solution. Check the scrap to see how many minutes of soaking are needed to produce the desired shade. The plastic workpiece is then placed in the dye for approximately the same number of minutes. Fig. 63-

63-3. *Plastic being placed in a tray of dye for coloring.*

63-4. *Dye being inserted into an internally carved flower.*

63-5. *Applying the coloring material with a brush.*

brushing paint onto the surface of the workpiece. Many plastics require special paints. For example, thermoplastics can be softened by certain thinners contained in some paints. Paint will not stick to polyethylene and polypropylene unless they are first specially treated. Choose the proper paint for the type of plastic you are using.

Lacquer can also be sprayed or brushed on plastic. Be sure the surface is clean, dry, and completely free of fingerprints.

Some very interesting effects may be obtained by glazing a plastic surface. Crystal-glazing liquids can be purchased in a wide range of colors and applied with equal success on plastic, glass, and metal. The liquid is brushed or sprayed on a clean, dry surface and allowed to dry. In drying, the beautiful crystals are formed. Fig. 63-6. No further treatment is necessary, other than wiping with a clean cloth. Plastic trays, bowls, wall plaques, and containers can be glazed to add to their beauty.

resin before molding the workpiece.

Certain acrylic plastics such as Lucite® and Plexiglas® can be colored by applying an aniline or special acrylic dye to a carved design. The plastic can be carved using a high-speed cutting tool. Flower designs as well as other designs can be carved into plastic this way. The carved design may be colored by using a medicine dropper or a hypodermic needle to apply the dye as shown in Fig. 63-4. Some coloring agents can also be applied with a brush. Fig. 63-5.

COATING

Most plastics can be decoratively coated by spraying or

63-6. *This glazed decorative jar shows some interesting crystal formations.*

3. Remove the plastic and rinse in tap water to stop the coloring action. Blot dry with a soft dry cloth to prevent water stains. If desired, apply a coat of hard-finish paste wax and polish with a soft cloth. Plastic castings and fiberglass can be colored by adding powdered dye directly to the

INDUSTRIAL DECORATIVE FINISHING

In the plastics industry, most of the decorating and finishing is done during the molding process. Coloring can be added to the plastic before molding. Designs or lettering can be part of the actual mold. Decorated plastic overlays can be placed in the mold with the product and the design fused onto the product during molding. (Children's melamine dinnerware decorated with cartoon characters is made in this manner.) However, there are many other industrial processes to further decorate plastics.

Hot stamping is often used to label plastic products. A pigmentized or a metallized plastic film is placed between the product and a die on which the label's design has been etched or engraved. The heated die is pressed against the film and the design is transferred to the plastic product. Hot stamping can also be done in the school shop.

Irregular surfaces can be decorated with any type of lettering or design by *electrostatic printing*. There is no actual contact between the product and the printing equipment. A design is cut in reverse in a stencil and the stencil is placed in front of the product. Powdered ink is charged with negative electricity. A plate behind the part is charged with positive electricity. Since negative charges of electricity flow to positive charges, the negatively-charged ink goes through the stencil openings to the product. The plastic product is then sent through an oven to set the design.

Another industrial process for decorative finishing of plastic products is *vacuum metallizing*. This is done in a vacuum chamber in which metal strips (usually aluminum) are heated until the strips vaporize onto the cool, rotating plastic parts. A protective lacquer finish is applied to the metal-coated plastic. This process gives the plastic a chromelike coating. It is used on chromelike parts in model car kits, radio and television control knobs, and dashboard instrument panel components.

QUESTIONS AND ACTIVITIES

1. What should be done to plastics before applying a decorative finish?
2. Define coloring.
3. Define coating. Give two examples of coating finishes.
4. What steps do you follow to color plastic by soaking it in a dye solution?
5. How can you color fiberglass?
6. Name two industrial processes of decorating plastic that has already been molded.

PART 4

Production, Manufacturing and Construction

CHAPTER 64

Introduction to Woodworking

From earliest times wood has served people in one form or another. In most of the areas of the world, wood has been plentiful and easy to obtain. In the United States wood has always been one of the most plentiful of our raw materials.

Today wood is used for a variety of purposes, from building homes to making fine furniture. The raw material, lumber, is changed to many forms of material for the woodworker. Plywood, particle board, and hardboard are all made from the basic material wood. In addition, wood is the raw material for thousands of other products, such as paper, rayon, charcoal, and turpentine. Fig. 64-1.

What you learn in this section about wood and how to work with it may lead to an interest in one of the many occupations related to woodworking. Understanding of the materials and the tools will also provide you with information and skill you may use around your home. Working with wood is one of the oldest crafts in this country and one of the most interesting.

OCCUPATIONS IN THE WOOD INDUSTRY

There are many jobs in the woodworking industries. Some require years of college education. These include professional careers, such as forestry, wood research and technology, and teaching woodworking. Other jobs require less education. Some

jobs are around the lumber mills where the trees are cut and made into lumber. Some people work in factories, also called mills, where the lumber is made into things such as doors and cabinets. These

workers may have to know how to run such woodworking machines as the planer, the jointer, and the circular saw. In these mills there are also maintenance workers who keep the machines

Wrather Corp.

A Flying Boat

Wood is widely used in building construction, from framework to trim, but we don't usually think of it as a construction material for airplanes. Yet the early airplanes were built of wood, and today the single-engine craft are often built of plywood over a metal frame. One of the most unusual wooden airplanes was the *Spruce Goose,* built during World War II as a cargo plane and troop carrier.

The *Spruce Goose,* so named because it was designed to take off and land on water, is the largest airplane ever built. It has a wing span of over 319 feet—longer than

a football field. There are four engines on each wing, and the tail is as tall as an 8-story building. This gigantic airplane weighs 200 tons.

The craft was built of wood because of the metal shortages caused by World War II. It took 5 years to build and cost $25 million. By the time it was finished (1947), the war was over. No cargo or troops were ever transported by this huge "flying boat." In fact, the *Spruce Goose* was flown only once—by its builder, Howard Hughes. Today it rests in a museum in Long Beach, California.

Forest Products	
Mechanically Derived	**Chemically Derived**
Lumber	Paper
Plywood	Plastics
Particle board	Charcoal
Hardboard	Rayon
Fiberboard	Dyes
Veneers	Turpentine
	Ethyl alcohol
	Cellophane

64-1. *Trees are a valuable resource, not only for building materials but also for a variety of chemically derived products.*

running and the saws and cutter knives sharp.

Many people earn their living as carpenters, cabinetmakers, patternmakers, and boat builders. They must be able to use all the hand and power woodworking tools. Each of these skilled crafts requires some training beyond high school, such as vocational school or apprenticeship. Perhaps you will find you have an interest in woodworking and want to take more courses in high school. Your instructor can give you information about jobs in woodworking and the training that is required.

QUESTIONS AND ACTIVITIES

1. Name four products that can be mechanically derived from trees.

2. Name four products that can be chemically derived from wood products.

3. Name two careers related to woodworking that require a college education.

4. Name at least three other jobs related to the woodworking industry.

CHAPTER 65

Lumbering

Lumbering is one of the oldest industries in America. Working with simple hand tools, the early colonists used trees to make their houses, furniture, wagons, bridges, and many other items for daily use.

The lumber industry grew as the country grew. Today the logger makes use of many machines. Lumbering methods vary depending on the part of the country and the size of the company, but the basic process remains fairly standard. It involves cutting trees, taking the logs to a sawmill, cutting them into lumber, and seasoning or drying the lumber.

LUMBERING PROCESSES

The first step is to select and mark the trees to be cut, or harvested. The trees are then partially cut, using power saws. Wedges are driven into the cut to make the tree fall in the right direction; then the cut is finished. After the tree is felled, the limbs and branches are trimmed off. The trunk is cut into logs of the desired length. The logs are moved to a loading site and stacked. Today some trees are harvested with automatic machines. Figs. 65-1 through 65-7.

65-1. *An automatic tree harvester in operation.*

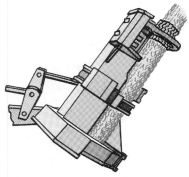

65-4. *Harvester head raises the tree and tilts it toward the ground. At the end of this step the tree is parallel with the ground, ready for delimbing. For high production efficiency, the operator can tilt the tree and delimb it while backing toward the pile.*

In years past, logs were floated to the sawmill in rivers and streams. There are a few places where this is still done, but today most logs are carried to the sawmill on boats, trucks, or railroad cars.

Modern sawmills use computers to keep track of operations. Fig. 65-8. When the logs arrive at the sawmill, they are *scaled* for species and quality and then stacked in a *log deck area* to await processing.

When the logs enter the mill, they go first to a *barking center.* Here machines remove the bark, dirt, and rocks from the logs. The logs are then moved to a computerized *bucking station,* where they are cut into the best lengths for use. Next the logs go to the

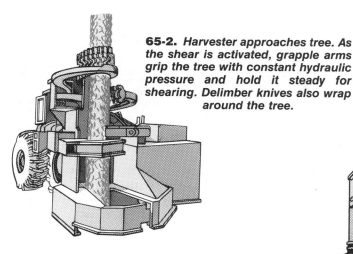

65-2. *Harvester approaches tree. As the shear is activated, grapple arms grip the tree with constant hydraulic pressure and hold it steady for shearing. Delimber knives also wrap around the tree.*

65-3. *Shear cuts tree. The 1" thick steel shear powered by a vane-type hydraulic pump cuts through the trunk of the tree. The cut is made within 4" of ground level.*

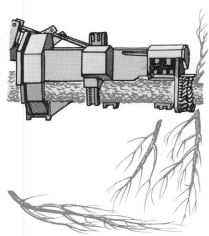

65-5. *Hydraulically driven chain pulls the tree through the harvester. Cross beams with sharpened grousers engage the trunk and drive the tree as grapple arms guide and hold it against the chain. Hard steel knives cut the limbs flush with the tree trunk.*

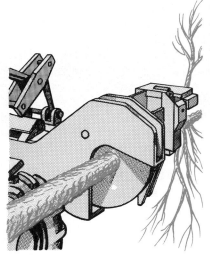

65-6. *Shear snips off tree top. Delimbing is effective down to a 2½" diameter top. The felling shear removes the top at any given length. Operator can also use the shear to cut any desired shorter lengths during delimbing.*

65-7. *Operator moves the harvested tree to a bunched pile. In a minute or less, the trimmed and topped tree is bunched and the operator moves to the next tree.*

head rig, where they are cut into boards by huge band saws.

The *chipper-canter,* controlled by the computer, makes sure that the maximum amount of lumber is obtained from the log. These operations are monitored at the *control console.* Then the lumber is *resawn* into desired sizes. The *chipper-edger* removes rough edges from the lumber. At the *trimming/sorting center* the lumber is cut to exact lengths. From there the lumber goes to large kilns where it is *dried.* (Some lumber is air dried.) The lumber is then *planed* to standard sizes. It is *graded* and *stored* to await *shipment.* The lumber is moved by truck, railroad, and ship all over the world for building houses and furniture and for thousands of other uses. The waste produced by the cutting and processing of logs can also be

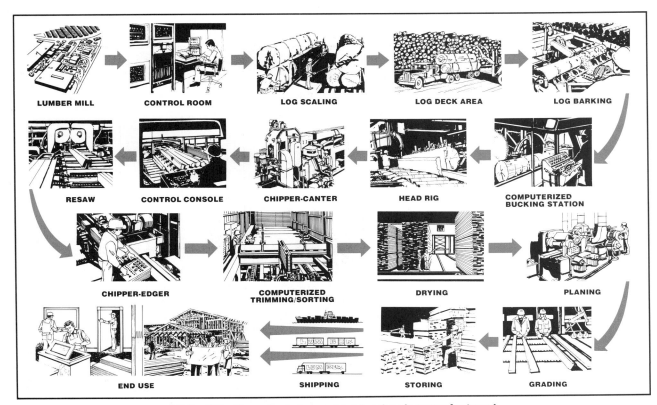

LUMBER MILL CONTROL ROOM LOG SCALING LOG DECK AREA LOG BARKING

RESAW CONTROL CONSOLE CHIPPER-CANTER HEAD RIG COMPUTERIZED BUCKING STATION

CHIPPER-EDGER COMPUTERIZED TRIMMING/SORTING DRYING PLANING

END USE SHIPPING STORING GRADING

65-8. *This flow chart shows how lumber is manufactured.*

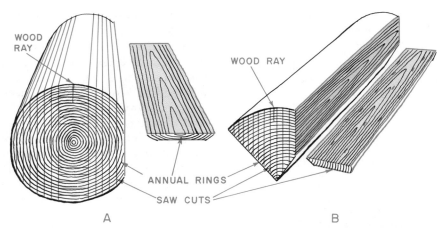

65-9. *Two ways of cutting lumber: (A) Plain-sawed. (B) Quarter-sawed. (Wood rays are horizontal passageways for food. Annual rings, or growth rings, are the new growth added each year.)*

made into useful products. Wood chips, for example, are made into particle board or paper.

Methods of Cutting Boards

Boards are cut from logs in two common ways. Logs cut by the cheaper method are said to be *plain-sawed* (if they are hardwood) or *flat-grained* (if they are softwood). In this method, the logs are squared and sawed lengthwise from one end to the other. Fig. 65-9.

Logs cut by the more expensive method are called *quarter-sawed* (for hardwood) or *edge-grained* (for softwood). In this method, the lumber is first cut into quarters and then sawed lengthwise so that the annual rings are at or nearly at right angles to the surface of the board. Fig. 65-9. Quarter-sawed lumber warps less than plain-sawed lumber. It also shows a better grain.

WOOD IDENTIFICATION

Each kind of wood has its own color, working qualities, and other properties. You should select the wood for your projects carefully. To do this you must know something about the more common kinds of wood. Much information is available from the government and from commercial sources to help you in recognizing and choosing woods.

Woods are divided into softwoods and hardwoods. The names really have little to do with how soft or hard the wood

65-10.
Common Hardwoods and Softwoods

Kind	Color	Working Qualities	Weight	Strength	Lasting Qualities (Outside Use)
HARDWOODS					
*Ash	Grayish brown	Hard	Heavy	Strong	Poor
Basswood	Lt. cream	Easy	Light	Weak	Poor
Beech	Lt. brown	Hard	Heavy	Medium	Medium
Birch	Lt. brown	Hard	Heavy	Strong	Fair
Cherry	Dk. red	Hard	Medium	Strong	Fair
Gum	Red-brown	Medium	Medium	Medium	Medium
*Mahogany (true)	Gold-brown	Easy	Medium	Medium	Good
*Mahogany (Philippine)	Med. red	Easy	Medium	Medium	Good
Maple, sugar	Red-cream	Hard	Heavy	Strong	Poor
*Oak, red	Flesh-brown	Hard	Heavy	Strong	Fair
*Oak, white	Gray-brown	Hard	Heavy	Strong	Fair
Poplar	Yellow	Easy	Medium	Weak	Fair
*Walnut	Dk. brown	Medium	Heavy	Strong	Good
*Willow	Brown	Easy	Light	Low	Fair
SOFTWOODS					
Fir, Douglas	Orange-brown	Medium	Medium	Medium	Medium
Pine, ponderosa	Orange to red-brown	Easy	Light	Weak	Poor
Pine, sugar	Creamy brown	Easy	Light	Poor	Fair
Redwood	Dk. red-brown	Easy	Light	Medium	Good

Woods marked with (*) are open grain and require a paste filler.

is—some softwoods are actually harder than some hardwoods. *Softwoods* are obtained from cone-bearing trees that stay green all year. *Hardwoods* are obtained from trees that have broad leaves which they lose every year.

Softwoods

Common softwoods are Douglas fir, ponderosa pine, sugar pine, and redwood. Douglas fir is used for construction lumber and for plywood. It is fairly hard, heavy, and stiff. Ponderosa pine is also used for construction lumber. It is fairly light and soft. Sugar pine is used for making patterns for foundries and for millwork. It is of a light color, soft, and straight grained. It is easy to work. Redwood is used for home construction, fences, and outdoor furniture. It is lightweight and strong. Redwood is very resistant to decay.

Hardwoods

Common hardwoods are basswood, birch, cherry, mahogany, maple, oak, and walnut.

Basswood, one of the softest hardwoods, is white with a few black streaks. It is fuzzy, good for bending, and has almost no grain. Basswood is used for drawing boards. It is also good for burned designs and thin lumber for jigsaw work. It is not good for use outside. Basswood is usually painted.

Birch has fine texture and close grain. It grows in the north central states. It has a fine wavy figure and is used for paneling and some furniture. Birch is a difficult wood to work, but will take almost any type of stain.

Cherry grows in most parts of the United States. Very desirable for furniture, it is a reddish brown, durable wood that does

not dent or mar easily. It is often given a lacquer finish.

Mahogany is a fine cabinet wood. Its color ranges from dark red to brown. True mahogany grows in Honduras, Africa, and South America. Mahogany is tough, strong, and easy to work. It polishes well and has a very distinctive grain pattern. Lauan, while not a true mahogany, is often called Philippine mahogany. It is excellent, inexpensive wood for furniture, cabinetwork, and boats.

Maple is a hard, tough, strong wood that wears well. The grain is usually straight. Maple is used a great deal for flooring and for furniture. A hard wood, it does not work too easily. Maple can be stained or given a natural finish.

Oak trees grow in all parts of

the United States. There are almost 300 kinds of oak, but for woodworking only the white oak and the red oak are important. Oak is used for flooring, furniture, and construction. White oak is a very popular cabinet wood. It is hard to work.

Walnut is one of the most beautiful woods that grow in the United States. A dark brown wood, strong and durable, walnut is used for furniture, cabinetwork, veneers, and gunstocks. Walnut is usually given a natural finish with varnish, lacquer, or oil. Figure 65-10 shows some common woods and their characteristics. Figure 65-11 shows common wood properties and how these are tested.

65-11.
Wood Properties and Tests

Property	Description	Test
Tensile strength	Resists stretching.	A specimen is pulled apart in a direction perpendicular to its grain.
Compressive strength	Resists squeezing.	Specimen is compressed (pushed together) until it fractures. Tests are done both along and across the grain.
Shear strength	Resists slipping of one segment in relation to another along the grain.	Various tests are done to determine the shear stress which may occur because of tension, compression, or torsion (twisting).
Bending strength (stiffness or rigidity)	Resistance to bending stress, which is a combination of shear, compressive, and tensile loading forces. Material returns to original size and shape after load is removed.	Bending tests (also called transverse flexure tests). Specimen is horizontally supported near its ends. A load is applied to the middle until specimen fractures. Tensile tests are also used.
Hardness	Resists wear, marring, and denting. Note: the terms *hardwood* and *softwood* are not an indication of relative hardness.	Brinell, Rockwell—a metal ball or diamond-shaped point is pressed against a test specimen.

QUESTIONS AND ACTIVITIES

1. Briefly describe the job of an automatic tree harvester.

2. What is plain-sawed lumber? What term is used to describe softwoods cut in this manner? What is the main advantage of plain-sawing?

3. Describe how lumber is quarter-sawed. What is this method called when it is used on softwoods? What are two advantages of quarter-sawed lumber?

4. What is the main difference between hardwoods and softwoods?

5. Name three common softwoods and tell what each is often used to make.

6. Name three common hardwoods and tell what each is often used to make.

CHAPTER 66
Wood Products

Forests are very valuable, not only as a source of wood but for other reasons as well. They reduce the washing away of soil during heavy rainstorms and melting snows. They help to prevent dust storms, protect wildlife, and provide recreational areas for hiking, fishing, and hunting. Trees also provide shade and protection from the wind.

Lumber (sawn timber) is very important, but there are many forest products besides lumber. This chapter discusses some of these.

VENEERS AND PLYWOOD

One very important use of wood is for the making of veneers and plywood. *Veneer* is a thin sheet of wood that has been cut from a log. Veneer is cut in one of three ways: *rotary-cut, plain-sawed* (or *plain-sliced*), and *quarter-sawed* (or *quarter-sliced*).

Fig. 66-1. Hardwood veneers are usually plain-sawed or quarter-sawed. Softwood veneer is rotary-cut.

To make plywood, several sheets of veneer are glued (laminated) together. There are three ways to make plywood:

● In *lumber-core plywood*, the sheets of veneer are glued to a thick middle layer of solid wood. This kind of plywood is the one most often used for fine furniture.

● To make *veneer-core plywood*, sheets of veneer are glued to each other. The veneer sheets in each layer are placed with the grain at right angles to the grain of the layer above and below. This gives the plywood great strength.

● *Particle-board plywood* has a

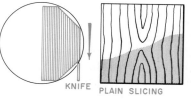

66-1. *Veneer is cut in one of three ways—rotary-cut, plain-sliced, or quarter-sliced.*

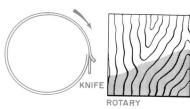

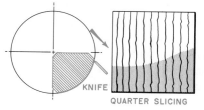

66-2. *Rotary cutting of veneer to make plywood.*

middle layer of particle board. (Particle board is discussed later in this chapter.) The other layers are veneer.

The following describes how veneer-core plywood is made:

The logs are cut into 8′, 10′, or 12′ lengths and the bark is removed.

The logs are placed on giant lathes which turn them against a razor-sharp knife. A thin, continuous sheet of wood is peeled off the log. This is rotary cutting. Fig. 66-2.

The veneer is cut into specific widths. It is then sorted for grade. Sheets of green (unseasoned) veneer are put through driers to remove the excess moisture.

Sheets of various sizes are joined together to make full-size sheets.

Glue is applied to the sheets, or *plies*. Sometimes two plies are glued with their grains parallel to form one layer. Then the layers are stacked with their grains at right angles to each other.

The glued sheets are placed in large hydraulic presses in which the glue is set under heat and pressure. When dry, the panels

are sanded, cut to certain lengths and widths, and then inspected and sorted.

Plywood is always made with an odd number of layers such as three, five, or seven. But the number of plies may be three, four, five, six, or seven, since one layer may have two plies. The plies or layers are held together with glue that is either moisture resistant or moisture proof. For outside use plywood that has been put together with moisture-proof glue should be selected.

Plywood has great strength and is fairly light. Because of these properties, plywood is often used for building construction, for small boats, and for furniture. Plywood can be worked with ordinary woodworking tools.

Some plywoods have surfaces that have been treated or decorated. These are usually used for paneling or furniture.

HARDBOARD

Hardboard is made by smashing wood into small chips. The chips are further refined into fibers which are then processed and compressed in heated hydraulic presses.

Hardboard has no grain. It will not split, splinter, or crack. It has a very hard surface. Hardboard panels can be worked with ordinary woodworking tools.

Hardboard panels are used for construction. They can be used for walls and ceilings, and also for cabinetwork. Some types of hardboard panels have wood-grained surfaces to make them look like natural wood.

PARTICLE BOARD

Particle board is made from wood shavings. The shavings are combined with adhesives or other chemicals and spread out on a steel plate. The layer is placed in a hydraulic press where it is heated and squeezed. After the board is cooled, it is trimmed and sanded.

Particle board is used for plywood, furniture, subflooring, paneling, counters, shelving, and concrete forms. It is also used for toys, Ping-Pong tables, cabinets, and signs. Particle board can be worked with ordinary hand and power woodworking tools.

OTHER WOOD CONSTRUCTION MATERIALS

Fiberboard

Fiberboard, also called wallboard or insulation board, is made from fibers of wood, cane, or other plants. It is softer and less dense than hardboard.

Fiberboard is commonly used as sheathing (covering) for the frames of buildings. It is also used as insulation, as siding, and as paneling.

Sandwich Structures

Sandwich structures consist of thin facings bonded to a thick core. The facings may be veneer, hardboard, or plastic laminate. The core is some light material such as paper, balsa wood, or plastic foam. The core material is usually arranged in a honeycomb pattern.

MODIFIED WOOD

Modern technology has developed a number of ways to treat wood to change its structure, improve its properties, and increase its range of applications. This is called *wood modification*. Materials with properties quite differ-

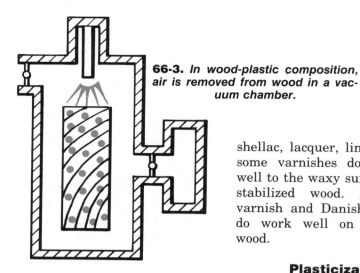

66-3. *In wood-plastic composition, air is removed from wood in a vacuum chamber.*

ent from those of the original wood are obtained by chemically treating wood, compressing it, irradiating it, or combining these and other treatments.

PEG Treatment

Wood can be modified chemically to help increase its *dimensional stability,* in other words to resist change when its environment changes. A chemical process to prevent wood from swelling, shrinking, or warping as it gains or loses moisture is the *PEG process.* In this treatment, green (freshly sawed and unseasoned) wood is soaked in a polyethylene glycol (PEG) solution for several days to several weeks, depending on the type of wood used. During soaking, the large molecules of the waxlike PEG displace the natural moisture in the wood cells. When dried, the wood is resistant to changes in humidity. PEG does not resist decay or insects; so wood in contact with moist soil should be treated with actual preservatives such as creosote or pentachlorophenol.

PEG-treated wood can be sanded, stained, or glued using the proper techniques. However, many common finishes such as

shellac, lacquer, linseed oil, and some varnishes do not adhere well to the waxy surface of PEG-stabilized wood. Polyurethane varnish and Danish oil finishes do work well on PEG-treated wood.

Plasticization

Some wood modification processes involve a pretreatment of the wood to soften, or plasticize, it. Such a process, called *plasticization,* acts on the structure of wood to loosen the fibers so the wood can become pliable and more easily shaped and molded.

Plasticization is followed by the application of heat and pressure, generally between forming dies, to produce a desired contour. The resultant product is excellent for such applications as forming rounded corners on desk legs.

Wood-Plastic Composition

Another important modification treatment is *wood-plastic composition* (WPC). In this process, wood is placed in a vacuum chamber where air is removed from the cells of the wood. Fig. 66-3. A liquid plastic monomer such as methacrylate is then introduced into the chamber. The

66-4. *The radiation causes polymerization. In polymerization, groups of identical small molecules (monomers) join together to form large molecules (polymers).*

vacuum is then released and atmospheric pressure forces the plastic monomer into the wood cells. Next the wood is bombarded with radioactive isotopes, causing polymerization to take place. (In polymerization, the small molecules—the monomers —link up to form large molecules.) Fig. 66-4. This process in effect converts the wood into plastic. The moisture resistance, strength, dimensional stability, and beauty of the wood are all improved greatly.

Finished wood-plastic composites include bowls, picture frames, salt and pepper shakers, door and tool handles, and furniture parts.

Compreg

Improved properties can also be obtained by applying chemical and compression treatments to wood. Wood treated with a ther-

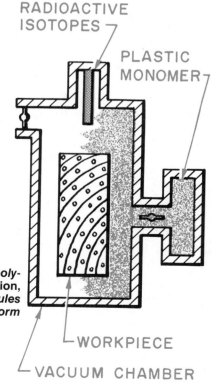

RADIOACTIVE ISOTOPES

PLASTIC MONOMER

WORKPIECE

VACUUM CHAMBER

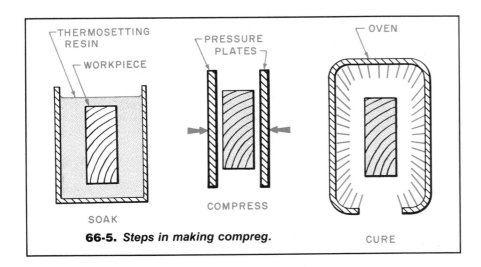

66-5. *Steps in making compreg.*

mosetting resin, such as a phenolic, then compressed and cured is known as *compreg*. Fig. 66-5. Through this process, a penetrating, bulking agent is deposited in the fibers of veneers, improving their stability. Additional advantages are improved appearance and moldability. Typical applications include antenna masts, small airplane propellers, and products for which electric insulation characteristics are important. Wood treated in a similar manner but without compression is called *impreg*.

QUESTIONS AND ACTIVITIES

1. Define veneer.
2. In what three ways is veneer cut from a log?
3. Name and briefly describe the three ways in which plywood can be made.
4. How is hardboard made? What is hardboard commonly used to make?
5. How is particle board made? Name at least four uses of particle board.
6. What is the purpose of PEG treatment of wood?
7. Briefly describe how wood is modified in wood plastic composition. In what ways does this process improve the wood?

CHAPTER 67

Estimating and Laying Out Stock

In Chapters 65 and 66 you studied some important facts about wood. The best way to learn more about wood is to work with it. You may do some wood projects in class or on your own. This chapter will tell you how to get started.

The first step is to plan your project. Review Chapters 7 and 9, which tell about product design and planning. Next, figure the amount of material you will need. The bill of materials (See Fig. 9-1 for an example) will list the dimensions of the lumber, plywood, or other wood products that are required for the project. However, if you were making the bookrack in Fig. 9-1, you wouldn't go to a lumberyard and order two pieces of lumber, one 6¼″ × 6¾″ and the other 1½″ × 5¼″. Lumber and other wood products come in standard sizes. You order these standard sizes and then cut out the size pieces you need.

LUMBER

Lumber is sold by the board foot. In simplest terms, one *board foot* is a piece of wood 1″ thick, 12″ wide, and 12″ long. (Thinner boards are usually figured as if they were 1″ thick.)

A simple formula for figuring board feet is:

$$BF = \frac{T \text{ (in.)} \times W \text{ (in.)} \times L \text{ (in.)}}{144}$$

Suppose you wanted to find the board feet of four pieces, each of which is 1″ thick, 6″ wide, and 24″ long. First determine the board feet in one piece:

$$\frac{1 \times 6 \times 24}{144} = 1 \text{ board foot}$$

Since there are four pieces, the total comes to 4 board feet. You can figure the cost of the lumber by multiplying the price of one board foot by the number of board feet needed.

Lumber Sizes

Lumber comes in standard sizes, called *nominal* sizes. For example, 1″ is a nominal thickness for hardwood boards. The actual size of lumber is always smaller than the nominal size. This happens because the lumber is seasoned (dried) and (usually) surfaced (run through a machine to smooth it) before it is sold. Thus a piece of hardwood with a nominal thickness of 1″ has an actual thickness of $^{13}/_{16}$″ if it has been surfaced on both sides.

Softwoods such as pine and fir are cut to standard length, thickness, and width. Hardwoods are cut to standard thickness only. They are more expensive and there would be too much waste if they were cut to width and length.

Lumber Grades

Lumber is graded according to its usefulness and freedom from defects. The best grades of softwood lumber are A and B. These are called "select" grades. The best grade of hardwood is FAS. This means "firsts and seconds."

Ordering Lumber

The bill of materials lists finished sizes. You will have to add to these sizes to make allowances for the wood that will be removed during sawing, planing, etc. Also, the thickness of lumber will depend on whether it is rough or surfaced. Lumber comes rough cut from the sawmill. If you want smooth lumber, order S2S or S4S. S2S means the lumber has been surfaced on two sides. S4S means all four sides have been surfaced.

To determine the cutout sizes you need:

• Add $^3/_{16}$″ to $^1/_4$″ to the finished thickness.
• Add $^1/_8$″ to $^1/_4$″ to the width.
• Add $^1/_2$″ to 1″ to the length.

Once you know the cutout sizes, you can decide which standard sizes to buy. You should order the sizes which will produce all the pieces you need with the least amount of waste.

When ordering lumber, you should specify:

• The number of board feet needed.
• Thickness. The standard (nominal) thickness is the rough one. Order nominal thickness and then state whether the lumber is to be rough (Rgh), surfaced two sides (S2S), or surfaced four sides (S4S).
• Grade.
• Kind of wood, such as white oak or hard maple.
• Seasoning (air or kiln dried).

• Widths and lengths (if ordering softwood; for hardwoods, it is less expensive to order "random widths and lengths").

Figure 67-1 shows the options you have when ordering lumber.

SHEET MATERIALS

Plywood, hardboard, particle board, and other sheet materials are sold by the square foot. For example, a sheet of plywood 4′ × 8′ contains 32 square feet. To figure its price, multiply the cost per square foot by 32.

Sizes

Plywood comes in standard thicknesses, such as $^1/_4$″, $^3/_8$″, $^1/_2$″, or $^5/_8$″. The nominal thickness of plywood and other sheet materials is the same as the actual thickness since these materials are not surfaced and they are made from seasoned wood. The most common plywood sheet size (width and length) is 4′ × 8′.

Hardboard is made in thicknesses from $^1/_{16}$″ to $^3/_4$″. The most common sheet size is 4′ × 8′.

Particle board comes in thicknesses from $^1/_4$″ to 2″. Widths range from 2′ to 5′, and lengths range from 4′ to 16′.

Grades and Types

The best grade of hardwood plywood is the premium grade, but it is very expensive. Grades 1 and 2 will be adequate for most projects. For softwood plywood (commonly called construction and industrial plywood), grades A-A or A-B are good choices. Be sure to specify exterior or interior plywood, depending on use. A doghouse, for example, would have to be built of exterior plywood.

Hardboard can be standard, tempered, or service. Standard hardboard has been given no additional treatment after manu-

67-1.
Lumber Selection Guide

Standard Sizes of Softwood			Standard Thickness of Hardwoods		Grade	
					Softwood	Hardwood
Nominal or Stock Size	Actual Size		Rough	S2S	1. Yard Lumber *Select*—Good appearance and finishing quality. Includes: Grade A—Clear. Grade B—High Quality. Grade C—For best paint finishes. Grade D—Lowest Select. *Common*—General utility. Not of finishing quality. Includes: Construction or No. 1—Best Grade. Standard or No. 2—Good Grade. Utility or No. 3—Fair Grade. Economy or No. 4—Poor. No. 5—Lowest. 2. Shop Lumber—For manufacturing purposes. Equal to Grade B Select or better of Yard Lumber. Includes: No. 1—Average 8″ wide. No. 2—Average 7″ wide. 3. Structural Lumber.	FAS—Firsts and Seconds. Highest Grade. No. 1 Common and Select. Some defects. No. 2 Common. For small cuttings.
	Green	Dry				
1″	25/32″	3/4″	3/8″	3/16″		
2″	1 9/16″	1 1/2″	1/2″	5/16″		
3″	2 9/16″	2 1/2″	5/8″	7/16″		
4″	3 9/16″	3 1/2″	3/4″	9/16″		
5″	4 5/8″	4 1/2″	1″	13/16″		
6″	5 5/8″	5 1/2″	1 1/4″	1 1/16″		
7″	6 5/8″	6 1/2″				
8″	7 1/2″	7 1/4″				
9″	8 1/2″	8 1/4″				
10″	9 1/2″	9 1/4″				
Surface	Method of Drying		Method of Cutting			
Rgh. or Rough—as it comes from the sawmill. S2S—surfaced on two sides. S4S—surfaced all four sides.	AD—Air dried. KD—Kiln dried		Plain-sawed or Flat-grained Quarter-sawed or Edge-grained			

facture. It is used for furniture and cabinets. Tempered hardboard has been treated to improve its hardness, stiffness, and finishing qualities. Service hardboard has less strength than standard, but it is lightweight. Hardboard is made with one side smooth (S1S) or two sides smooth (S2S).

Particle board is available in various forms, from unfinished to laminated.

Ordering Sheet Stock

When ordering plywood, state the species of wood for the face (surface) plies; the number of pieces; their width and length; the number of plies; the type (interior or exterior); the grade; and the thickness.

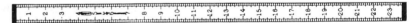

67-2. *Two-foot wooden bench rule. The common markings for this rule are inch, half-inch, quarter-inch, eighth-inch, and sixteenth-inch.*

For other sheet materials, list the number of pieces; their width and length; the type; and thickness.

MAKING A LAYOUT

After you have the plans and the materials, you need to measure and lay out the pieces for your project. To make a layout you must know how to measure accurately. (See Chapter 8.) The common tools for measuring and laying out stock are the 2′ wooden rule, the try square, the zigzag rule, the T bevel, the

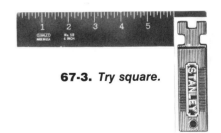

67-3. *Try square.*

marking gauge, and the framing square. Figs. 67-2 through 67-7. Other tools for marking the layout on the stock are a pencil or short-bladed knife, a combination square, a scratch awl, dividers, and a pencil compass.

When making a layout, keep the following in mind:

● Most lumber has some defects, such as knots. You will have to lay out the pieces for your project around them.

● Be sure to allow extra material for cutting, planing, and shaping.

● Wood is stronger in the direction of its grain. When laying out support parts such as legs, the length should be in the same direction as the grain, not across it.

Using Metric Measurements

You may be required to use metric measurements for some of your projects. Metric measurements in woodworking are rather simple. The important SI base unit for woodwork is the metre. The derived units of area and volume will also be used.

The main metric measuring tools for the woodworker are the 150, 300, 450, 600, and 1000 millimetre bench rules. These are

67-6. *Marking gauge.*

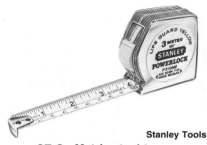

Stanley Tools

67-8. *Metric steel tape.*

roughly the same as the 6, 12, 18, 24, and 36 inch customary rules. Steel tapes are used for longer measures. Fig. 67-8.

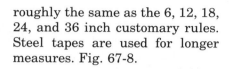

METRIC STANDARDS

Metric lumber sizes (or standards) have not yet been determined for the United States. We are not sure what these metric standards will be. For example, the standard thickness of a ¾″ board may become 20 mm. If you measure a ¾″ board, you will find that it comes closer to 19 mm, but this is an odd metric size. In order to make the dimensions more convenient, we will probably accept a 20 mm size to replace the ¾″ board thickness. A ½″ piece of lumber may be 12 mm thick, and a 1″ thick board will probably be 25 mm.

The standard 4′ × 8′ sheet of plywood will probably be 1200 × 2400 mm. It is interesting to note

that this size will have a bearing on the spacing of the studs and joists in house construction. The standard 4′ × 8′ (48″ × 96″) sheets of plywood are sized so that they will join at the center of a stud. Studs are spaced 16″ (or sometimes 24″), center to center. A 1200 × 2400 mm sheet of plywood would be 46.8″ × 93.6″. It could not be used with the present stud spacings. Thus the spacing of the studs, and other framing members, will have to be changed. Plywood sheathing, insulation board, plaster board, and a number of other sheet construction materials will come in this 1200 × 2400 mm size.

Marking Stock for Length

The first step is to make sure one end of the board is square (at 90 degrees to the edge). To do this, place the blade of the try square against the edge of the board and mark a line across the

67-4. *Zigzag rule. This rule is used mostly by carpenters and cabinetmakers.*

67-5. *Sliding T bevel. This tool can be adjusted to any angle.*

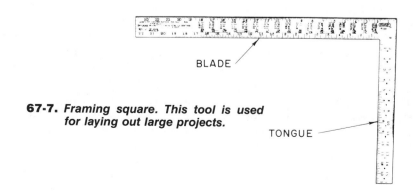

BLADE

TONGUE

67-7. *Framing square. This tool is used for laying out large projects.*

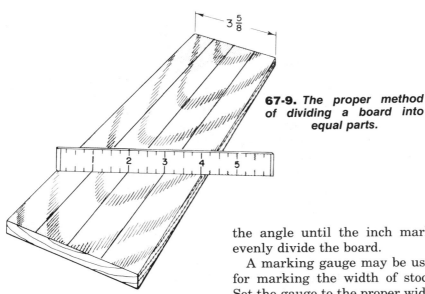

67-9. *The proper method of dividing a board into equal parts.*

67-11. *Using a marking gauge. Push the gauge away from you.*

end. Be sure to miss cracks or other imperfections in the end of the board. On wide boards you may use a framing square for this.

Now lay out the length with a rule. Mark with a sharp pencil. Square off the length with a square as you did for the first end.

Marking for Width

Measure the desired width with a rule as you did for length. If you need to divide a board into several equal parts this may be done as shown in Fig. 67-9. Place a rule across the board and shift

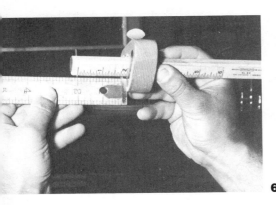

67-10. *Setting a marking gauge.*

the angle until the inch marks evenly divide the board.

A marking gauge may be used for marking the width of stock. Set the gauge to the proper width with a rule. Fig. 67-10. Mark the board by pushing the gauge forward with the head held firmly against the edge of the board. Fig. 67-11. Do not mark the board too deeply. The thickness can be marked in the same way.

Angles

Angles may be marked with a T bevel. Set the T bevel to the desired angle and tighten the blade. Then hold the handle against the edge of the wood and mark along the edge of the blade with a pencil.

Circles

Before drawing a circle, you should be familiar with the different measurements of a circle. The *circumference* is the distance completely around the outside of the circle. The *diameter* is the length of a straight line that passes through the center of the circle and divides the circle into equal halves. The *radius* is equal to one-half the diameter. The following directions describe how to draw a circle using a compass.

Dividers can be used in a similar manner.

1. With the legs of the compass together, the leg with the metal point should be $\frac{1}{32}''$ longer than the leg with the pencil point.

2. Spread the legs of the compass until the distance between them equals the distance you want the radius of the circle to be.

3. Place the metal point of the compass where you wish the center point of the circle to be. (You might want to put a piece of masking tape over this center point or a small rubber eraser over the metal tip of the compass. Either one will help protect the wood from scratching.)

4. Hold the compass between your thumb, forefinger, and third finger. Tilting the compass slightly toward yourself, turn the compass clockwise (counterclockwise if you are left-handed) to make a complete circle. Apply just enough pressure so that the circumference is marked by a sharp, clear line. Too much pressure can cause the legs to spread, ruining the circle.

Rounded Corners

Often the corners of projects are rounded to improve their ap-

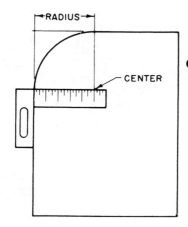

67-12. *Marking stock for a rounded corner.*

pearance. Corners can be marked for rounding using a compass (or dividers), a try square, and a pencil. Fig. 67-12.

1. Spread the compass for the desired distance. (The further the legs are spread, the wider the curve, or arc, will be.)

2. Place the metal point of the compass at the corner of the board. Use the pencil point to mark the distance at the end and again along the side of the board.

3. Holding the try square against the side of the board, draw a line from the pencil mark on this side across the board.

4. Holding the try square against the end of the board, draw a line from the pencil mark down the board until this line intersects (crosses) the first line. The point where the lines intersect will be your center point.

5. Place the metal point of the compass on this center point. Put the pencil point leg at the side of the board. (Do not change the radius set on the compass.) Turn the compass from the left to the right to mark the rounded corner.

Enlarging a Design

Sometimes you may want to develop a pattern from a picture or print in a book or magazine. These are often too small, but can be enlarged by the use of grid systems. Fig. 67-13. Directly over the print draw a grid system with small squares. Letter the vertical lines and number the horizontal ones, as shown in the top part of Fig. 67-13.

Perhaps you are planning to triple the size of the pattern. If so, make another grid system with squares three times as large. Letter and number the lines as before. If a line on the pattern crosses the grid system where lines H and 4 come together, make a small dot at point H4 on the large grid. Transfer enough of these points to show the outline of the pattern. Connect the points with lines to complete the pattern.

Transferring a Pattern

There are several ways to transfer the pattern to the workpiece.

● Cut out the pattern with scissors and tape it to the wood. Trace around it with a pencil or knife. Then remove the pattern.

● Cut out the pattern with scissors and glue it to the wood. Leave the pattern on the wood as you cut the piece.

● Put carbon paper between the pattern and the wood. Trace the design and then remove the pattern.

If a pattern is to be used many times, it is a good idea to make it from thin wood or metal. Such a reusable pattern is called a *template.*

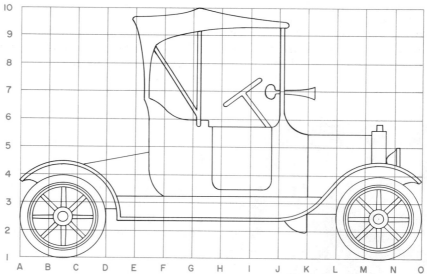

67-13. *Enlarging a pattern by using a grid system. Patterns may be reduced in a similar manner.*

QUESTIONS AND ACTIVITIES

1. In simple terms, what is a board foot? What is the formula for figuring board feet? How many board feet are in a board 1″ thick, 12″ wide, and 60″ long?

2. Why are the nominal size and the actual size of lumber different? What is the actual size of a piece of dry pine with the nominal thickness of 2″ and the nominal width of 4″?

3. What does the label "S4S" tell you about a piece of lumber?

4. What six things should you specify when ordering lumber?

5. What three points should you keep in mind when making a layout?

6. Describe how to mark stock for length.

7. What is the circumference of a circle? The diameter? The radius?

8. Describe two ways to transfer a pattern to a workpiece.

9. What is a template?

CHAPTER 68

Wood Cutting Principles

There are several ways of cutting wood. You may use a saw, a chisel, or a piece of abrasive paper. The common kinds of cutting are shown in Fig. 68-1.

Some woodworking tools are used for more than one kind of cutting. Drills and crosscut saws really cut with a shearing action. The lathe tool also cuts by shearing as it is held against the revolving piece. Probably the most commonly used tool in woodworking is the plane, and it too cuts with a shearing action.

68-1.

Cutting

Cutting is the process of removing or separating pieces of material from a base material.

KIND OF CUTTING	DEFINITION	EXAMPLES
Sawing	Cutting with a tool having pointed teeth equally spaced along the edge of a blade.	Operations on the circular saw, crosscut saw, ripsaw, coping saw, band saw, jigsaw.

(Continued)

Cutting (Continued)

KIND OF CUTTING	DEFINITION	EXAMPLES
Shearing	Cutting usually between two cutting edges crossing one another, or by forcing a single cutting edge through a workpiece.	Cutting with a wood chisel, knife, plane.
Abrading	Cutting by wearing away material, usually by the action of mineral particles.	Hand sanding, belt sanding, disc sanding.
Drilling	Cutting with a cylindrical tool usually having two spiral cutting edges.	Operations using the drill press, auger bit, twist drill.
Milling	Cutting with a tool having sharpened teeth equally spaced around a cylinder or along a flat surface.	Jointer-planer operations; cutting with wood files.
Turning	Cutting by revolving a workpiece against a fixed single-edge tool.	Operations on the wood lathe.

QUESTIONS AND ACTIVITIES

1. Define cutting.
2. Name three kinds of cutting. Give a definition and at least one example of each.

3. What is the most commonly used cutting tool in woodworking? What kind of cutting action does this tool use?

CHAPTER 69

Sawing Wood

For cutting stock to size you need a ripsaw and a crosscut saw. The saw used for cutting across the grain is called a *crosscut saw.* The saw used for cutting with the grain is called a *ripsaw.*

CUTTING TO LENGTH WITH THE CROSSCUT SAW

The teeth of the crosscut saw look like little knife blades. The teeth are bent alternately to the right and the left. This bending is called the "set" of the saw. The outside edges of the teeth cut the wood fibers on either side, and the center of the cutting edges removes these fibers. The groove

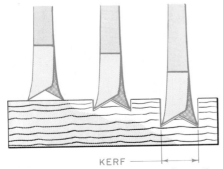

69-1. *This drawing shows how the teeth of a saw form a kerf that is wider than the saw.*

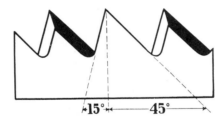

69-2. *Close-up of a crosscut saw blade.*

left by the saw is called the *kerf.* Fig. 69-1. The teeth have "set" to make the cut wider than the saw. This keeps the saw from binding in the cut.

Crosscut saws come in many sizes. The easiest to use is one about 18" to 20" long. A crosscut saw that has 8 to 10 teeth per inch is best for cutting dry wood. Fig. 69-2.

When sawing stock to length, first use a try square to mark the cutting line. If the board is long, lay it across two sawhorses. A short board may be clamped in a vise to hold it. Remember to place the cutoff line outside the supports, not between them. If you have the board on a saw-horse, place your left knee on the board to hold it. Grasp the saw in

the right hand. If you are left-handed, reverse this procedure. Place the saw in the waste stock, just outside of the layout line. This is where the saw kerf will be. Do not saw directly on the layout line or the piece could be too short. Place your free thumb against the smooth part of the saw to guide it. Fig. 69-3.

Starting with most of the blade below the board, pull up on the saw using the teeth close to the handle to start the kerf. Pull the saw up in this way two or three times before starting to saw. If

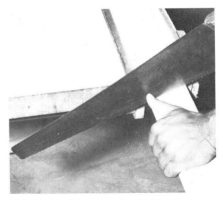

69-3. *Starting a cut. Guide the saw with your thumb until the cut is started.*

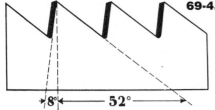

69-4. *Close-up of a ripsaw blade.*

8°   52°

you try to begin the cut on the down stroke, the saw may jump out of your hand and mar the wood or cut your hand. Hold the saw at an angle of 45 degrees to the stock. Cut with steady, even movement, using full-length strokes. The crosscut saw cuts on both the forward and the back strokes. Do not jerk the saw and do not force it.

Make sure you are cutting square with the board. While sawing, watch the line, not the saw. Blow the sawdust off the line so you can see it. If the saw starts to go away from the line, correct this by twisting the handle a little. Hold the end of the board for the last few cuts. This will keep the board from splitting before the saw kerf is complete.

CUTTING TO WIDTH WITH THE RIPSAW

Sometimes you need to cut boards to width. The ripsaw is used to cut wood with the grain. It has teeth that are like chisels. A ripsaw should be about 24″ long. Fig. 69-4. Short stock may be clamped in a vise. If the board is long, you will need to use saw-horses, and you may want to place a guide board just outside the layout line to help keep the cut straight. You may also need to place a wedge in the kerf to keep the saw from binding when cutting a long board to width. The ripsaw should be held at an angle of 60 degrees. The ripsaw cuts only on the forward, or down, stroke.

CUTTING PLYWOOD

Plywood should be cut with a fine-tooth crosscut saw. Place the plywood face up. Support the wood so that it will not sag. With a pencil, mark the line to be cut. Hold the saw at a low angle to the plywood to avoid splitting.

CUTTING CURVES

For making many projects it may be necessary to cut curves or irregular shapes from wood. For such cuts the saw must have a thin blade. The two handsaws most commonly used for irregular cuts are the *coping saw* and the *compass* or *keyhole saw*.

The coping saw has a U-shaped frame into which a blade is fastened. Most coping saws have a screw handle to tighten the blade. This kind of a saw is easy to use because the blade can be turned at any angle to the frame.

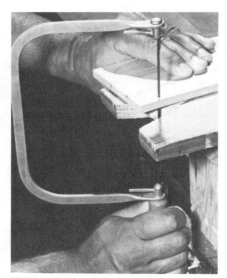

69-5. *Using a coping saw to cut a piece of wood held on a saw bracket.*

69-6. *Using a compass saw to cut a circle in a large sheet of plywood.*

However, you must make sure that the pins holding the blade are turned the same amount so the blade does not twist or bind. Blades for the coping saw have teeth like those of a ripsaw. For most work a blade with 15 teeth to the inch will be satisfactory.

You can use a coping saw with the work held either in a vise or over a saw bracket. Fig. 69-5. If you are using a saw bracket, the teeth should point *toward* the handle. If the work is held in a vise, the teeth should point *away* from the handle.

Grasp the handle of the saw in your hand; move the saw up and down or back and forth depending on where you are sawing. Cutting is done on the down or forward stroke, so apply light pressure as you push forward and release the pressure as you pull back. Keep the blade moving at a steady, even pace. The blade may break if you jerk it or put too much pressure on it. At sharp corners keep working the saw and turn it slowly in the direction of the line. Twisting or bending the blade at the corners will also usually break it.

Compass and keyhole saws are used when you cannot use a coping saw, as for cutting curves in large boards or in plywood. The compass saw looks like a regular

Rockwell International Power Tool Division
69-7. *A jigsaw.*

handsaw except that it is much smaller and almost comes to a point at the end. A keyhole saw is smaller than a compass saw and is used for cutting keyholes and for similar work. Figure 69-6 shows a compass saw in use.

The *jig* or *scroll saw* is a machine for cutting curves. Fig. 69-7. The saw blade moves up and down and is used for the same kind of cutting as the coping saw. The jigsaw is a safe machine if you follow the rules. You will probably use a jigsaw in making some of your projects. Special blades are made for this saw, but a coping saw blade without pins can be used successfully.

Cutting with the jigsaw requires the same care as cutting with a coping saw. Adjust the guide so that the spring tension holds the work firmly against the table. Hold the work with the thumb and fingers as shown in Fig. 69-8. Don't force the stock

69-8. *The correct way to hold the work when cutting with a jigsaw.*

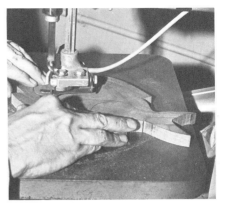

into the saw. Use an even pressure. Turn the stock slowly when cutting a curve; if it is turned too sharply or rapidly, the blade will break.

If you are going to cut an internal curve or design, drill a hole in the center of the waste stock first. Remove the throat plate from the jigsaw. Unfasten the blade from the plunger. Slip the work over the blade. Fasten the top of the blade back to the plunger. Replace the throat plate. Adjust the guide to the correct pressure. Then take a cut from the hole to the layout line.

CUTTING WITH THE BAND SAW

The *band saw* is also used for cutting curves, circles, and irregular designs. This saw can be used for straight crosscutting and ripping too. The band saw is used for heavier wood than the jigsaw; it has a wider blade and cuts faster. It does not make as smooth a cut as the jigsaw. The

band saw has a blade that runs over two large wheels. One of the wheels is on the top of the machine and one on the bottom. The upper wheel is adjustable in order to tighten the blades.

To cut on the band saw, the wood is placed on the table and pushed into the blade. The saw cuts rapidly; so be sure your fingers are out of the way. The band saw should not be used without permission of the instructor.

CIRCULAR SAW

In Fig. 69-9 you see a *circular saw*. This is used for power sawing. Your instructor may use this to do the same work as is done with a hand saw.

Rockwell International Power Tool Division
69-9. *A circular saw.*

QUESTIONS AND ACTIVITIES

1. What is the saw kerf? Where should the saw kerf lie in the board to be cut?

2. What is the "set" of a saw?

3. Describe the teeth and the set of the crosscut saw. At what angle do you hold a crosscut saw when cutting?

4. How do you place and then move the saw to start the kerf?

5. At what angle is the ripsaw held when cutting? What are some things to remember when cutting long boards with a ripsaw?

6. What saw do you use to cut to width? To cut with the grain? To cut plywood? To cut a circle in a large board?

7. What direction should the teeth of a coping saw point when using a saw bracket to hold the workpiece? When using a vise?

8. Name three differences between the use and cutting action of a bandsaw and a jigsaw.

CHAPTER 70

Planing and Chiseling Wood

After pieces have been sawed, they are rough. Even surfaced wood still has small knife marks when it comes from the mill. If these are not removed, they will show up when you apply a finish. The hand plane is used to remove marks and smooth rough edges.

The plane is a tool which you must learn to handle correctly. It is the most complicated hand woodworking tool you will use. It takes more care and adjustment than any other tool.

PARTS AND TYPES OF PLANES

Figure 70-1 shows the major parts of the plane. The *body,* or *bed,* of the plane is made of steel. The *base,* or *bottom,* is smooth or sometimes ribbed. Behind the opening in the base is a *frog.* This is the support for the *double plane iron.* A brass *adjustment*

nut sets the depth of cut. The *lateral adjustment lever* is used for making a sideways adjustment of the cutter. The double plane iron consists of the *plane iron* itself, which is the cutting edge, and the *plane iron cap,* which breaks

the chips (or shavings) and forces them up and out. The double plane iron fits over the frog and is held in place by the *lever cap.*

There are four common types

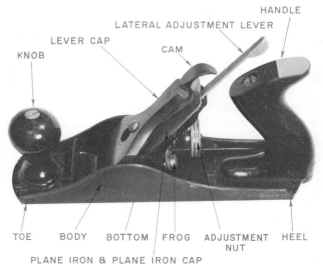

70-1. Major parts of a plane.

HANDLE

LATERAL ADJUSTMENT LEVER

LEVER CAP

CAM

KNOB

TOE   BODY   BOTTOM   FROG   ADJUSTMENT   HEEL
                                  NUT

PLANE IRON & PLANE IRON CAP

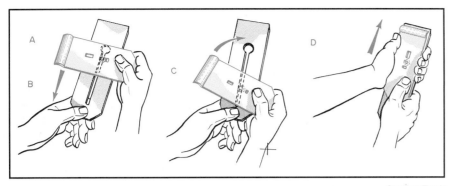

Stanley Tools

70-2. *The correct way to assemble a double plane iron.*

of bench planes. They are all much alike. The big difference is in their length. The *jack plane* is either 14″ or 15″ long. The *smooth plane* is 9¼″ or 9¾″ long. These planes are used for general-purpose planing. The *fore plane* and the *jointer plane* are much longer. These are used for planing long edges such as doors.

SHARPENING THE PLANE IRON

The plane must be sharp. One way to check for sharpness is to sight along the edge. A sharp edge will not reflect much light. A dull edge looks shiny.

If your plane iron is very dull or has nicks, it will have to be ground. Grinding is the shaping and forming of the cutting edge. Your instructor will grind your plane iron or show you how.

If your plane iron just needs touching up or if it has been ground, it must be honed. Honing makes the cutting edge razor-sharp. (The plane iron must of course be removed from the plane for grinding and honing.)

To hone, select an oilstone with a flat, true surface. Apply a few drops of oil to the face of the stone. Place the bevel (the angled side) of the plane iron flat on the surface of the oilstone. Raise the end of the plane iron so just the

cutting edge rests on the stone. Move the iron back and forth or in a circular pattern on the face of the stone. A wire or feather edge will form on the cutting edge. To remove this, turn the plane iron over and lay it flat on the stone. Move it back and forth a few times. *Be sure to hold the iron flat.* Then turn the plane iron over and give it a few strokes. Turn it over and stroke the back again. Repeat this until the wire edge is gone. The cutting edge should now be sharp.

ASSEMBLING THE PLANE

The first step in assembling the plane is putting together the double plane iron, as shown in Fig. 70-2. Holding the plane iron in one hand with the beveled edge of the blade down, place the plane iron cap across it. Fig. 70-2(A). Put the cap screw through the hole. Then slide the plane iron cap down from the plane iron's cutting edge. Fig. 70-2(B). Turn the plane iron cap 90°, so that it is straight with the plane iron. Fig. 70-2(C). Slide the cap forward until it is ⅟₁₆″ from the cutting edge. Fig. 70-2(D). (Never slide the plane iron cap over the cutting edge of the blade or you could nick the blade.) Holding the plane iron and the plane iron cap together, tighten the cap

screw securely with a screwdriver. If the two parts do not fit tightly, shavings can get between the cap and the plane iron and the plane will not cut properly.

Now the assembled double plane iron must be placed into the plane. Holding the double plane iron with the beveled edge down, carefully place it on the frog and over the lever cap screw. Be careful not to hit the cutting edge on the side of the plane. The long slot in the plane iron fits over the roller of the lateral adjustment lever. The small slot in the plane iron cap fits over the lever which adjusts the depth of cut. (This lever is controlled by the adjustment nut shown in Fig. 70-1.) Slip the lever cap in place. Using thumb pressure, push the cam (Fig. 70-1) down to fasten the double plane iron securely in place. If the lever cap is too tight, you will have difficulty adjusting the plane. If it is too loose, the plane won't stay properly adjusted. You can use a screwdriver to tighten or loosen the lever cap screw until the cam will close with a push of your thumb.

ADJUSTING THE PLANE

Before starting to plane, you should adjust the tool for the correct depth of cut. Hold the plane upside down with the bottom at eye level. Turn the brass adjusting nut until the cutting edge of the plane iron just appears beyond the bottom of the plane. Then move the lateral adjustment lever to the right or left until the blade is parallel with the bottom of the plane. Turn the brass adjusting nut again until the cutting edge just appears beyond the bottom of the plane. Test the plane on a scrap piece of wood. Continue adjusting it until shavings are smooth, silky, and even. For fine work or smoothing a surface, a fine cut is best.

FOR YOUR SAFETY . . .

● Make sure the cutting edge of the plane is sharp. A dull blade tends to jam and stick.

● Clamp the workpiece onto the workbench or in a vise to hold it steady.

● Stand facing forward and balanced on both feet so that you will be able to plane with steady, even strokes.

● Hold the knob of the plane with one hand and the handle with your other hand. Keep your fingers away from the cutting edge at the underside of the plane.

● Don't try to cut thick shavings. It is safer to cut off thin shavings, and you will get better results.

PLANING A SURFACE

First check the board to make sure there is no metal that will come in contact with the cutting edge of the plane. Planing over metal objects such as nails or screws can ruin the cutting edge of the blade.

Lock the board to be planed in the vise or between the dog of the vise and the bench stop. The work must be centered or it will slip when you start planing. If the wood is rough, you may need to take a few cuts with the plane to see which way the grain runs. The board should be locked so that you are planing with the grain. Planing against the grain will make the board rougher.

Grasp the knob of the plane in your left hand and the handle in your right hand. Stand with your feet apart, your left foot forward, and your right side near the bench. The body, hand, and foot positions should be reversed if you are left-handed. Use a back-and-forth motion with your body. Apply pressure to the knob on the front of the plane at the start of the stroke. Apply even pressure on both the knob and the handle as the whole plane comes onto the board. Apply pressure to the handle at the rear of the plane as it begins to leave the surface. Lift the plane off the board on the return stroke. If you take a shearing cut, with the tool at a slight angle to the direction of cut, the plane will work more easily. Work across the board gradually. Plane down the high spots first, as these will require more planing than the rest of the surface. After the surface of the board begins to get smooth, check it with a straightedge in several directions (edge to edge, corner to corner) to see if it is true. Light will show under the straightedge where there are low spots. It is best to check the entire length and width of the board.

The first surface you have planed is called the face surface. This is the surface you use to start squaring up stock.

PLANING END GRAIN

Planing end grain is harder than planing the surface. When you plane the end of a board, you must cut off the fibers of the wood. The block plane is used for planing end grain. Fig. 70-3. It is much smaller than other planes and can be held in one hand.

Mark a sharp line across the surface and the edge to indicate how much stock should be removed. Clamp the board in a vise with about one inch of the end grain showing. If the board sticks out too far from the vise, the piece will not be held securely enough to plane. The cutting edge should be very sharp, and the depth should be set for as thin a cut as possible.

The block plane is held in one hand. Begin from one edge and work toward the middle of the board. Then begin from the other edge and do the same thing. Be sure to hold the block plane square with the work. It is wise to take a shallow cut to keep the plane from jumping. By planing halfway across the end, then starting from the other side, you are not likely to split the wood. Another way to prevent splitting is to take a piece of scrap wood of the same thickness as the stock and place it against the edge of the stock. Still another way is to cut a bevel on the waste edge of the stock, then begin to plane from the other edge all the way across the wood.

SQUARING UP STOCK

Squaring up stock means to make all opposite sides (surfaces, edges, and ends) flat and parallel with each other. There are various ways to do this. One method is described here.

1. Plane the best surface. This is called the "face surface." Follow the procedure described in "Planing a Surface." After planing, mark it so you will know it is the face surface.

2. Plane the best edge. This will become the "face edge." Fasten the stock in a vise, with the edge extending 2 or 3 inches

70-3. *A block plane.*

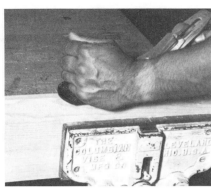

70-4. *When planing an edge, fasten the stock in the vise with the edge extending 2 or 3 inches above the vise. Plane with the grain.*

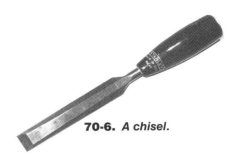

70-6. *A chisel.*

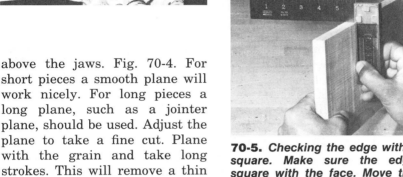

70-5. *Checking the edge with a try square. Make sure the edge is square with the face. Move the try square along the entire edge of the board, checking for squareness in several places.*

above the jaws. Fig. 70-4. For short pieces a smooth plane will work nicely. For long pieces a long plane, such as a jointer plane, should be used. Adjust the plane to take a fine cut. Plane with the grain and take long strokes. This will remove a thin shaving all along the stock. Do not remove much wood. With a try square check the edge against the planed face. Fig. 70-5. The edge and face surface should be square with each other. It is also a good idea to hold a straightedge along the edge from one end to the other to check for straightness. Mark the edge so you will know it is the face edge.

3. Plane one end. Follow the procedure described under "Planing End Grain."

4. Cut the board to length. (See Chapter 69.) It should be about 1/16" longer than the desired finished length. Then plane this end to the finished length.

5. Set a marking gauge to the correct width. Mark the stock along its entire length with the gauge. Remember to push the gauge away from you. If there is much stock to be removed, the board should be ripped to within 1/8" of the finished measurement. Lock the stock in the vise, and plane this second edge just as the first. Be sure to check it against the face with a try square.

6. Mark the stock for thickness, again using a marking gauge. Mark a line on both edges. Check the lines to see if there are any high spots that need more planing than the rest of the board. Lock the stock between the bench stop and the vise dog. Then begin to plane the length of the board. Work from one side to the other. Planing this surface is like planing the face surface except you have to watch the lines.

As you near the lines, keep checking the piece with a straightedge and try square. Check this surface against both edges with the try square.

CHISELS AND GOUGES

Some cutting jobs that cannot be done with a saw or a plane are done with a chisel or a gouge. A chisel has a flat blade with a bev-

eled end. Fig. 70-6. It is used to cut flat pieces from wood in order to shape and fit parts. A gouge has a curved blade with a beveled end. It is used for cutting grooves and holes.

FOR YOUR SAFETY . . .

Chisels and gouges should be used with great care. They cause more injuries than any other woodworking tool.

• Always carry chisels and gouges with the pointed end down.

• Clamp the work securely before using a chisel or gouge.

• Use a wooden or plastic mallet, not a metal hammer, to hit the handle of a chisel or gouge. Do not use any mallets with a pointed tang chisel; the tang could split the handle.

• Keep your hands away from the front of the cutting edge. Move the tool away from your body, not toward it, as you work.

Using a Chisel

To use a chisel, fasten the work securely in a vise. Fasten it so you can cut with the grain. You can also chisel across the grain, but never against the grain (sloping down into the wood). When going against the grain, the chisel tends to dig into the wood, splitting rather than cutting it. For rough cutting hold the chisel with the bevel against the wood. For light cuts, turn the chisel over. A shearing cut is easier to make than a straight cut. When cutting across the grain, it is best to work first from one side and then the other. If you try to cut all the way across the piece from one side, the opposite side will split. Curves can be cut on boards by starting at a corner and taking several cuts.

Using a Gouge

Mark the area you wish to remove with the gouge. Fasten the work in a vise or use a hand clamp to clamp it to the bench. Hold the handle in your right hand and use your left hand to guide the blade. (Reverse this if you are left-handed.) Start near the center of the waste area. Holding the gouge at a 30-degree

70-7. *The spokeshave.*

angle, make long strokes with the grain, removing long, thin shavings. As you near the layout lines, take thin shavings in all directions from these lines toward the center.

Many bowls are made by using gouges. When gouging out a bowl, you may need a mallet to make heavy cuts. It is better to take heavy cuts across the grain. The gouge is less likely to dig in.

Gouges are also used for wood carving on bowls and model boat hulls.

SPOKESHAVE

For some projects it is necessary to form a curved surface. Curves can be formed with a bladed tool such as the spokeshave. Fig. 70-7. The spokeshave was originally used for shaping the spokes of wooden wheels. It is used for finishing the edges of curves and making irregular shapes. The cutter blade is sharpened much like the plane iron. It has two adjusting nuts on the top. The spokeshave may be either pushed away from you or pulled toward you. Files and rasps may also be used for shaping curved surfaces. (See Chapter 73.)

QUESTIONS AND ACTIVITIES

1. What is the hand plane used for?
2. What two parts make up the double plane iron? What is the function of each part?
3. What is the function of the frog on the plane? The function of the lateral adjustment lever?
4. What are at least three safety rules to remember when using the plane?
5. Why shouldn't you plane against the grain? Why shouldn't you chisel against the grain?
6. What type of plane is used for planing end grain? For planing long edges such as doors? For general-purpose planing?

7. Describe two ways to keep from splitting the wood when planing end grain.
8. Define squaring up stock.
9. Name the six basic steps to follow in squaring up stock.
10. Describe a chisel. What are chisels used for?
11. Describe a gouge. What are gouges used for?
12. What are at least three safety rules to remember when using chisels and gouges?
13. What is a spokeshave used for?

CHAPTER 71

Abrading Wood

Sanding is the process of cutting the wood fibers with some type of abrasive. *Abrasives* are hard materials that grind or wear away a softer material. The main reason for sanding is to smooth the wood surfaces before the finishing operation. Sanding is very important because finishes make defects show up more. Scratches you can barely see will be noticeable after you apply the finish to your project.

Sanding on either wood or plastic projects is usually not started until all the other work is done. There are times when some shaping can be done with abrasive paper. However, a general rule is never to try to make abrasive paper take the place of a chisel or a plane.

Abrasive papers are sold in many forms. (See Chapter 37.) Coated abrasive (sandpaper) is most commonly used in the school shop. The type of abrasive grains you use will depend on the kind of wood used and whether you are sanding by hand or with a machine. The grade of abrasive paper you select will make a difference in your work. A carefully planed surface can be sanded with a fine paper (150 or 180) and be ready to finish. If tool marks show on the wood, it will be necessary to use a coarser paper first. The coarser grades of paper are usually used for shaping edges of wood.

Each piece of a project should be sanded before assembly. Always brush off the surface after sanding. After the project is assembled, light sanding should be done again before the finish is applied. Use a tack cloth to remove any dust left by sanding before applying any finish.

SANDING BY HAND

For hand sanding flat surfaces, a sanding block should be used. The sanding block shown in Fig. 71-1 was made by tightly holding a piece of sandpaper over a block of wood to which heavy felt or leather has been glued as a backing. (This backing is needed between the paper and wood block because, if a sliver of wood gets directly between the paper and wood block, it creates a hard spot. This hard spot can tear the paper or cause uneven sanding.)

71-1. *Sanding a flat surface with a sanding block. Be sure to sand with the grain of the wood.*

71-2. *Sanding the edge of a board. Be sure to keep the sanding block square with the surface.*

You can also use commercially-made sandpaper holders.

Fasten the piece to be sanded in a vise or hold it firmly on a bench. Apply an even pressure to the block, and sand the surface *with the grain*. Move the block back and forth and work from one side to the other. Don't sand the edges of the surface too much. Start with a medium paper and finish with a fine paper.

When sanding an edge, fasten the piece in the vise so that the edge is showing. Notice how the small sanding block is held in Fig. 71-2. Use two hands on the sanding block. Keep both thumbs on top of the block. Use your fingers to support and guide the block to keep it from rocking. Be sure to keep the sanding block square with the face.

To sand an end, use the same procedure as for sanding an edge. Sand in one direction only. Sand the corners and arrises lightly. (An *arris* is the edge formed by two surfaces.)

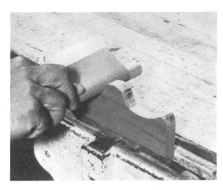

For sanding convex, or outside curves, you can simply hold the sandpaper in your hand. For sanding inside curves, or concave surfaces, the abrasive paper can be wrapped around a stick such as a large dowel. Fig. 71-3.

Round pieces such as stool legs can be sanded by using the paper like a shoeshine cloth. Turned parts that are straight can be sanded with the grain.

Very small parts may be sanded by first fastening a piece of abrasive paper to a board. This is clamped in a vise or held on a bench top. Hold the small pieces in your hand and rub them back and forth over the paper. Fig. 71-4.

POWER SANDERS

Some shops may have one or more kinds of power sanders. The most popular are belt sanders and disk sanders. With stationary sanders such as the one in Fig. 71-5, the work is placed on the table and held against the belt or disk.

The disk sander is best used for coarse sanding such as edge work. Hold the work flat on the table of the sander. (The table itself can be tilted for angle and chamfer sanding.) Hold the work lightly against the half of the disk that is moving downward. Move the work smoothly back and forth across the disk. If you

71-3. *Sanding an inside curve with the sandpaper wrapped around a dowel.*

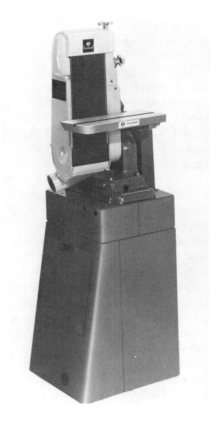

71-4. *Small pieces may be sanded on a sanding board.*

do not keep the work moving, the wood will be burned and the abrasive paper ruined.

The belt sander can be used for medium or fine sanding. Belts of different grit can be used on the belt sander. Be sure that the belt is installed properly. Belt sanders can be used in the horizontal position for sanding surfaces, and in the vertical position for sanding angles, edges, and ends.

When you cannot take the work to the sander, a portable sander can be used. The portable belt sander, for example, uses a sanding belt like the floor machine. The belt revolves on two wheels. The sander is placed over the work and moved back and forth in a straight line with the wood grain. Never apply pressure when using a portable belt sander because it cuts very rap-

idly. This sander is hard to use because the belt makes the machine run away from you. The finish sander has a more gentle action. Straight-line action leaves the smoothest finish. Orbital (circular) action is better for rough sanding.

FOR YOUR SAFETY . . .

● When using portable power sanders, always turn the power on before placing the sander on the workpiece. When you are done sanding, always lift the sander from the workpiece before turning off the power. Make sure the sander has come to a full stop before setting it down.

● You should wear goggles and a dust mask when using power sanding tools.

● Never use any power sanding tool without the permission of your instructor.

Rockwell International
Power Tool Division
71-5. *A belt sander.*

QUESTIONS AND ACTIVITIES

1. Define sanding.
2. Define abrasives.
3. What is the main reason for sanding? Why is this so important?
4. What should be done to a freshly assembled and sanded workpiece before applying any finish?
5. Briefly describe how to sand a flat surface by hand.
6. Describe how to hold a sanding block when sanding an edge.
7. Describe how to hand sand inside curved surfaces.
8. What type of sanding is the disk sander best used for? The belt sander?
9. What are some rules to remember when using portable sanders?

CHAPTER 72

Drilling and Boring Wood

Many woodworking projects involve the drilling or boring of holes. Holes ¼″ or smaller in diameter are *drilled;* larger holes are *bored.*

Figure 72-1 shows some common drilling and boring tools. The *twist drill* (A) is used in hand drills, portable electric drills, and drill presses to cut holes ½″ or less in diameter. For cutting holes that are ¼″ to 1″ in diameter, an *auger bit* (B) is used. For still larger holes, an *expansion bit* is needed (C). This tool can be adjusted with different cutters to bore holes from ⅞″ to 3″ in diameter. A *Foerstner bit* (D) is used to bore holes nearly all the way through a piece of wood without splitting the other side. It can also be used to even out the bottom of a hole made by an auger bit or to enlarge a hole. Foerstner bits commonly range from ¼″ to 2″ in diameter. All these bits must be installed in a hand or power tool before they can be used.

FOR YOUR SAFETY . . .

When using drilling or boring tools, follow these rules for safe operation.
- Fasten the work in a vise or clamp it securely.
- Make sure the bit is in the chuck straight. Tighten the chuck firmly.

When using power tools. . .
- Wear eye protection or a face shield.
- Avoid loose clothing. For example, long baggy sleeves could get caught in the machinery.
- Be sure to remove the chuck key before turning on the power. Otherwise, the chuck key could be thrown out at you.
- Always wait for the portable electric drill to come to a complete stop before setting it down.
- When using a drill press, wait for it to come to a complete stop before removing waste material from the table.

- If a workpiece gets caught in the drill, turn off the power. Never try to stop the movement with your hands.

BORING HOLES

As already mentioned, one tool for boring holes is the auger bit. Auger bits come in sizes ranging from No. 4 to No. 16. The number

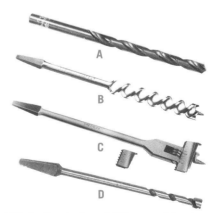

72-1. *Tools for drilling and boring: (A) Twist drill. (B) Auger bit. (C) Expansion bit. (D) Foerstner bit.*

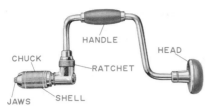

Stanley Tools

72-2. The brace is used with an auger bit for boring holes.

72-3. Small pieces can be clamped in a vise. Be sure to keep the bit square with the work by lining it up with a try square.

stamped on the tang of the bit indicates the size in sixteenths of an inch. For example a No. 4 would be 4/16″ or 1/4″. A No. 7 would be 7/16″. A set of auger bits has all the numbers from 4 through 16.

Auger bits are used in a *brace*. Fig. 72-2. Most braces have a ratchet that allows boring holes in a corner even though there isn't room for a complete swing of the handle.

To install a bit in a brace, hold the shell of the chuck in your left hand and turn the handle to the left until the jaws are open. Insert the bit and turn the handle to the right to fasten it. When placing the bit in the brace, be sure to have the corner of the bit in the U grooves of the jaws.

Measure and mark the position of the holes by drawing two intersecting lines. Punch the center of the hole with a scratch awl.

Place the stock in a vise so that the brace can be used in a horizontal position. Fig. 72-3. The punch mark should be near the top or side of the jaws of the vise. Guide the bit with your left hand and start it in the hole made with the scratch awl. Hold the head of the brace with your left hand. Turn the handle with your right hand. (If left-handed, reverse these instructions.) Be sure to keep the bit square with the work. You can sight along the top of the bit. Have another student sight the bit to see if it is straight up and down.

Do not press too hard on the brace. A properly sharpened auger bit will almost feed itself into the wood. Continue to bore until the point of the bit just comes through the stock. Turn the han-

72-4. A depth gauge fastened to an auger bit.

72-5. Using a T bevel as a guide for boring a hole on a slant.

dle in the opposite direction to back the bit out of the hole. Now turn the wood around and bore from the other side. If this is not done, the wood will split when the bit goes through. Another way to keep the wood from splitting is to clamp a scrap block of wood to the back of the stock. Then you can bore all the way through.

Sometimes you need to bore only partway through a board. For this you need a *depth gauge*. Fig. 72-4 shows a commercially-made depth gauge. You can make your own depth gauge using a piece of dowel rod. Bore a hole in the piece of dowel so it will fit over the bit like a sleeve. Cut away the dowel until the desired bit length is exposed.

To bore a hole at an angle, set a sliding T bevel at the proper angle to use as a guide. Start the auger bit as you did for straight boring. When the screw feeds into the wood, tilt the bit to the proper angle. The bit should line up with the T bevel. Fig. 72-5.

To bore a large hole use an expansion bit. If you look closely at the expansion bit in Fig. 72-1(C),

you will notice there is a graduated scale on the cutter. This scale will help you set the proper diameter to be cut. Set the distance from the feed screw to the spur to equal one-half the diameter of the hole. This equals the radius of the hole. Be sure to lock the cutter. It is a good idea to bore a hole in a scrap piece to check the size of the hole. When the feed screw comes through the wood, finish the hole from the other side. It is especially important that the work is held tightly in a clamp or vise when using expansion bits.

As mentioned earlier in this chapter, Foerstner bits are used for boring holes that go only partway through a board. The sizes range from ¼" to 2". They are numbered the same as auger bits. Draw a circle where you want the hole.

DRILLING HOLES

Holes ¼" or smaller are usually drilled with a hand drill. (See Chapter 39 for more information on the hand drill, electric drill, and drill press.) Twist drills have straight shanks. These drills may be used for making holes in either wood or metal. Twist drills in a set range in size from ¹⁄₆₄" to ½" in intervals of ¹⁄₆₄".

The *push drill* is used for boring small holes in wood. Fig. 72-6. The special drill points are carried in the handle. To use this tool, select the desired drill point, place it in the chuck, and tighten. When you push down on the handle, the point turns, thus drilling the hole.

The *hand drill* is used with the twist drill. Fig. 72-7. It has three jaws in the chuck. These jaws hold the round shanks of the twist drills. To insert the twist drill, turn the crank counterclockwise until the shank of the drill will slip into the jaws. Make sure the drill is in the chuck straight. Turn the crank clockwise to tighten the jaws. Clamp the work in a vise. Mark the hole to be drilled with a scratch awl or sharp nail. Place the point of the

72-8. Portable electric drills. The one on the bottom has a side handle.

drill on this mark and turn the crank evenly. Apply light, even pressure to the handle, pressing straight down (if drilling vertically) or straight ahead (if drilling horizontally). The twist drill will break if the hand drill is tilted after you begin to drill. When removing the drill from the hole, continue cranking in the same direction as you back out of the hole.

An electric hand drill is a useful tool for drilling and boring holes, as well as many other jobs. Fig. 72-8. These drills are sized according to the largest drill bit they will hold. The most common size is ⅜". Electric drills come in single-speed, two-speed, and variable-speed models. (On variable-speed drills, the speed is changed as you change, or vary, the pressure on the trigger.) Before inserting a twist drill or bit, be

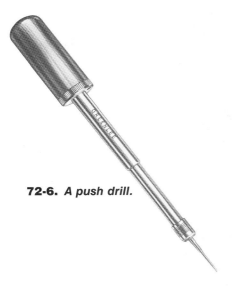

72-6. A push drill.

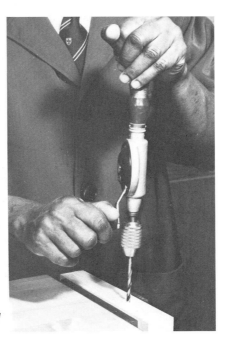

72-7. Using a hand drill for drilling a small hole in a piece of wood.

Fred W. Gillman

72-9. *Using a drill press. Notice that the stock is held firmly by a clamp.*

sure the electric drill is un-plugged. Open the chuck, insert the twist drill or bit, turn the chuck clockwise to close the jaws, and tighten the jaws with a

chuck key. Be sure the drill bit is centered in the chuck, and be sure to remove the chuck key before beginning to drill. After marking the exact center of the hole to be drilled with a center punch or scratch awl, place the drill over this mark. Guide the drill by placing one hand on the housing or the side handle (if the drill has one). Hold the drill

steady; do not force it into the wood. Do not let the revolving chuck touch the surface of the workpiece or it will mar the wood.

A drill press, Fig. 72-9, is especially good for drilling several holes in a small piece. Before using a drill press, be sure you have no loose clothing that might catch in the drill or spindle. To use this machine, place the workpiece on the table of the drill press over a piece of scrap wood. Adjust the height of the table until the workpiece just clears the twist drill or bit. Adjust the speed for the size of the cutting tool and the type of wood. Faster speeds are used for smaller diameter tools and softwoods. Slower speeds are used for larger bits and hardwoods. Do not operate a drill press until your instructor has given a thorough demonstration of how it is used.

QUESTIONS AND ACTIVITIES

1. What are drilled holes? Bored holes?

2. Name four common drilling or boring tools and tell what type or size hole each is used to cut.

3. Name one safety rule to follow when using all types (hand and power) of drilling and boring tools. Name four safety rules for using power drilling or boring tools.

4. What is a brace used for?

5. What does the number on an auger bit indicate? What number auger bit would you use to bore a ½″ diameter hole?

6. When is a depth gauge needed?

7. Name two kinds of tools used to drill holes by hand. Name two kinds of power tools used to drill holes.

CHAPTER 73

In *milling*, wood is removed by a cutting tool that has sharpened teeth equally spaced around a cylinder or along a flat surface. Milling is done to smooth and shape wood. For example, rough boards are smoothed and cut to uniform thickness on a machine called a planer. The edges of wood are often rounded or cut to some other curved shape by milling tools. In this chapter you will learn of some methods of hand and machine milling.

FILES AND RASPS

Many times, small curves will require the use of a rasp or a file. Fig. 73-1. Files are available in many sizes and shapes. The most often used files are the half-round cabinet and flat files. The rasp removes large amounts of stock quickly, but leaves a rough surface. Be sure to use a handle on a file or rasp. See Chapter 40 for additional information on files.

The Surform® tool is for shaping. Fig. 73-2. Its tool steel blade has rasplike teeth that make it easy to cut wood. Holes between the teeth keep the tool from becoming clogged with wood shavings. This tool is used like a rasp, but it really cuts the wood instead of scraping it. It produces a smooth, flat surface. It is good for

73-2. *One type of Surform® tool.*

Rockwell International Power Tool Division
73-3. *The jointer is used mainly to smooth the surfaces and edges of wood after it has been cut with a saw.*

shaping odd-shaped projects such as canoe paddles and gun stocks. It is also a good repair tool for smoothing an edge or end that has splintered or chipped.

FOR YOUR SAFETY . . .

When using files or rasps, follow these safety rules:
● Always put a handle over the tang of the file or rasp before using it.
● Use the right tool for the job.
● Never use a file or rasp for prying or hammering. It could break, sending sharp chips in all directions.
● Keep the teeth of the file or rasp clean. A clogged tool could slip off the workpiece and injure you.
● Fasten the work firmly in a vise or clamp it to the workbench.

JOINTER

Your shop may have a jointer for smoothing the edges and faces of boards. The jointer is a surfacing machine. Fig. 73-3. It has a base, two tables, and a cutter head. When used improperly, it is a dangerous machine. Probably your instructor will operate the jointer for you. The jointer is usually used for smoothing an edge and making it square with the face.

73-1. *A wood rasp.*

ROUTER

The portable router is a hand-held machine that is used for many cutting and shaping jobs. It can cut grooves, round out edges, make bead and cove edge cuts, and do much more. Fig. 73-4. It consists of a high-speed motor mounted on an adjustable base. The chuck at the bottom of the motor shaft can hold a number of differently shaped cutters. The router can be a very dangerous tool in inexperienced hands. **Check with your instructor before using it.**

Porter-Cable Corporation
73-4. A portable Router.

QUESTIONS AND ACTIVITIES

1. Define milling. What is the purpose of milling?

2. What are at least three safety rules for using files or rasps?

3. Describe the Surform® tool and tell what it is used for.

4. What jobs does a jointer do?

5. What kinds of jobs does the portable router do?

CHAPTER 74

Wood Turning

Many projects require turned parts for their construction. The wood lathe is the machine used to make these parts. Fig. 74-1. Long pieces of wood are supported at both ends of the lathe. The stock is held between the live (moving) center and the dead (stationary) center. This is called *spindle turning*. The posters for a bed would be made by spindle turning. Bowls and similar pieces are made by *faceplate turning*, in which the stock is mounted on only one end of the lathe.

As the workpiece turns, the operator holds cutting tools against it. The common cutting tools include a 1″ gouge, a ½″ skew, a roundnose tool, a spear, and a parting tool. Fig. 74-2.

There are two methods of turning wood—namely, cutting and scraping. In *cutting*, the tool is held so that the cutting edge digs into the revolving wood. In *scraping*, the tool is held at right an-

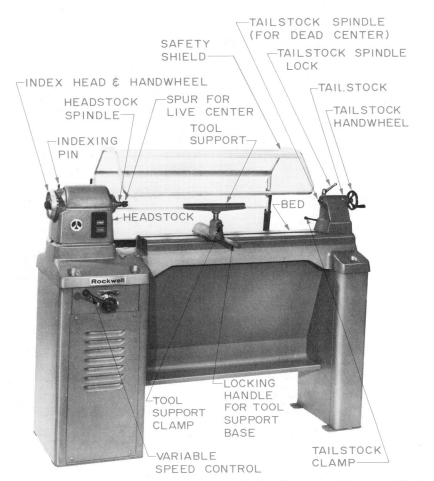

INDEX HEAD & HANDWHEEL

HEADSTOCK SPINDLE

INDEXING PIN

SAFETY SHIELD

SPUR FOR LIVE CENTER

TOOL SUPPORT

HEADSTOCK

TAILSTOCK SPINDLE (FOR DEAD CENTER)

TAILSTOCK SPINDLE LOCK

TAILSTOCK

TAILSTOCK HANDWHEEL

BED

TOOL SUPPORT CLAMP

LOCKING HANDLE FOR TOOL SUPPORT BASE

TAILSTOCK CLAMP

VARIABLE SPEED CONTROL

Rockwell International Power Tool Division

74-1. *A lathe for wood turning.*

gles and fine particles are worn away. Scraping is easier than cutting.

FOR YOUR SAFETY . . .

● Wear goggles or a face shield when using the lathe.

● Examine the workpiece before you place it on the lathe to make sure it is not split and is free of nails.

● Be sure you keep proper tension on the belt. It should be just tight enough to keep it from slipping.

● Do not run the lathe at excessive speeds.

● Be sure the lathe tools are sharp, and be sure you keep a firm grip on the tools while turning.

● Keep the tool support securely locked in the proper position. Never try to adjust the support while the lathe is running.

● When spindle turning, be sure both the tailstock and the tailstock ram are locked securely.

● To remove the faceplate, use a wood wedge between the spindle pulley and the headstock. Never use the pulley index pin to lock the pulley.

SPINDLE TURNING

The piece to be turned should be about 1″ longer and ⅛″ to ¼″ thicker than the finished piece.

1. At each end of the wood, mark the center.

2. Select one end to be the headstock end. This is the end that will be mounted on the live center of the lathe. Make ⅛-inch deep saw cuts on this end. These should be through the center point and at right angles to each other.

3. Place the stock on a solid surface. Center punch a hole at the exact center of each end.

4. Place the spur center over the headstock end of the wood. Tap it with a mallet to drive it into the wood. The center must be in the hole, and the spurs must enter the saw kerfs.

5. Place the spur center in the spindle of the headstock.

6. Loosen the tailstock clamp and bring the tailstock to within about 1½″ of the end. Use the tailstock clamp to lock the tailstock to the bed. Then turn the tailstock handwheel to bring the tailstock up to the wood. Force the cup (dead) center into the center hole of the wood about 1/32″. Back the tailstock off and

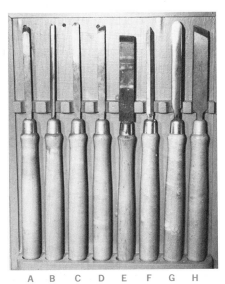

A B C D E F G H

74-2. *A set of lathe turning tools: (A) Round-nose tool. (B) Small gouge. (C) Diamond-point tool. (D) Small skew. (E) Flat skew. (F) Parting tool. (G) Large gouge. (H) Large skew.*

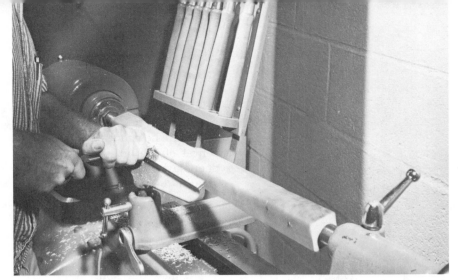

74-3. *Turning between centers on the lathe. This student is making a rough cut with the large gouge.*

74-5. *Faceplate turning. Scraping tools such as a round-nose tool are especially good for removing material from the inside of a bowl.*

rub some wax on the wood to prevent it from burning. Put the wood back on the dead center. Then tighten the spindle lock.

7. Adjust the tool rest to clear the stock by about ⅛″ and slightly above center. Rotate the stock by hand to make sure the stock clears the tool rest.

To begin, the lathe should be set on the slowest speed. Use the large gouge. Hold it against the wood as shown in Fig. 74-3. The gouge is used for rough turning. For finish turning, use a skew and increase the machine speed. Fig. 74-4. Be careful with the skew so as not to catch it in the stock.

Always make sure to have a good hold on the lathe tools. Keep the tool rest close to the work. If there is too much space between the tool rest and the work, the tool may catch and be thrown out of your hands.

FACEPLATE TURNING

Lamp bases, bowls, and trays are turned on the faceplate. Fig. 74-5. Faceplate turning is done by fastening a piece of wood to the faceplate with screws. The faceplate is mounted on the headstock spindle. The tool rest is turned around so that it is in front of the work. Be sure the wood is held securely. Begin turning with the gouge the same as for turning between centers.

Do not run the lathe at excessive speeds. This is especially important in the beginning before the work is trued up. Unbalanced pieces running at a high speed may fly out of the lathe and injure you or someone else. Never use the wood lathe without the instructor's permission.

James L. Shaffer

74-4. *Finish turning on the wood lathe.*

QUESTIONS AND ACTIVITIES

1. Define spindle turning. Give an example of an object made by spindle turning.

2. Define faceplate turning. Name two things that can be made by faceplate turning.

3. What is cutting? What is scraping?

4. Name at least four safety rules to remember when turning wood on the lathe.

5. What is the gouge generally used for on the lathe? What is the skew used for? Name a job the roundnose tool might be good for.

As shown in Fig. 75-1, steaming and laminating are the main methods used to bend wood. Skis, toboggans, and certain furniture parts are made this way. Wood can also be bent without heat. All of these methods are described below.

STEAM BENDING

Certain kinds of wood bend more easily than others. These are ash, hickory, birch, and oak. Before bending, wood should be steamed or boiled in hot water. This softens the wood cells so that they can be stretched or compressed; then the wood will bend much more easily than when dry.

To soften the wood, a heating tube is often used. This tube is closed on the lower end and has a cover on the upper end. Water is poured into the tube and heated.

The wood is inserted into the tube and the cover placed on the upper end.

After the pieces have been heated, they are removed and placed in the bending form. This must be done slowly and carefully because wood will split when pulled too rapidly. As the wood is bent, it is clamped to hold it until formed.

The wood should dry for at least 24 hours in the form. After forming, it is sanded in the usual manner.

BENDING WITHOUT HEAT

Flat stock can be bent without heating. Cut a saw kerf in one end of the stock. Figure 75-2. Next, cut pieces of veneer wider than the stock to fit in the saw kerf. Put waterproof glue on both sides of the veneer and slip it into the kerf. Clamp the stock in the bending form and let the glue dry. Spruce and mahogany are good woods to use for this process,

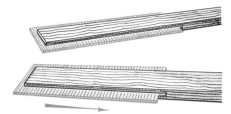

75-2. Veneer inserted in saw kerfs of flat stock. Spruce and mahogany work well for this process.

which is a variation of the laminating process explained next.

LAMINATING WOOD

Another way of forming wood products is by laminating. This is the process of building up thickness by gluing several layers of wood together. The grain runs in the same direction. (In plywood the grains of alternate layers run at right angles to each other.)

Laminating is done to produce the attractive beams found in buildings such as schools and churches. Fig. 75-3. These beams are very strong and resist fire better than solid pieces. Laminated beams can be made from short pieces.

Laminating can be done in the school shop to make small projects such as salad servers. First, make a full size pattern of the curve to be bent; then decide on the number of thicknesses to use in the project. Usually an odd number is best—three, five, or seven.

Next, make the form from hard maple or birch. It must be wide enough to allow at least 1″ on ei-

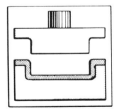

75-1.

Forming

Forming is the process of shaping wood without adding to or removing any of the material.

KIND OF FORMING	DEFINITION	EXAMPLES
LAMINATED SHEETS CLAMPS FORM **Bending**	Forming by uniformly straining wood around a straight axis.	Steam bending, bending without heat, and laminating.

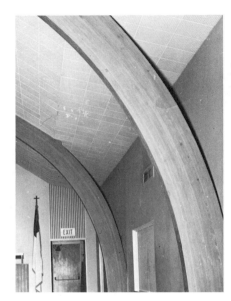

75-3. *Laminated wooden beams used in a church building.*

75-4. *This kind of form is needed for laminating in the school shop. Salad servers could be made using forms like these.*

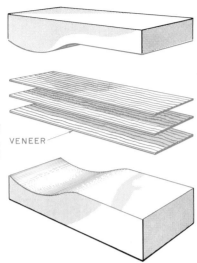

VENEER

ther side of the veneer. Lay out the curve on the block; then carefully cut the curve. This is best done on the band saw. **Be very careful when using the band saw. Follow your teacher's instructions.** The two halves of the form must fit perfectly. Sand the

form lightly and apply a sealer, such as shellac, to make a better surface. Fig. 75-4.

Cut several pieces of veneer large enough for the project. You can use one or several kinds of wood for the different layers. Spread glue evenly on all the layers. Do not put glue on the outside of the top and bottom pieces.

Place a piece of waxed paper on one half of the form and lay the veneer pieces on it. Lay a piece of waxed paper over the top veneer and set the top half of the form in

place. Clamp the two halves together with wood clamps. Allow the piece to dry under pressure. When dry, remove the piece from the form. Saw out the design with the jigsaw, and sand.

If you are making a project that will be used around food, finish with salad oil or light mineral oil.

QUESTIONS AND ACTIVITIES

1. Define wood forming.
2. Define wood bending.
3. Briefly describe the steps to steam bend wood.
4. Describe how to bend flat wood without heat.

5. What kinds of woods work well in steam bending? In bending without heat?
6. Briefly describe how to form wooden objects by laminating.

CHAPTER 76

Wood Fastening Principles

Mechanical fasteners most frequently used on wood are screws and nails. Machine bolts, special rivets, and spring clips are also used. Figure 76-1 shows the fastening methods used for wood.

To join wood by adhesion, a substance which is different from the materials being joined is used. This is usually some kind of glue. Many kinds of glue are used for wood products. Waterproof glues are often used on wood that will be exposed to the weather. Other kinds may be used where the wood will not be exposed to moisture.

Two different materials may be fastened together by adhesion. For example, plastic can be cemented to wood by using epoxy cement or contact cement.

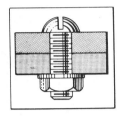

76-1.
Fastening

Fastening is the process of joining materials together. The materials may be joined permanently or semipermanently. Different materials require different types of fasteners. Wood materials used in the shop may be fastened in one of two ways—with mechanical fasteners or by adhesion.

KIND OF FASTENING	DEFINITION	EXAMPLES
Mechanical	Permanent or semipermanent fastening with special locking devices.	Nails. Screws. Corrugated fasteners. Machine bolts. Rivets.
Adhesive WOOD / WOOD / GLUE	Permanent fastening by bonding like or unlike materials together with glue or cements.	Glue. Contact cement. Epoxy cement.

QUESTIONS AND ACTIVITIES

1. Define fastening.
2. Define mechanical fastening. Give three examples of metal fasteners.

3. Define adhesive fastening. Give two examples.

There are many kinds of wood joints. Some are used more often than others, but all joints must be laid out, cut, fitted, and assembled. Joints are held together with glue or with glue plus nails, or screws. The strength of a joint is determined largely by how much surface of one piece touches the other. Joints which do not have much surface touching are reinforced with dowels or a spline. A *spline* is a thin piece of wood inserted in a groove in both parts of the joint.

The pieces to be joined should first be squared and cut to size. Lay out the parts carefully. Use a sharp pencil to mark the two pieces of a joint so that they will not get mixed up. This is especially important when gluing several pieces edge to edge for a tabletop.

The *edge joint* is shown in Fig. 77-1. This is not a very strong joint; so it is often reinforced with dowels. Dowels are hardwood rods, usually of birch or

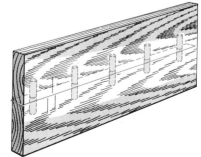

77-1. *Two pieces glued together edge to edge. Dowels are used to make the joint stronger.*

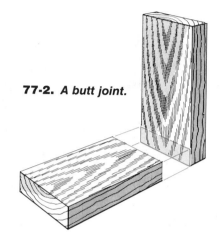

77-2. *A butt joint.*

maple. They are available in a variety of sizes from ⅛″ to 1″ in diameter, with or without grooves. Dowel rods are particularly good for reinforcing the joints of tabletops or wide boards built up from several narrower boards.

Butt joints are made by fastening the end of one piece to the surface or edge of the other. Fig. 77-2. Dowels or corner blocks may be used to strengthen this joint. Carpenters use butt joints in house construction and in the construction of boxes and crates.

Lap joints are made by overlapping the ends or edges of two pieces of wood. One-half the thickness of each piece is cut away so that when the two pieces are joined, the surfaces on the wood are flush with each other. There are several types of lap joints. The cross-lap joint is made when two pieces of wood cross each other, usually at right an-

gles. The middle-lap joint is used to make a T-shaped joint. The end-lap joint is used to join the ends of two boards so that they will be at right angles to each other. Fig. 77-3. Many frames are made using the end-lap joint.

The lap joint is laid out by marking the width and depth of the wood to be removed. The backsaw is used to make the depth cut. The wood between the cuts is removed with a chisel. The sawing should be done inside the marking lines so as to make a good tight fit.

The *rabbet joint* is used for fitting panels into a frame, as when making doors for cabinets. A groove is made in the edge, or end, of the wood used for the frame. This groove is the same size as the thickness of the piece that will be fastened to it. Fig. 77-4.

The *dado joint* is commonly used for supporting shelves in bookcases. Fig. 77-5. The width and depth of the dado are marked on the board to be cut. The width

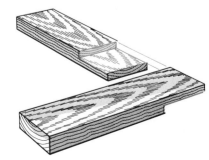

77-3. *A lap joint. This particular one is an end lap.*

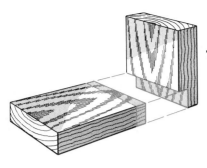

77-4. Rabbet joint. This joint is used in furniture doors.

depends on the board that are to fit into the joint. The backsaw is used to cut the sides to the proper depth. A chisel is used to remove the excess wood. Be sure to chisel from both sides to avoid splitting the wood. On rough work, the dado joint may be fastened with nails. On finish work the joint is usually fastened with glue.

The *miter joint* is used for cutting and fitting rafters and moldings. It is also commonly used for picture frames. Fig. 77-6. This joint helps eliminate end grain, but it is a rather weak type of construction. Miter joints can be strengthened with splines or dowels.

A T bevel or a miter square is commonly used to lay out the miter cut on the two pieces of wood to be joined. Then the stock may be clamped in the vise and sawed. If the pieces to be joined are small, they can be sawed with a backsaw. However, the easiest method of cutting miters is to use a miter box. The best miter boxes are adjustable for sawing at many different angles. Fig. 77-7.

Miter joints are usually nailed and glued. The first nail is driven from the outside edge, at a right angle to the miter cut, until the point comes through and makes an impression on the opposite piece. The two bevels of the miter joint are then given an even coating of cabinet glue. One half of the joint is securely fastened in a

77-5. Dado joint. This joint is used for shelves in bookcases.

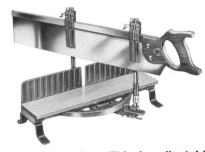

77-6. Miter joint. This is a common joint for making picture frames.

vise. Then the half which contains the nail is placed in position and the nail driven in. The joint is then removed from the vise and the opposite side is clamped in the vise. Another nail is driven in from the opposite side. This will counteract the slippage that occurred when the first nail was driven in, and will help line up the joint and make an even and square contact. All

surplus glue should be wiped from the edges of the joint with a damp cloth before the glue sets.

The *mortise-and-tenon* joint is one of the strongest joints in woodworking. These joints are commonly used to fasten rails to the legs of tables and chairs and to join parts in many other types of better-quality furniture. When properly made and fitted, the mortise-and-tenon joint is strong. In the mill, mortise-and-tenon joints are rapidly made by machines. They can be made successfully with hand tools, but they require more time and more careful workmanship than some other joints. The mortise is the opening, and the tenon is the part which fits into it. Fig. 77-8.

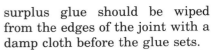

77-7. Miter box. This is adjustable to any angle for making miter cuts. The most common miter joint is made by cutting each of the two pieces to be joined at an angle of 45 degrees.

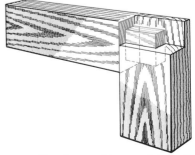

77-8. Mortise-and-tenon joint. This joint is used on legs and rails of tables and chairs.

QUESTIONS AND ACTIVITIES

1. What determines the strength of a joint? What can be done to a joint to make it stronger?

2. Define an edge joint and tell what might be made using this joint. How can edge joints be strengthened?

3. Sketch a butt joint. What types of things are made using butt joints? How can butt joints be reinforced?

4. How are lap joints made? Name and describe two types of lap joint.

5. Describe a rabbet joint and tell what types of things are made using this joint.

6. Sketch a dado joint. What is a common use of this joint?

7. Sketch a miter joint. What are these joints commonly used for?

8. What are mortise-and-tenon joints commonly used for? Which part is the mortise? The tenon?

CHAPTER 78

Mechanical Fasteners for Wood

NAILS

The most common mechanical fastener for wood is the nail. Almost everyone has had some occasion to drive and pull nails. It seems simple, but some skill is needed to drive a nail straight without bending it.

The most common tool for driving nails is the claw hammer. Hammer size is indicated by weight. A 16-ounce hammer is good for most work. The face of the hammer head should be slightly rounded.

The *nail set* is another tool used for nailing. It is a short metal punch with a cup-shaped end used to drive the head of a nail below the surface of the wood.

There are many kinds and sizes of nails. Nails are made of aluminum, brass, copper, or mild steel. Some mild steel nails are galvanized (coated with zinc) to keep them from rusting.

The length of nails is indicated

by their penny size. (The letter "d" means penny.) Nails range from 2d to 60d (1″ to 6″). The larger the number, the bigger the nail.

Nails also come in various diameters. The diameter of a nail is indicated by a gauge number. The smaller the gauge number, the bigger the diameter. The diameters range from 0.03″ (No. 20 gauge) to 0.28″ (No. 1 gauge).

The kinds of nails that you will use most are *common, box, casing,* and *finishing.* Fig. 78-1. Box nails are thin with flat heads. They were designed for nailing together boxes. The common nail looks like a box nail except it is heavier. Casing nails have small heads. They are used for finish carpentry or sometimes for projects. The finishing nail is the finest of all the nails and is used for cabinetwork.

Brads are really small finishing nails that are used for fastening thin stock. They are indi-

cated by length in inches. The corrugated fastener is used for holding some joints and is good for repair work. Fig. 78-2.

FOR YOUR SAFETY . . .

Follow these safety rules when using a hammer:

- Wear eye protection.
- Make sure the hammer is in good condition before using it. The head should not be loose or damaged.
- Keep the hammer free of grease and oil so that it won't slip when you use it.
- Watch the nailhead, not the hammer. Then you will be less likely to hit your fingers.
- Never use a claw hammer on hardened metal, such as chisels and punches, or on another hammer.
- Never put nails in your mouth.
- Follow the correct nailing procedure, as described in this chapter.

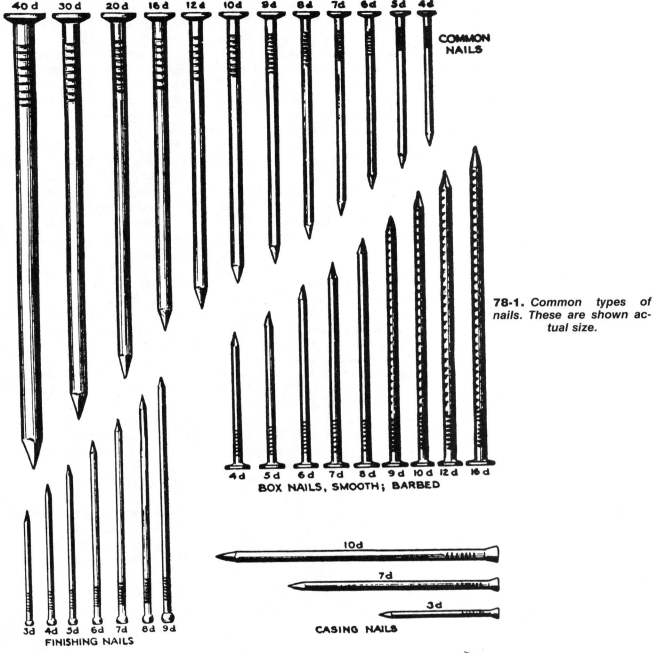

78-1. *Common types of nails. These are shown actual size.*

COMMON NAILS

40d 30d 20d 16d 12d 10d 9d 8d 7d 6d 5d 4d

BOX NAILS, SMOOTH; BARBED

4d 5d 6d 7d 8d 9d 10d 12d 16d

3d 4d 5d 6d 7d 8d 9d
FINISHING NAILS

10d
7d
3d
CASING NAILS

Driving Nails

Before driving nails, be sure to choose the correct kind and size for the job. Small nails are chosen for thin stock, heavier ones for thick stock. To make a tight joint, nails are sometimes driven at an angle.

To start a nail, hold it in one hand between your thumb and fingers. Hold the hammer in your other hand and tap the nail lightly once or twice. Fig. 78-3. Remove your fingers from the nail once it is started. To drive the nail, use your wrist, elbow, and arm to swing the hammer. Watch the head of the nail, not the hammer. Try to drive the nail with a few sharp blows

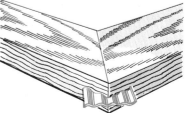

78-2. *This corner is held together with a corrugated fastener. The fastener should be driven flush with the wood surface.*

293

78-3. *The proper way to start a nail.*

rather than many light taps. If a nail bends, pull it out and drive a new one.

Do not place several nails along the same grain marking. This may split the wood. Do not drive casing or finishing nails completely in with the hammer. Use the nail set. Hold the nail set as shown in Fig. 78-4. Then drive the nail until it is about $\frac{1}{16}''$ below the surface.

If you are nailing hard wood, you may need to drill holes for the nails. This will keep the wood from splitting and the nails from bending over. The holes should be slightly smaller than the size of the nail. A little wax on the nail will make it drive easier.

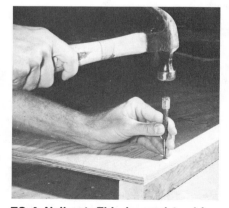

78.4 *Nail set. This is used to drive the head of a nail about $\frac{1}{16}''$ below the surface of the wood.*

If the nails you are using are so long that they will go all the way through the pieces being nailed, you may need to clinch the joints. Drive the nails completely through the pieces and bend the nail point over in the direction of the grain.

Sometimes you may need to pull some nails. Force the claw of the hammer under the head of the nail. Pull on the handle. When the nail is pulled partway out, slip a scrap of wood under the hammer head. This helps keep the nail straight and gives better leverage. Fig. 78-5.

SCREWS

Screws are another type of mechanical fastener used to assemble wood projects. It takes longer to install screws than nails, but they make a stronger joint. Also, joints made with screws are easy to take apart and reassemble. A few screws will do the work of several nails.

Screwdrivers come in many sizes and shapes. The size is given as the length of the blade. There are two common types of screwdrivers: the plain and the Phillips head. The plain screwdriver is used for slotted-head screws. The tip of the plain screwdriver should be the same width as the head of the screw. The blade should be flat at the tip and fit properly into the screw head. This is very important. If a screwdriver is ground or worn to a sharp edge, it may slip out of the slot and mar the wood. The Phillips head screwdriver is made for driving screws with recessed (Phillips) heads. Fig. 78-6.

There are several things you need to know when choosing wood screws. These are: the kind of head, the diameter or gauge, the length, the kind of metal, and the finish.

● There are three kinds of

screw heads you may use in woodworking. These are the flathead, the ovalhead, and the roundhead. All three are available with either slotted or Phillips heads. Fig. 78-6.

● The common screws come in almost any length from $\frac{1}{4}''$ to over 6″.

● Wood screws are available in gauge numbers from 0 to 24. The smaller the number, the thinner the screw. For example, a number 9 screw would be larger than a number 7.

● Most screws are made of steel. Some are made of aluminum and brass, for use where moisture might rust the steel screws. Most flathead screws have a bright finish. Roundhead screws are usually finished dull blue.

To get the most holding power, you should choose a screw long enough to go into the second piece of wood almost the entire length of the screw threads. This will be about two-thirds the length of the screw. End grain wood does not hold very well; so select a longer screw for this.

Drilling Holes for Wood Screws

To fasten two pieces of wood with screws, you need to drill two

78-5. *Using a claw hammer to pull a nail. When the nail is partway out, slip a block of wood under the head.*

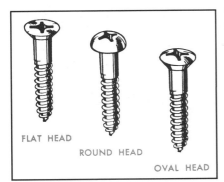

FLAT HEAD

ROUND HEAD

OVAL HEAD

78-6. *Common head shapes of wood screws. Note that all of these have the recessed (Phillips) head.*

78-7. *Two types of countersinks: the one on the top is for a brace, and the one on the bottom is for a hand drill or drill press.*

78-8. Drill Guide

No. of Screw	For Shank Clearance Holes	For Pilot Holes*		No. of Auger bit to Counterbore for Sinking Head (by 16ths)
		Hard Woods	Soft Woods	
0	1/16	1/32	1/64	
1	5/64	1/32	1/32	
2	3/32	3/64	1/32	3
3	7/64	1/16	3/64	4
4	7/64	1/16	3/64	4
5	1/8	5/64	1/16	4
6	9/64	5/64	1/16	5
7	5/32	3/32	1/16	5
8	11/64	3/32	5/64	6
9	3/16	7/64	5/64	6
10	3/16	7/64	3/32	6
11	13/64	1/8	3/32	7
12	7/32	1/8	7/64	7
14	1/4	9/64	7/64	8
16	17/64	5/32	9/64	9
18	19/64	3/16	9/64	10
20	21/64	13/64	11/64	11
24	3/8	7/32	3/16	12

*Sometimes called "anchor holes."

holes for each screw—one for the screw shank (the part of the screw between the head and the threads) and the other, called the pilot hole, for the screw threads. NOTE: When drilling wood, observe the safety rules listed in Chapter 72.

The shank hole is drilled in the first piece of wood. It should be large enough so that the screw can be pushed in with the fingers. Select a drill bit with the same diameter as the screw shank.

The pilot hole is drilled in the second piece. Select a bit that is equal to the smallest diameter of the threaded part of the screw. The depth of the pilot hole should depend on the hardness of the wood. Pilot holes in soft wood only need to be about half the length of the screw. In hard wood

they should be drilled the entire length of the screw thread.

Flathead and ovalhead screws should be countersunk. *Countersinking* is a way of enlarging the top of the hole so that the top of the screw will be level with or below the level of the wood surface. A countersink should be used in a brace or a drill press. Fig. 78-7. Cut just deep enough so that the screw will be level with the surface of the wood. The table in Fig. 78-8 shows the sizes of bits needed for shank and pilot holes and for countersinks.

To install the screws is a simple job if you have drilled the correct shank and pilot holes. Place the screw in the hole and push it

as far as you can without forcing. Grasp the handle of the screwdriver in one hand. Use the other hand to guide the blade of the screwdriver. Turn the screw until it is set. Be careful not to let the screwdriver slip out of the slot and mar the wood. Don't tighten too much, or you might strip the threads or break off the screw.

QUESTIONS AND ACTIVITIES

1. What is a nail set and what is it used for?

2. Which is longer: a 5d nail or an 8d nail? What does gauge number indicate? Which is smaller: an 18-gauge nail or a 5-gauge nail?

3. What are the three most often used nails other than the "common nail"? Describe each and tell what each is used for.

4. Name three safety rules to remember when using a hammer.

5. Briefly describe how to start and then drive a nail in an ordinary nailing job.

6. Name three advantages of using screws instead of nails.

7. Which three screw heads are commonly used in woodworking?

8. What is the pilot hole for? What determines the size of the pilot hole? How deep should pilot holes be?

9. What is countersinking?

10. Describe how to install a screw after the proper holes have been drilled.

CHAPTER 79

Adhesive Fastening of Wood

To fasten wooden pieces together permanently, use some kind of glue. This type of fastening is done for several reasons. Boards can be glued together edge to edge to make larger surfaces as, for example, a tabletop. Boards can also be glued together face to face to make them thicker, as in making a lamp. Joints that are to be fastened together permanently are also glued.

KINDS OF GLUE

You should be familiar with six kinds of glue: animal hide glue, polyvinyl, resorcinol, casein, contact cement, and epoxy glue. The ones most commonly used are animal hide glue and polyvinyl.

Animal hide glue, which is usually brown, is made from the hooves, hides, and bones of animals. It is a good, general-purpose glue for furniture and other wood projects. However, it is not waterproof; so it cannot be used for gluing pieces that will be exposed to the weather.

Polyvinyl glue is odorless, dries colorless, and does not stain. It is good for gluing furniture. It is always ready to use; no mixing is required. However, it is not waterproof and therefore should not be used on projects for the outdoors.

Resorcinol is very strong, and it is waterproof. This is the glue to use for outdoor furniture and for boats. It has a dark color and requires mixing. Therefore it is not often used unless a waterproof glue is needed.

Casein glue must also be mixed before use. It is a good glue for oily woods, such as teak. Do not use it on redwood, as it will stain acid woods. Casein glue is not waterproof, although it is water-resistant.

Contact cement is used for bonding veneer or plastic to plywood. As the name implies, it bonds on contact. The pieces to be glued must therefore be positioned very carefully before being brought together.

Epoxy cement will stick to almost anything. It can be used on wood, metal, plastic, ceramics, and many other materials. It must be mixed before using.

CLAMPS

The two types of clamps that are most useful for woodworking are the *bar clamp* and the *hand screw.* Each clamp is tightened with a screw adjustment. The bar clamp is used for wide pieces. Fig. 79-1. The hand screw is used for smaller pieces. Fig. 79-2. Before starting to glue up a project, be sure to have plenty of clamps handy. When using bar clamps, place a piece of scrap stock between the clamp and the work. This will protect the work from the clamps.

GLUING UP STOCK

First assemble all the parts to be glued to make sure they fit correctly. You can also mark the pieces at this time so that they can easily be put in proper order for gluing and assembling. Adjust the clamps and try them on the pieces before applying the glue. There should be a clamp about every 12″ to 15″ when gluing edge to edge. When using glue on a bench top, be sure to protect the bench with wrapping paper or newspapers.

Using a brush, stick, or roller, apply glue to both the surfaces to be joined. Make sure that both pieces are completely covered. Do not apply too much glue, or it will squeeze out of the joint when the clamps are tightened.

When gluing edge to edge, place glue on all the edges to be

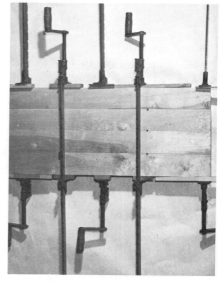

79-1. These boards for a tabletop have been glued edge to edge. They are being held in place by bar clamps. Notice the scrap pieces to protect the edges of the boards.

79-2. *Using hand screws in assembling a project.*

glued. Then lay the pieces on the bar clamps. Tighten the outside clamps slightly. Tap the ends of the boards until they are even. Then tighten the clamps. *Don't tighten too much.* You cannot pull a poor joint together by using extra pressure on the clamps. Remove the surplus glue before it hardens. The more glue you remove now, the less scraping you will have to do later.

For gluing stock face to face, hand clamps are used. First arrange the pieces the way they are to be glued. Set the clamps to the proper opening. Apply the glue to all surfaces to be glued. Clamp the pieces together. Be careful to keep the jaws of the clamps parallel. This will keep the pressure even.

The time it takes glue to dry depends on the kind of glue used. White polyvinyl glue takes only about one-half to one hour to set. It is best to let glue dry overnight with the clamps in place. Any excess glue can be removed with a scraper or chisel.

Many of the joints used in making wood products are fastened with glue. Glue is applied to both parts of the joint. The joints are assembled and either clamped or nailed to hold the pieces until the glue dries. (Nails will of course stay in; clamps are removed after the glue dries.) When projects are assembled with glue, they cannot be taken apart.

DIELECTRIC HEATING

Dielectric heating is based on the fact that molecules can be excited, or made to move rapidly. These moving molecules cause friction, which causes heat. The movement is caused by very high frequency radio waves. The radio waves move the glue molecules so quickly that they generate heat. The heat dries the glue. This heating action is similar to what happens to food when it cooks in a microwave oven.

With this method of drying, glued boards can be ready for working in minutes instead of days, as required by ordinary methods. The dielectric heating machine consists of a high frequency generator and a hand gun. Fig. 79-3. The high frequency waves from the generator are transmitted to the wood by the hand gun. As the hand gun is moved along the glue joint, the high frequency waves excite the glue molecules, causing heat which dries the glue rapidly. Since the heat is generated inside the glue, drying is much faster than when the glue is heated by placing the wood in an oven.

The process can be used for gluing in any position where the hand gun can be placed on the glue joint. It can be used for edge to edge gluing, on the edges of boards (Fig. 79-4), or even for repairs (Fig. 79-5).

Not every kind of glue is suitable for dielectric heating. The glues most commonly used are the urea resin glues, such as urea formaldehyde resin.

Workrite Products Co.

79-3. *A dielectric heating machine.*

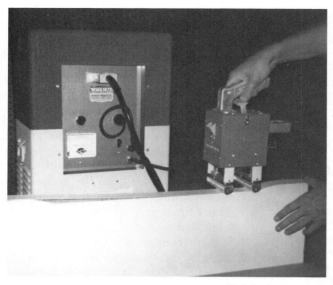

Workrite Products Co.

79-4. *Gluing a counter top with roller electrodes on the hand gun.*

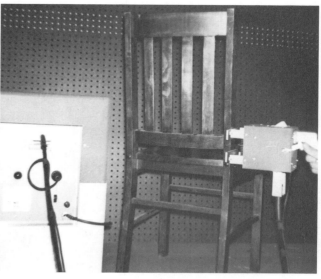

Workrite Products Co.

79-5. *Repairing furniture. The glued joint is being dried by dielectric heating.*

QUESTIONS AND ACTIVITIES

1. Name the six kinds of glue discussed in this chapter and tell what each might be used to fasten.

2. Name a good all-purpose glue. A good waterproof glue. Which glues require mixing?

3. What types of clamps are used when gluing edge to edge? When gluing face to face?

4. Give a brief description of dielectric heating. What is the main advantage of this type of gluing?

CHAPTER 80

Wood Finishing Principles

A well-constructed project deserves a good finish. More projects are ruined by improper finishing than by any other mistake during production. Notice the fine finishes that are applied to commercially constructed pieces of furniture. A properly applied finish brings out the beauty of the wood and gives the completed project a mark of quality.

Wood may be finished in one of several ways, as shown by the chart. Fig. 80-1. A knowledge of the materials used for wood finishing is important. Just as important is knowing the right order for using the materials. If the steps in the finishing process are not followed in the correct order, a well-constructed project may be

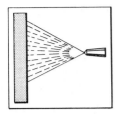

80-1.
Finishing
Finishing is the process of treating the surface of a material for appearance and/or protection.

KIND OF FINISH	DEFINITION	EXAMPLES
Coloring*	Applying penetrating chemicals to a material to change its color.	Staining.
Coating*	Applying a layer of finishing substance to the surface of a material.	Painting. Lacquering. Dipping.
Remove Finishing	Finishing by cutting the surface by abrasive action; finishing by charring.	Wire brushing. Sandblasting. Rubbing with coated abrasives. Charring.

*Coating could be considered a coloring process, since coating usually does change a product's color. However, it is customary to consider coating and coloring to be separate processes, as defined in the chart above.

not only protects a wood surface but also changes its appearance. It is important to choose the right kind of finish for each job and to apply it well. Otherwise appearance of the wood will not be changed for the better.

Before applying any finish to a product, you must prepare the surface. This is usually done by sanding, as explained in Chapter 71.

BASIC FINISHING STEPS

There are certain basic steps in finishing. Although you will not need to perform them all for every wood product you construct, you should become familiar with them:

1. *Bleaching* lightens the color of the wood.

2. *Staining* brings out the grain of the wood. It is also done to change the color of the wood.

3. *Filling* is required for some types of wood. Oak, mahogany, and walnut have large cells which form little troughs when the wood is cut. These must be filled with a paste filler to obtain a smooth finish. Birch, cherry, and maple have smaller cells; so they require only a liquid filler. Pine, cedar, and redwood require no filler.

4. *Sealing* is done after staining to keep the stain from bleeding. Sometimes a wash coat of shellac is used (seven parts alcohol to one part of four-pound-cut shellac). If a lacquer finish is to be applied, a lacquer sealer may be used.

5. *Applying the finish.* Lacquer, varnish, or synthetic finish is applied. These may be applied with a brush or sprayed. Usually more than one coat is needed. Each coat should be sanded before the next is applied. Use a 180-grit abrasive paper. Make sure the finish is dry before sanding.

ruined. Many times the finish makes the difference between professional and amateurish work.

Finishes are applied to wood to protect the surface and to change the appearance. Unfinished wood will soak up moisture from the air during humid seasons and dry out during dry seasons. Wood used outside is exposed to rain, sun, wind, snow, and cold. It will

last longer if protected by some kind of finish. Such wood is usually painted, but sometimes it is varnished. Telephone poles, fence posts, railroad ties, and other wood which is used on or in the ground will decay rapidly without some kind of protection. Wood used for these purposes is usually pressure-treated with creosote to help preserve it.

As mentioned earlier, a finish

6. *Rubbing, buffing, and waxing.* These are done after the last coat of finish is applied. The surface may be rubbed with pumice stone or with very fine 380 or 400 wet-or-dry sandpaper. Steel wool may also be used. The surface is rubbed until it is dull. The final step is to apply a coat of good wax and polish.

TEXTURING WOOD

Sanding is a removal process which is usually done to make the wood smooth. However, some types of removal finishing leave the wood rough, or, in other words, give it texture.

One method of texturing wood is by *sandblasting.* Sand under high pressure is forced against the wood surface, cutting away part of the wood fiber. The soft part of the wood is removed, leaving the hard parts higher (or in relief). Large panels for use in homes, offices, and restaurants are sometimes finished in this way.

Another texturing method is called *charring.* A blowtorch is used to burn away the soft fibers and blacken the wood. Rubbing with a wire brush removes loose material and leaves the wood dark brown or black. Great care must be taken not to burn the wood too much and start a fire.

Panels of certain woods, such as walnut, birch, cherry, maple, and mahogany, are sometimes *grooved* to improve their appearance. The panels are passed through machines which have cutters set at the desired intervals. The resulting grooves are about $\frac{1}{16}''$ deep.

Texturing of wood panels is mostly done in the mills. The equipment is too large and costly for home or school shops.

You can purchase wood panels with most finishes already applied. When choosing these materials it is best to consult a reliable dealer.

FOR YOUR SAFETY . . .

Most finishing materials are volatile (they evaporate easily) and flammable (they catch fire easily). Thus you must be careful of the fumes as well as the material itself.

• Always work in a well-ventilated area.

• Wear gloves, goggles, and protective clothing.

• When spray finishing, wear a respirator.

• Do not work near flames or other sources of heat or sparks.

• Always store finishing materials in tightly closed containers. Keep them away from heat, flames, and sparks.

• Rags or papers that have been soaked in finishing materials should be put in a metal can.

QUESTIONS AND ACTIVITIES

1. Define finishing.

2. Name and define the three kinds of finish and give an example of each.

3. List the six basic finishing steps.

4. Why do some woods need filling? Name three types of wood that require a paste filler before a finish is applied. Name three types of wood that require no filler.

5. Why is sealing done? What are two good sealers?

6. What is texturing? Name and describe three methods of texturing wood.

7. Name four safety rules to remember when using finishing materials.

STAINS

Stain improves the appearance of wood by adding color and bringing out the grain. It also helps preserve the wood. Sometimes stain is used to make a cheaper wood look like a more expensive kind.

There are many kinds of stains. Only oil stains and water stains are used in most schools. Before you apply stain to your project, test it on a scrap piece of the same wood to see if it is what you want.

Stains come in many colors and shades. They are usually labeled according to the type of wood they resemble. There are walnut, light oak, dark oak, mahogany, and many others. Oil stain is easier to apply, but water stain is cheaper, has a more even color, and is less likely to fade.

Applying Stain

All sanding should be complete before applying the stain. Wipe the surface of the project and be sure there is no grease or glue left on the wood.

A soft, clean brush is best for applying stain, but a rag or a sponge may be used. Fig. 81-1. Wear gloves to keep your hands clean. As soon as the wood is the color you want, wipe off the excess stain with a clean cloth. Fig. 81-2. It is better to apply two light coats of stain than one heavy coat. Allow the stain to dry for 24 hours before you apply any other finish.

It is easier to stain large surfaces if they are in a horizontal position. Be careful that no drops of stain fall on the work. Begin at the center and work toward the edges. Inside corners and recessed surfaces should be stained first.

Water stain will raise the grain of the wood. Before applying this stain, sponge the surface lightly with water. After the wood is dry, sand lightly with 5/0 sandpaper. This will help the stain flow on evenly.

End grain soaks up stain rapidly and becomes too dark. To prevent this, coat the end grain with solvent just before applying the stain. Coat with turpentine when using oil stain, with water when using water stain.

BLEACHES

Bleaching lightens the color of wood. This is done with chemicals that remove some of the color but do not injure the wood.

Household laundry bleach may be used on light-colored woods. Mix one-half pint of bleach with one gallon of water and apply with a brush or a rag. Let the wood dry, then sand.

For other bleaching, a solution of oxalic acid crystals in hot water can be used—12 ounces of crystals per gallon. Apply with a brush, and let the bleach stay on

81-1. *Applying stain with a brush.*

81-2. *Wipe the excess stain off the wood with a clean, soft rag.*

the wood 10 to 15 minutes. Be careful not to let the solution come in contact with your skin.

After the oxalic acid bleach has been on the wood for the required time, it must be neutralized. To do this, sponge the wood with a solution of three ounces of borax in a gallon of water. Let the wood dry, then sand it.

For large projects it is better to use commercial bleaches. With these it is possible to remove a little color or all of it.

Bleaching solutions are very strong chemicals. Special care should be used when working with them. It is best to wear rubber gloves and an apron.

PENETRATING FINISHES

Penetrating finishes soak into the wood to seal and protect it. They may be applied plain or mixed with color. No other finishing is needed, except for buffing.

Penetrating finishes are applied with a rag. Sealacell®, Minwax®, and oil finish are some penetrating wipe-on finishes commonly used in school shops as well as in cabinet shops.

QUESTIONS AND ACTIVITIES

1. What effects does stain have on wood?

2. What advantages does water stain have over oil stain? What is the main advantage of oil stain?

3. Describe the basic steps in applying stain to a surface.

4. What special treatment should be given to wood before applying a water stain? Why should these things be done to the wood?

5. How do you prevent end grain from becoming too dark when staining wood?

6. What is the purpose of bleaching wood? What are two solutions that can be used to bleach wood?

7. Describe the effects of penetrating finishes. Name two penetrating finishes. What other finishing is needed after applying this type of finish?

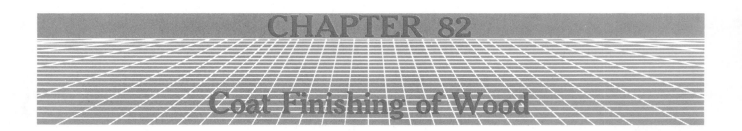

CHAPTER 82
Coat Finishing of Wood

SHELLAC

Shellac is one of the coat finishes applied with a brush. It is a good finish for many projects because it is easy to apply and dries quickly. However, it turns cloudy if it gets wet; so it is not good to use where there is water. It can be used by itself or as a sealer over a stain or filler. Shellac is often used to seal knots before they are painted.

Shellac is made by dissolving lac gum in alcohol. Most lac gum comes from India and Thailand where a tiny insect deposits it on trees. The gum is gathered, dried, purified, and ground.

Common mixtures are 3 or 4 pounds of lac gum to a gallon of alcohol. These are called 3- or 4-pound cuts. Natural shellac is orange. As a result it gives many light-colored finishes an unat-

tractive appearance. White shellac, a bleached form of the orange shellac, is better for general use, though it does not leave as tough a finish.

Keep shellac in a glass container; it will turn brown if kept in metal. Before applying shellac, wipe the project clean with a cloth that has been dipped in alcohol. It is better to apply several thin coats of shellac than one or

82-1. *Dip about one-third of the bristle length into the shellac and wipe off the sides on the container.*

two heavy ones. For the first coat, mix one part shellac with one part alcohol. For the other coats, mix one part shellac with three-fourths part alcohol. Thinned shellac sinks into the surface of the wood better.

A soft bristle brush is best for applying the shellac. Use a clean varnish brush about 1½" to 3" wide. Dip about one-third of the bristle length into the shellac and wipe off the sides of the brush on the top edge of the container. Fig. 82-1.

Begin at the center of a flat surface and work toward the edges. Work quickly using long, light strokes. Don't go over the same area twice. Shellac dries quickly and the brush will stick and leave marks.

It requires about 4 or 5 hours for a coat of shellac to dry completely. Before applying the next coat, go over the surface with steel wool or 5/0 sandpaper. Be sure to rub with the grain. Before applying the shellac, wipe the surface with a clean rag.

Be careful not to apply the shellac too thick. This will make the wood look yellow. Do not let

the shellac build up and run over the edges of the project. After the entire project has been covered, clean the brush in alcohol. The brush may be hung in shellac, as no scum forms on the surface. The container should be tightly covered to prevent the alcohol from evaporating and to keep out dirt.

VARNISH

Varnishes are mixtures of gum resins, vegetable oils, and various thinners and driers. When properly applied, varnish is an excellent finishing material. It produces a hard, bright surface. However, it is difficult to get a good varnish finish in the shop. This is because varnish dries slowly, and the sticky surface of the drying varnish becomes covered with tiny dust particles.

Modern research has eased the problem. Old style natural varnishes took up to 48 hours to dry. Today synthetic varnishes dry overnight and will not collect dust after about two hours. For most work a high gloss or satin finish varnish is used. These are easier to apply than others. For outside finishes where there is moisture, spar varnish is best. This is the kind used on boats. It is also used for tabletops and other pieces that will have hard wear or contact with water.

For the small shop, varnish should be bought in small cans. Scum forms on the surface of open cans and is difficult to remove. If you use the varnish from a can that has a scum, be sure it

is all removed, or it will mar a good finish.

To apply varnish, first find a dust-free place. If you have no finish room, wait until all the machines are turned off. Do not varnish on cold, damp days, as the varnish will not dry properly. The temperature in the room should be between 70 and 80 degrees F.

Wipe the project with a clean rag. Select a 1½" to 2" varnish brush. Stir the varnish (slowly, to avoid air bubbles) and then pour a small amount into another container. For the first coat, thin the varnish with the recommended solvent (check the label).

Dip the brush in the varnish about one-third the length of the bristles. Do not overload it. Apply the varnish with long, easy strokes. Brush first with the grain, then across the grain. When the brush is "dry" continue brushing easily with the grain. You can brush out varnish more than shellac because it dries more slowly. Work from a corner or edge toward the center. Be careful not to let the varnish build up and run over the edges.

When you finish the first coat, soak the brush in a can of turpentine. Cover the varnish can as soon as you are finished so that a scum will not form. Allow the first coat to dry about 24 hours. After it is dry, rub the surface lightly with 6/0 sandpaper. Use your fingers, not a sanding block, for holding the sandpaper. Be especially careful when sanding edges and corners. If you rub too hard, you will remove the varnish and the stain.

Wipe the surface clean before applying the second coat. Do not apply the second coat too soon. Make sure the first coat is completely dry.

The second coat is applied the

same as the first. If it gives good coverage, you will not need to apply any more. Allow it to dry and, if a third coat is needed, sand with 6/0 sandpaper just as after the first coat. The final coat may be rubbed with pumice stone and water. This will produce a dull (satin) finish. Sometimes rottenstone and oil are used for final rubbing. Pumice stone and rottenstone are finely ground abrasive powders. Pumice stone is available in several grades. Rottenstone is very fine.

After the varnish has dried and been rubbed, apply a good paste wax. Polish with a clean cloth. This should provide a professional looking finish that you will be proud of.

LACQUER

Lacquer finishes are available in both clear and colored formulas. Lacquer dries quickly and leaves a hard, durable, protective coating. Fig. 82-2. It is usually made from nitrocellulose, resins, and solvents. Lacquer can be brushed or sprayed on. With either type of application, three to

82-3. Painting wood with a brush. Be careful not to apply too much paint. You should brush with even strokes.

five light coats are needed, and there is no need to sand between coats.

To brush on lacquer, choose a brush with soft bristles, such as camel's hair or sable. A foam polybrush can also be used. Dip one-third of the bristle length into the lacquer. Load the brush heavily and do not wipe any lacquer onto the side of the container when removing the brush. Apply the lacquer with long, rapid strokes, lapping the sides of each stroke. Allow the finish to dry at least 30 minutes between coats. Lacquer thinner should be used for thinning purposes and for cleaning the brushes.

To spray on lacquer with an aerosol can, first place the project on a newspaper-covered bench in a well-ventilated room. Starting at the side nearest you, spray in back and forth strokes, moving toward the back of the project. Be sure to overlap each stroke. After the surface has been entirely coated in this way, turn the project a quarter of a turn and spray the surface again in the same manner. This is one coat; the drying time between coats is the same as for brushed-on lacquer.

After the last coat of lacquer has dried overnight, sand with 10/0 wet-dry sandpaper moistened with water. Then rub the surface with a rubbing compound. Wash the surface with water. When the project is thoroughly dry, polish it with lemon oil or wax.

PAINTING

Painting is a good way to finish certain projects, such as those which do not have an attractive grain. Many pieces of furniture and cabinetwork are painted. Paint seals the surface so that it is not affected by moisture. Paint also makes the surface better looking and easier to keep clean.

Many kinds of paint are available. There are exterior and interior house paints in gloss, semigloss, and flat finishes. There are special paints for floors and for masonry. All these paints are either oil-based or water-based.

American of Martinsville

82-2. Most commercially made furniture has a lacquer finish.

82-4. *Brushes may be cleaned and wrapped in waxed paper for storage.*

The paint to use depends on the job. Salespeople at paint stores can help you select the right type of paint.

Almost all paints have a vehicle, pigment, and driers. The *vehicle* is the liquid part of the paint. This might be linseed oil, turpentine, water, or some other liquid. The *drier* speeds up the drying of the paint. The *pigment* is the solid part of the paint. This provides the color.

Before painting, be sure the surface is clean and smooth. Go over any knots with a light coat of shellac. The first coat of paint may be a primer. The primer should be brushed into the pores of the wood.

To apply paint, brush a small amount at a time onto the wood. Fig. 82-3. (Beginners often apply the paint too thick.) Allow the first coat to dry about 24 hours. Go over the surface with a medium grade sandpaper. Apply the second coat. (In a three-coat process, this is called the undercoat.) If a third coat is to be applied, sand the undercoat with a fine sandpaper. Then apply the third coat, being sure to brush it evenly. On wood which has been painted before, one or two coats of paint are usually enough. On new wood, three coats may be needed.

Enamels are made from polyester plastics, usually with an oil base. Because they do not cover as well as other paints, it is usually best to apply an opaque primer coat. Enamels dry with a gloss finish. They are available in many colors. Enamels are usually used for small projects which need a colored finish. Enamel paint is applied with a brush the same way as other paint. Use a short, even stroke and work in a small area. Enamel should be allowed to dry 24 hours between coats.

BRUSHES

Good quality brushes are important for a good finish. For each type of finish you apply, you should have several brushes ranging from 1″ to 4″ in width. The bristles should have flagged (split) ends. In good-quality brushes, the bristles are set in rubber.

Cleaning and care of brushes is a very important part of finishing. The same solvents that are used to thin the finish are usually used for cleaning brushes. There are also special brush cleaners. For storing overnight, brushes may be placed in a container of solvent. Brushes should always be suspended, not made to stand on the bristles. When they are going to be stored longer than overnight, they should be cleaned in solvent, then washed with soap and water. (Brushes used in water-based paint can be cleaned with soap and water alone.) Then they should be rinsed, dried, and wrapped in waxed paper. This way brushes may be stored for long periods. Fig. 82-4.

AIRLESS SPRAYING

This is a method of spraying finish on a surface without using compressed air in the spray gun. Pressure is applied to the liquid in its container, such as a five-gallon paint can. This pressure forces the finish through the connecting hose to the spray gun, then out a slot which produces a fan-shaped spray pattern. Fig. 82-5. The slot may be turned for vertical or horizontal spray. In this method all the finishing solution settles on the object being sprayed. Paint, lacquer, and other synthetic liquid can be applied this way.

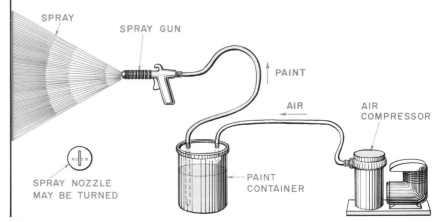

SPRAY

SPRAY GUN

PAINT

AIR

AIR COMPRESSOR

SPRAY NOZZLE MAY BE TURNED

PAINT CONTAINER

82-5. *Airless spraying. Compressed air forces paint from the container through the hose to the spray gun. The spray gun breaks the paint into a mist for painting.*

QUESTIONS AND ACTIVITIES

1. What are the advantages of using shellac as a finish? The disadvantages?

2. What must be done to the project before applying shellac? What type and size brush should be used? What type of container should shellac be kept in? Why?

3. Briefly describe how to apply one coat of shellac.

4. What are the advantages of varnish as a finish? The disadvantages?

5. Briefly describe how to apply a coat of varnish.

6. Describe how to brush on lacquer. How many coats are needed?

7. What three things do most paints consist of? Tell a little about each.

8. How do you treat knots that are to be painted?

9. How should brushes that are to be stored for long periods of time be cared for?

10. Describe airless spraying.

PART 4

Production: Manufacturing and Construction

CHAPTER 83

Introduction to Building Construction

Construction has always been an important part of life. Early human beings lived in caves to obtain shelter from the wind, rain, and snow. The first construction may have been digging to improve the cave. Later, people probably began to build shelters using other materials, such as wood, stone, and earth. As their knowledge increased and they developed better tools, they began to build larger, more comfortable structures. Eventually people learned to use natural materials to build roads, aqueducts (passageways for water), temples, and other structures as well as homes. Not only did construction serve the practical and spiritual needs of people; it helped shape civilization. Think, for example, how different history would be if people had never learned to build roads.

Today construction is a very important part of industry in the

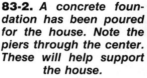 

83-2. A concrete foundation has been poured for the house. Note the piers through the center. These will help support the house.

United States. Millions of people make their living working in some part of the construction industry. The total value of buildings constructed in any one year amounts to billions of dollars. The building industry plays such a major role in our economy that it is considered an important indicator of the nation's economic state. A slump in the construction industry usually means a slowdown in the economy. An increase in new construction is considered a sign of economic recovery. Fig. 83-1.

The construction industry differs from the usual manufacturing industry in the location where the work is performed. In most manufacturing industries there is a factory to which materials are brought. The workers come to the factory to make something, such as an automobile or a radio. In the construction industry the workers travel to the location of the work. As

soon as one building is completed, they must go to a new site where another building is being constructed.

In the future more and more of the elements of construction may be done away from the site. Even today roof trusses, pipes for plumbing, and some wall sections are built or assembled at a central point or factory. They are then taken to the construction site and incorporated into the building. Sometimes entire buildings are made up of sections constructed in a factory. The sections are taken to the building site and assembled. Very little additional work is needed. Figs. 83-2 through 83-4.

As the population has increased, there has been an increase in the need for new construction. To meet the needs of people for shelter, millions of new homes and apartments have been built. Millions of older homes have been remodeled to

83-1. Construction is a major American industry.

83-3. *A crane is used to unload a section of the house. It is carried to the site on wheels and axles similar to those used for mobile homes.*

lions of people working in the industry itself and in the industries which supply goods for construction. Within the construction industry there are over two dozen skilled trades, such as carpenter, plumber, electrician, bricklayer, painter, and others. These workers follow the plans developed by architects, engineers, and interior designers. They use materials provided by the wood, metals, and plastics industries. Altogether, the construction industry directly or indirectly provides jobs for about 15 percent of our working population.

modernize and expand living space. Office buildings and stores have been newly built or remodeled to meet the needs of business and industry. New roads, bridges, dams, and other structures have also been built. The following three chapters will briefly describe the processes involved in building homes and small commercial structures. Heavy construction will be discussed in Chapter 87.

CAREERS IN CONSTRUCTION

The construction industry requires the skills and labor of mil-

83-4. *The completed house. It looks like a conventional site-built home, but it is less expensive and can be put up in less than a week.*

QUESTIONS AND ACTIVITIES

1. Describe the role of the construction industry in our nation's economy.

2. How does the construction industry differ from the usual manufacturing industry? In what way is this difference becoming less pronounced?

3. Discuss some advantages of using some preconstructed elements in building a house or other structure.

4. Name four skilled trades in the construction industry.

CHAPTER 84

Before a new home or other building is constructed, someone must want it. Someone must decide to build a particular building on a certain lot.

CHOOSING THE BUILDING SITE

There are several factors to consider in choosing a lot:

• Location. Is the neighborhood suitable? How close are the facilities that will be needed, such as transportation, schools, churches, or stores?

• Utilities. Are there water, electricity, and gas lines? Sewers? Telephone lines?

• Shape and contour of the lot. Is it big enough? Is it flat or does it curve up and down so that a special building design will be needed?

• Zoning. Communities are usually divided into zones. Each zone allows only certain types of buildings. Most communities have zones for single-family dwellings, multiple-family dwellings, apartments and condominiums, and for light commercial, heavy commercial, light industrial, and heavy industrial development.

• Deed restrictions. Within each zone there are usually specific rules about a building's size and its placement on the lot.

• Legal documents. There should be a *survey* showing the boundaries of the property. There must be a *deed,* which is the evidence of ownership. There must

be an *abstract of title,* a history of the deeds and other documents pertaining to the lot's ownership. These legal documents are needed to obtain a loan and a *building permit* (a paper from the city or county giving permission to construct the building). Real estate agents often help a buyer obtain these documents. Many buyers hire lawyers to examine the documents and make sure everything is in order. Fig. 84-1.

SELECTING THE BUILDING PLANS

When choosing a building design, the following must be considered:

• Local building codes. The building codes specify standards for types of materials and construction techniques. In addition to the community's building

codes, the lending institutions list standards which a building must meet in order to qualify for a mortgage loan.

• The purpose of the building. Perhaps the building will be a family dwelling. If so, how many people will be living in the house? What is their life style? A retired couple may want a single-story home that is small and easy to maintain. A young family may want a home with room for expansion.

• Costs. Some buildings are much more expensive to build than others. The owner usually tries to strike a balance between what is desirable and what is affordable.

Besides building costs, other costs must be taken into consideration when choosing a design. Maintenance and real estate

James L. Shaffer

84-1. *Before you buy property, it's a good idea to have a lawyer check the legal documents.*

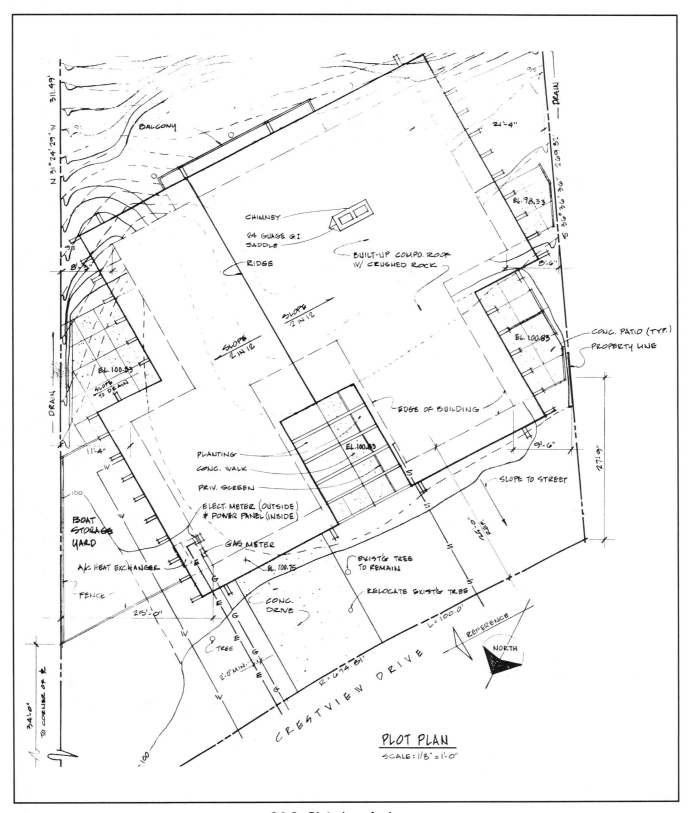

84-2. *Plot plan of a house.*

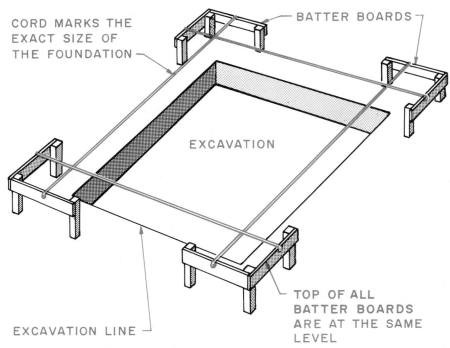

CORD MARKS THE EXACT SIZE OF THE FOUNDATION

BATTER BOARDS

EXCAVATION

TOP OF ALL BATTER BOARDS ARE AT THE SAME LEVEL

EXCAVATION LINE

84-3. Batter boards for laying out building.

taxes are two examples. Energy costs are another. Today many buildings are being designed to use less energy for heating and cooling. Energy-efficient buildings may have such features as extra insulation, energy-saving sheathing, storm windows and doors, weather stripping around doors and windows, caulking, attic ventilation, and insulated duct work. The size and placement of windows also affect energy costs. For example, some buildings in the northern states have a sun porch that works like a greenhouse to trap the sun's heat.

There are various ways to obtain plans for a building. The least expensive is to buy stock plans from a company which specializes in them. Local building contractors may also have plans, which they can alter somewhat to suit their customers' needs. A building design that is completely original requires an architect.

CONTRACTORS AND FINANCING

Once a lot has been bought and a building plan chosen, a contractor must be hired. The usual practice is to ask several contractors to make bids (price estimates) on the job. The plans should be thoroughly discussed with the contractors, and prices and schedules determined.

After a contractor has been chosen from among those who made bids, financing must be arranged. Savings and loan institutions are the most common source for loans to finance building projects.

Once financing has been obtained, a contract (legal agreement) with the contractor can be signed. The contractor will now be responsible for getting the

84-4. Position of foundation in relation to batter boards and cord.

building constructed. The contractor obtains the building permit, has the lot cleared and graded, and hires the various workers who will do the actual construction. Often, the contractor specializes in one type of work, such as carpentry. In such cases, the contractor's employees will do the framing and other carpentry. Other kinds of work—masonry, electrical, plumbing, etc.—will be subcontracted out to companies specializing in those areas.

LAYING OUT THE BUILDING

After the site is prepared, the building must be laid out (outlined) on the ground. The *plot plan* shows the building in relation to the property. Fig. 84-2 (p. 311). The actual layout of the building is done by using *batter boards*. These are usually 2″ × 4″ stakes driven into the ground with a board fastened to the top of them. Batter boards are located near each corner of the building. A nail is driven into the top of the boards and a cord is stretched between nails. The cord outlines the size and shape of the building. Fig. 84-3.

As you can see in Fig. 84-4, the batter boards themselves are placed outside the construction area in order to allow room for construction machinery to operate without knocking them over.

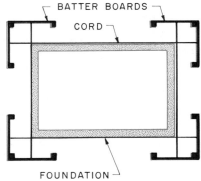

BATTER BOARDS

CORD

FOUNDATION

QUESTIONS AND ACTIVITIES

1. What are six factors to consider when choosing a building site?

2. What is the difference between a deed and an abstract of title?

3. What are three things to consider when choosing a building design?

4. Describe how to choose a contractor to construct a building.

5. Describe how the actual layout of a building is done.

CHAPTER 85

Foundations, Framing, and Roofing

FOUNDATIONS

Once the building has been laid out, the construction can begin. If the building is to have a basement, this must be dug. Usually this is done with large earth-moving machinery. If the building is to be built on a slab, the land must be leveled.

Next the foundation is built. The foundation supports the weight of the building. Foundations are usually made from concrete. (Some are constructed of wood on a gravel base.)

For firm support, foundations must have enlarged bases, called *footings*. Local codes specify type and size of footings, as well as their depth. In cold climates, footings must be below the frost line. Footings are usually poured concrete.

The foundation wall may also be poured concrete, or it may be built of concrete blocks. Poured walls require *forms*. These are usually built on the site from wood. Fig. 85-1. The height of the foundation wall depends on whether the house will be built over a basement (in which case the foundation walls become the basement walls), a crawl space, or simply a concrete floor slab. Fig. 85-2. Some slab houses have

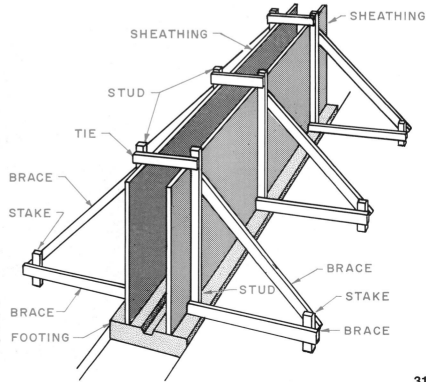

85-1. *Foundation wall form. The concrete will be poured into this form and allowed to harden. Then the form will be removed. (Note the footing at the bottom.)*

SHEATHING

SHEATHING

STUD

TIE

BRACE

STAKE

BRACE

FOOTING

BRACE

STUD

STAKE

BRACE

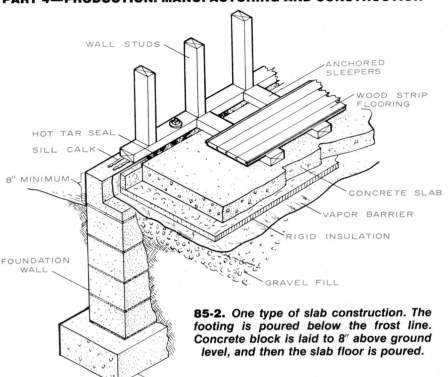

85-2. *One type of slab construction. The footing is poured below the frost line. Concrete block is laid to 8" above ground level, and then the slab floor is poured.*

85-4. *Slab foundation with plumbing in place.*

a unified slab and foundation; the footing and the floor slab are one poured-concrete unit. Fig. 85-3.

If the house is to be built over a basement or crawl space, the utilities, such as plumbing and electrical equipment, can be placed in position after the foundation is completed. If the house is to be built on a slab, such as shown in Fig. 85-4, the utilities must be placed before the foundation is poured.

If the foundation is poured, the concrete must be allowed to set at least 2 days before the forms are removed. The concrete must then be allowed to cure another 14 to 25 days before the rest of the construction may begin. A concrete block foundation wall must be allowed to cure at least 24 hours before further construction is done.

FRAMING

Generally, wooden frames are used for small buildings, such as houses. Usually the framing for these buildings is divided into three major parts. These are the floor framing, the wall framing, and the roof framing. Floor framing, where wooden floors are utilized, is made up mainly of horizontal members called *joists*. (Joists are also used for ceiling framing.) Most wall framing is made up of vertical members called *studs,* and most roof framing is made up of sloped members called *rafters.* Fig. 85-5.

To begin framing a house, a piece of lumber called a *sill* or sill plate is laid down on top of the

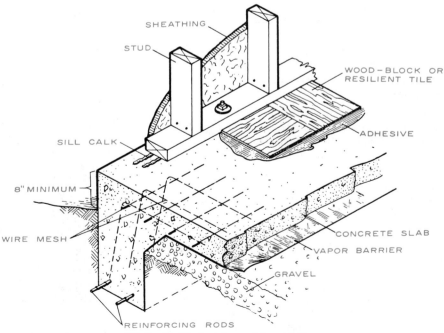

85-3. *Unified slab and foundation.*

foundation. The sill is held to the foundation by *anchor bolts* which were placed in the concrete every 5 or 6 feet while the concrete was moist. The threaded end of the bolts points up. Holes are marked and drilled in the sill so that the sill can be placed down over the anchor bolts. Fig. 85-6. After the sill is leveled, it is held by tightening the anchor bolts.

In some areas of the United States, termites (wood-eating insects) are a problem. Termites can eat their way through framing and make tunnels which weaken the structure. In areas where there are termite problems, a *termite shield* is placed

under the sill. The shield is a metal strip which stretches along the full length of the sill. It is wider than the sill and therefore sticks out on each side. This shield helps keep the ground-dwelling termites from burrowing through holes or cracks in the foundation and into the wood frame.

Floor Framing

The floor framing supports the load on the floor and also gives support to the walls. The floor framing in houses with basements usually includes sills, posts, girders, joists, and subflooring. Fig. 85-7. *Posts* are

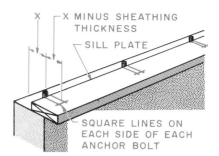

85-6. *Marking the sill for anchor bolts.*

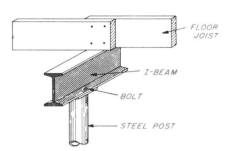

85-7. *Floor framing members.*

upright wood or steel members which support girders. *Girders* are horizontal wood or steel members which support the floor joists. A typical house will have a single steel I-beam running the length of the building, through the center. *Floor joists* are planks set on edge. One end rests on the sill and the other on the girder. Generally the floor joists are spaced 16″ on center. That is, the center of one joist is 16″ from the center of the next. Fig. 85-8. At the sill, the ends of the joists are usually held in place by nailing them to a header. A *header* is a piece of lumber with the same dimensions as the joist. It is set flush with the outside of the sill. To keep the centers of the joists from twisting and turning, *bridging* is used between the joists. Fig. 85-8. Bridging is made from 1″ × 4″ boards or pieces of metal.

A subfloor is nailed to the top of the joists. The *subfloor* may consist of boards laid diagonally

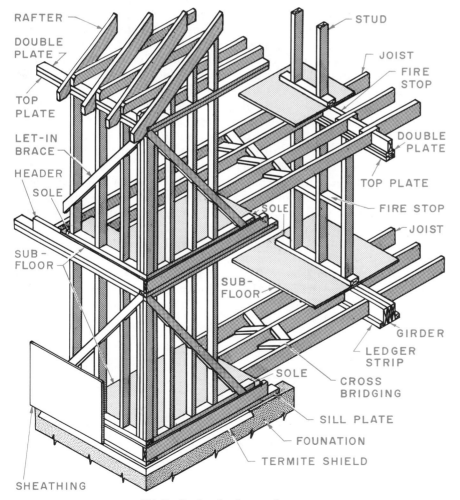

85-5. *Parts of a house frame.*

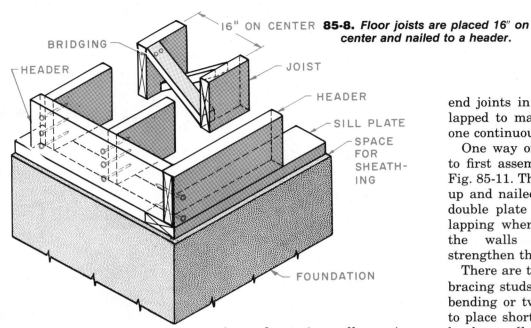

16" ON CENTER **85-8.** *Floor joists are placed 16" on center and nailed to a header.*

BRIDGING

HEADER

JOIST

HEADER

SILL PLATE

SPACE FOR SHEATHING

FOUNDATION

over the joists or of plywood panels. Later, a second floor, the "finish floor" will be laid over the subfloor.

Houses built on a concrete slab do not need floor joists. Instead the flooring is installed over wood "sleepers" which lie flat on the concrete slab. Fig. 85-2. Some types of flooring, such as asphalt tile, can be installed directly on the slab.

Some homes today are being built with prefabricated floor trusses instead of joists. Floor trusses are longer than joists, and they don't require bridging. Since the trusses are put together in a factory, they can be quickly installed at the building site. The "open" type of floor truss construction allows ductwork, electrical wiring, and pipes to run through the trusses. The basement ceiling can then be nailed to the bottoms of the trusses. Fig. 85-9.

Wall Framing

Wall framing supports the roof and serves as a base for the inte-

rior and exterior wall coverings. Fig. 85-10. The main framing members are the *studs,* upright pieces of nominal 2" × 4" lumber which are usually spaced 16" on center. The bottoms of the studs are nailed to a *sole plate.* This is a piece of 2" × 4" lumber laid flat and nailed to the subflooring (or fastened to the concrete slab) exactly where the walls are to be. Exterior walls are placed over the foundation. The interior (partition) walls are usually over an inside foundation wall or over a girder. Across the tops of the studs, a *top plate* of 2" × 4" lumber is nailed. A second top plate (called a double plate or rafter plate) is nailed over the first. The

end joints in the two plates are lapped to make them more like one continuous piece.

One way of building a wall is to first assemble it on the floor. Fig. 85-11. The wall is then tilted up and nailed into position. The double plate is nailed on, overlapping where walls join, to tie the walls together and to strengthen them.

There are two common ways of bracing studs to keep them from bending or twisting. One way is to place short lengths of 2" × 4" lumber, called fire blocks, horizontally between the studs. These *fire blocks* serve as braces as well as draft stops to prevent the spread of fire in a building. These blocks are nailed in line, or they may be staggered above and below a line for ease in nailing. Another common way to brace studs is to use diagonal *let-in braces,* usually of 1" × 4" lumber. A brace is set to run from one top corner of the frame down to some point on the sole plate at an angle of about 45°. This makes a triangular frame within the wall section and makes the frame rigid and solid. In the studs, notches are cut to the thickness of the diagonal brace so that they do not interfere with

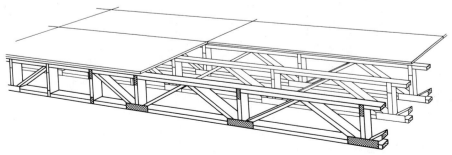

85-9. *One type of open truss. (Closed trusses have a solid center instead of webbing.)*

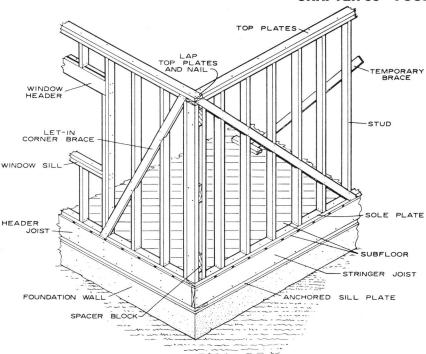

85-10. *Typical wall framing.*

85-12. *Assembled wall sections ready to be nailed in place. Note the let-in brace at the center of the picture. The other braces are temporary.*

the installation of the sheathing material. Fig. 85-12. (Sheathing usually consists of boards or plywood nailed to the outside of the frame. The finish siding will be applied to it.)

Extra framing is needed around the openings for doors, windows, heating ducts, plumbing pipes, and other openings. Fig. 85-10.

After one story of the frame is completed, a second story may be built in the same way. Joists and headers are placed on the double plate of the first story and covered with a subflooring. Walls are built on top of this to form the second story.

CEILING AND ROOF FRAMING

Most wood frame buildings have sloped roofs. The top of the slope is called the *ridge*. The bottom of the slope forms the *eaves* (overhang).

One method of roof framing uses rafters. *Rafters* are inclined members that support the roof loads. The rafters rest on the top plates and meet at the ridge. Where the lower end of the rafter hits the top plates, a notch (called a bird's-mouth) is cut in the rafter so that there will be a flat edge resting on the plate. Usually the rafter goes a foot or more past the face of the wall. This holds the roofing that goes beyond the outside wall to form the eaves. Rafters are set in pairs, one sloping down each side of the roof. Fig. 85-13.

When rafters are to be used *ceiling joists* are required. These are horizontal members to which the ceiling is fastened. They also serve as floor joists for an attic or second story. The ceiling joists are supported by beams, girders, or bearing walls.

Trusses may be used in place of conventional rafters. These usually combine the rafters and ceiling joists into one triangular piece. Fig. 85-14. Trusses may be

85-11. *Wall sections being constructed on the floor.*

317

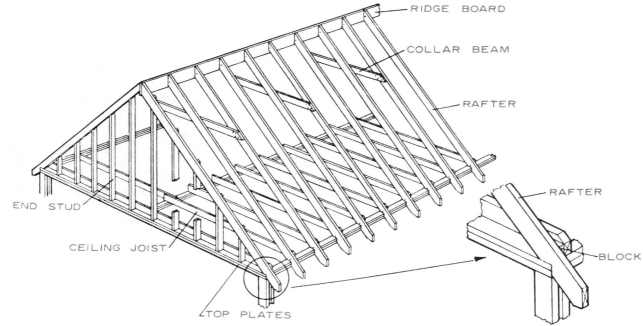

85-13. *Roof framing using rafters, a ridge board, and joists.*

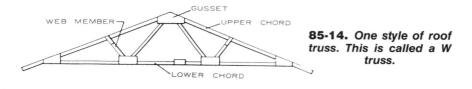

85-14. *One style of roof truss. This is called a W truss.*

made on the site, or they may be made in a factory.

Roof frames are covered with *roof decking,* which serves the same purpose as sheathing and subflooring. The roof decking consists of boards or sheet material nailed over the framing.

Usually the decking completely covers the framing. Sometimes, when wood shingles are used, spaces are left between roof decking to save material and to allow better air circulation to dry the shingles after wet weather.

Usually, an underlayment is placed over the decking. (This may be omitted when wood shingles are used.) Flashing—usually of sheet metal—is installed around chimneys, soil stacks, and other areas to prevent water seepage.

Various materials are used for the final covering. Shingles of wood or asphalt are common. So are tile and slate. Sometimes sheet materials are used. Flat roofs are often covered with layers of asphalt-saturated felt, mopped with hot asphalt or tar, and topped with gravel.

QUESTIONS AND ACTIVITIES

1. What is the function of a foundation? Describe what supports the foundation. In what two ways are most foundation walls constructed?

2. Describe a sill plate and tell how it is held in place.

3. Describe a termite shield and its purpose.

4. Name and describe the main elements of floor framing.

5. Name and describe the main members of wall framing.

6. Name and describe two common ways of bracing studs.

7. Name and describe the main members of roof framing.

8. Describe the following: header, roof truss, eaves, bridging.

During the construction of a house, the utilities must be considered. Utilities include heating pipes or ducts, lines for electricity and gas, and pipes for water and sewage. Even before the foundation or slab is constructed, outside utilities must be installed. The water pipe must be placed from the city main to the house. The gas line from the main to the house must also be installed. The soil pipe to carry the waste from the house to the sewer must be put in. Once all these utilities are "roughed in," the foundation or slab is poured.

HEATING

The utilities in the building itself are installed as the house is being framed. One of the very important utilities is heat. Before installing the heating system, it must be decided what type would be best for the location and the type of house.

There are several ways of heating a building. One uses steam or hot water going through pipes and radiators. Another uses hot air forced through sheet metal ducts. Whatever type of heating system is selected, the piping or ductwork must go in as the house is being framed and before the outside sheathing goes on or the interior walls are finished.

Another heating method uses electricity in resistance wires located in the ceiling. This method is very clean and noiseless. It also eliminates the need for pipes or ductwork. However, the building needs to be well insulated if electric heat is used, as the heating is more gradual than with other methods.

More and more homes and other buildings are using solar energy to provide at least part of the heat. See Chapter 98 for a discussion of solar heating.

PLUMBING

Once the heating pipes or ducts have been installed, the plumbers begin to install the water and sewage pipes. A plumber generally runs the soil pipe, which is the large drain pipe, and the smaller drains first. To run soil pipe, one has to start at the sanitary sewer connection, which is just inside the foundation wall. All drain lines have to slope toward the sanitary sewer. These lines run to the fixtures and continue up through the roof where they act as vents.

Vents allow the fixtures to drain properly and prevent the siphoning of water up from the traps under each fixture. If there were no vents, the effect would be the same as turning a bottle full of liquid upside down. The liquid just gurgles out because air can't get into the bottle easily. The same principle applies to a drainage line.

A trap is installed under each fixture to keep sewer gas and odors from entering the room.

Water lines generally start from the water meter, which also is just inside the foundation wall, and run to all the fixtures. To prevent accidental burns, the hot water is always installed on the left side, as you face the fixture. The hot water line starts from the water heater and runs to all the fixtures where hot water is needed. The outlets of water lines are generally shut off with test plugs or nipples and caps until the inside walls and floors are completed. They are removed when the fixtures are installed. Fig. 86-1.

Permanent fixtures such as bathtubs and shower bases are installed at the time the roughing in takes place because they must be built into the walls.

ELECTRICITY

The electrician generally hooks up the service to the house very

86-1. *The plumbing pipes have been installed and capped, ready for the finish plumbing.*

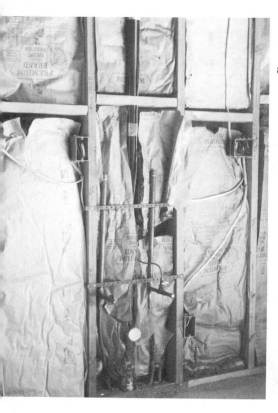

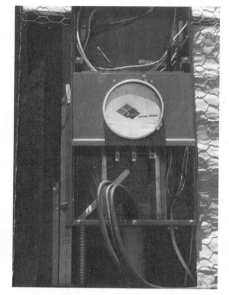

86-2. *Wiring roughed in, along with some plumbing.*

86-3. *Main power panel being wired in. Cables run to outlets and to underground utilities.*

early in the construction so that other people working on the house can have electric power.

After the heating and plumbing have been roughed in, the electrician will locate the number of outlets as set by specifications and attach the outlet boxes to the walls. He or she will then drill all the holes (through the framing) needed to run the wires to these boxes. Fig. 86-2. The wires are attached to the boxes and left coiled up in the boxes. The feed end is attached to the service panel box, but it is not hooked up to the terminals. This leaves the circuit dead, and no electricity flows through it until the time comes to finish the installation. Fig. 86-3.

There are many other appliances and fixtures that may have to be roughed in. Examples of these are communications lines and ductwork for kitchen fans.

Today many areas specify that all utilities be underground. Communication lines, such as telephone and electrical cables, are run through plastic conduit buried underground. Figure 86-4 shows conduit through which power cables will be run.

FINISHING

Once the utilities are in, the finishing operations may be com-

86-4. *Underground power cables will be run through these plastic pipes. The tall metal object in the illustration is the terminal panel for telephone cables.*

pleted. In many areas of the country, insulation is installed in the outside walls (the walls that enclose the house). Insulation comes in many forms. One common form is blanket insulation, which comes in rolls. It is easy to cut to length and staple to the framing. Fig. 86-5. Besides outside walls, insulation should also be placed in the ceiling and in any floors that are above unheated areas, such as crawl spaces.

Inside the house, the walls and ceilings are covered with a coating of plaster or with dry wall. When plaster is to be applied to the interior, lath must be placed over the studs to hold the plaster. Lath may be made of gypsum, wire mesh, or other materials. In the past most walls were of lath and plaster.

In many homes today, sheets of dry wall, such as gypsum board, are nailed over the studs. The nail holes and joints are covered with a plasterlike material to make a smooth wall. Holes are cut in the dry wall for electrical outlets and heating ducts. Figs. 86-6 & 86-7.

Other parts of the interior are then finished. Doors are hung

86-6. *Holes in dry wall for electrical and communication outlets.*

86-5. *Insulation stapled to an outside wall along the stairway.*

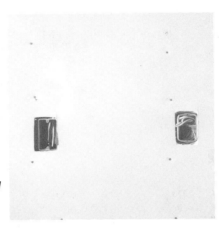

wallcovering. The cabinet work is completed in the kitchen and bathrooms, and the faucets and other plumbing fixtures are installed. The light fixtures, outlets, switches, and covering plates are installed. When all this work has been completed, the house is ready for the owner to move in.

LANDSCAPING

The grounds around the house must be made attractive and useful. This is called landscaping.

First the ground is graded to make it slope away from the house. This is done to make sure that rainwater will drain away from the house. Next topsoil is spread. To make a lawn, grass seed is sown or strips of sod laid. Usually, trees and shrubbery are planted to make the lot more attractive and to provide shade and privacy.

RENOVATING AND REMODELING

With the high price of new homes, many people are finding it more economical to buy an older home and renovate or re-

and the door casings constructed. The trim is placed around the windows and the windows put in place. The finish floor is installed. This might be hardwood, which is nailed to the subfloor, or asphalt or vinyl tile, which is put down with an adhesive. Carpeting may also be installed.

At the same time the interior is being finished, the outside is receiving its finishing material. There are several ways of finishing the exterior of a house. Wood siding called clapboard or lapped board siding may be used. These are usually boards 6″ to 10″ wide placed horizontally around the house and lapped over one another to shed water. Another method is to use 4′ × 8′ textured plywood panels. This method of siding a house is fast, and when the panels are painted, they present a pleasing appearance.

Another exterior finish is aluminum or plastic siding. It has the advantage of not needing to be painted.

Still another method of finishing exteriors uses a masonry material called stucco. *Stucco* is usually made from portland cement, sand, and lime. It is applied over

a felt paper and chicken wire reinforcement. Fig. 86-8. A stucco finish has the appearance of rough cement. Fig. 86-9.

The final work on a house consists of finishing the interior. The walls and ceilings are painted the desired colors or covered with

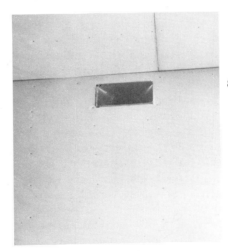

86-8. *House partially wrapped for stucco. Panels of felt paper and chicken wire are fastened to the studs, and stucco is applied to the outside like plaster.*

86-7. *Hole in dry wall for heating duct.*

86-9. *Finish coat of stucco. House may be painted any color.*

model it. To *renovate* a house means to restore it to its former state by cleaning, repairing, and/or rebuilding. To *remodel* a house means to change its structure, as when adding a room or modernizing a kitchen.

Some older homes are a true bargain; others are so run down that they are not worth repairing. Here are some things to check before buying a "handyman's special."

• The surroundings. Are other homes in the neighborhood in good condition? If not, are they being repaired, or are they being allowed to deteriorate?

• The lot. Trees are good for shade and as windbreaks, but a house surrounded by many trees and bushes may be constantly damp. The dampness is not only unpleasant; it can lead to damage. Damp wood is prone to dry rot (a decay caused by fungi). Moisture also encourages the growth of mildew and causes paint to blister and peel. Another source of moisture problems is damp or wet earth. Check to make sure the ground slopes away from the house so that rainwater does not run towards the foundation.

• The foundation. Check for cracks and dampness. Large cracks in the foundation indicate a major structural problem, but even small cracks can be a source of water leaks. If there's a basement, look for evidence of flooding. In basements without a closed ceiling, check the framing. Joists, beams, etc., should be straight and solid. You can check for dry rot and insect damage by poking an awl into the wood. Decayed or damaged wood will be soft and crumbly. It's a good idea to have a professional exterminator inspect the house for termites, carpenter ants, and other insects that damage wood.

• The exterior walls. Problems with the walls can indicate uneven settling of the foundation. If the house has horizontal siding, stand at each corner and sight down the lines of the siding. A dip towards one end indicates a sinking corner. Check masonry walls for crumbling mortar and for bulging. Also check stucco walls for bulges and cracks. Another potential problem: paint peeling. It could just be age or two coats of incompatible paint, but it could also mean that moisture from inside the house is moving out through the walls because of inadequate vapor barriers. If the paint is only blistering or peeling outside the kitchen or bathroom, the problem is probably caused by moisture.

• The roof. Look at the ridge. If it is not straight, the foundation may have settled unevenly, and you should look for cracks in the foundation. Check shingles for curling and breaking. Examine the chimney for loose mortar, broken bricks, or damaged flashing. Inspect the attic. Stains on the rafters could indicate leaks in the roof or poor ventilation. Check the framing members for dry rot. Also note the amount and type of insulation. Check eaves and soffits for leaks and damage, and don't forget the gutters and downspouts.

• Electrical wiring. There should be three wires entering the house. This indicates 240-volt service. The old two-wire, 120-volt system is not enough for today's needs. Inside the house, check the fuse box. A lot of 20- and 30-ampere fuses indicates those circuits are being overloaded. In each room of the house, note the number and location of outlets. Use a plug-in receptacle analyzer to check whether outlets work and are properly grounded.

• The plumbing. Turn on the faucets; first the hot, then the cold. Flush the toilets. Inspect the plumbing for leaks, clogs, and rust. Check the water heater for rust on the outside and for water leaks or stains underneath. Make sure the water heater has a pressure-relief valve.

• The heating system. Turn on the furnace to make sure it works. On an oil-fired burner, look for leaks and rust stains on the floor. Heavy soot inside the burner is a sign of poor combustion. In a forced-air system, *after the burners light,* hold a lighted match near the flue. The flame should bend up into the flue, indicating good draw. When the blower starts, check the flames.

They should be blue, not yellow. Wobbling or slanting flames indicate a cracked heat exchanger. Such a condition allows poisonous fumes to get into the house. If there is a fireplace, examine the flue to check the condition of the liner. Also measure the flue. Fireplaces built for coal or gas have flues that are too small for wood fires, which require a flue at least 8″ × 12″.

● Interior walls. Cracks and holes in plaster can indicate problems in the wall framing.

● Floors. Drop a marble on the floor several times. If it always rolls in one direction, the floor is not level. This can be a sign of rot in the floor framing.

● Windows and doors. These should fit snugly but open and close easily. Wood-framed windows are more energy-efficient than aluminum windows. In cold climates, there should be storm windows and doors.

Even severe problems with a house can often be repaired, but the cost may exceed the benefits. By carefully inspecting a house before buying it, you can determine what needs to be done and obtain estimates of the costs. Then you can decide whether to buy or to keep looking.

QUESTIONS AND ACTIVITIES

1. Explain why vents are needed in plumbing.

2. What are two ways of finishing inside walls? What are three types of finishing materials for the exterior of a house?

3. Describe stucco.

4. Describe how the grounds should be landscaped.

5. What is the difference between renovating and remodeling a house?

6. Describe how to check the foundation of a house before buying it.

7. Describe how to check a house for possible problems with the roof.

CHAPTER 87

Heavy Construction

Homes, small apartments, and small shopping centers are usually constructed of wood or brick or a combination of wood, brick, and stucco. Fig. 87-1. For large buildings, bridges, dams, and other large structures, stronger construction is needed. Because such structures must withstand

87-1. *Construction of a small shopping center. Note the large wooden beams spanning the openings for each store.*

NASA

NASA

Construction in Space

One of the purposes of the Space Shuttle is to build large structures in space—communications antennas, solar power collectors, and space platforms. Some of these huge structures will fold into compact containers on Earth, travel in the Shuttle cargo bay, and then unfold (or deploy) in space. One hoop-column type would open up much like an umbrella does. A cylinder no larger than a school bus could be deployed in one hour to form a giant antenna 100 metres across. This would cover an area of nearly 0.78 hectare, or about two acres. Other structures will be sent up in separate pieces, unfolded, and then erected (or assembled) into a single gigantic structure in orbit. Deployable structures will, in a sense, build themselves. They will unfold with the push of a button. Erectables will not. Someone or something will have to snap the separate pieces together.

87-2. *These workers are installing steel reinforcing bars in a nuclear power plant. Later, concrete will be poured over the bars to form the floor and walls.*

Southern California Edison

tremendous forces, the construction is usually done using steel and concrete. Fig. 87-2. Freeway or interstate highway bridges are good examples of construction using steel reinforcing rods and concrete. Fig. 87-3.

Large office buildings are constructed using a steel frame. The steel beams are either bolted or welded together to form the framework. The walls and windows are then bolted or welded to the frame. Fig. 87-4.

Sometimes buildings are constructed using poured concrete pillars. First wooden forms for the pillars are built. Reinforcing steel is placed in the forms. Electrical conduit and plumbing pipes are also placed in the forms. Concrete is then poured

87-3. *Bridge construction for a freeway.*

87-4. *The crane is lifting steel beams into place for the framing of a high-rise office building. The steel is placed in position by steelworkers and either welded by welders or bolted by the steelworkers.*

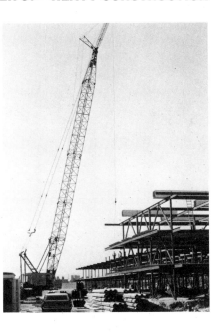

87-5. *A hotel under construction. Note the steel studs, plumbing pipes, and placement of electrical conduit.*

into the forms. After the concrete has set, the forms are removed. Once the concrete pillars and floors are completed, metal studs are fastened in place. Wherever necessary, the electrical conduit and plumbing pipes are fastened to the metal studs in ways similar to wood construction. Fig. 87-5.

QUESTIONS AND ACTIVITIES

1. What materials are used to build large structures that must withstand tremendous forces?

2. How are concrete pillars constructed?

3. What is the difference between deployable structures and erectable structures used in the space program?

PART 5

Power and Energy

Section 1—Electricity

CHAPTER 88

Introduction to Electricity

This is an age of electricity. Without it the efficient factories, the comfortable homes, and the swift transportation and communication of this era could not exist. Fig. 88-1. Electricity is almost always available when you want it. A switch can turn it off and on just as water can be turned off and on with a faucet. Electricity can be changed into other forms of energy. There are appliances to turn it into heat, sound, motion, and light. These appliances include toasters, stoves, bells, motors, electric fans, and light bulbs.

Electricity is as important in industry as it is in the home. Electric motors provide power for almost all industries. For example, changing electrical energy to heat for welding metal is a very important part of industry.

The generation (production) of electrical energy is an industry in itself. With a few exceptions, most of the electrical generators are turned by steam turbines. The steam is usually produced by heat supplied by coal, oil, or atomic power. Today, with the emphasis on conservation of resources, engineers are looking for other heat sources. One possible source is heat from inside the earth. This is called *geothermal steam*.

Electricity is used for transportation such as the diesel electric trams. A diesel engine powers a generator which produces electricity. This electricity is used to run powerful electric motors attached to the wheels of the engine. These electric motors actually provide the propulsion power for the engine.

Electricity in one form or another is used in information technology. The radio, television, telephone, telegraph, and computer all depend on electricity. Thus you can see that it is a very important part of our life.

CAREERS

There are hundreds of jobs related to electricity. Perhaps you will become interested in one of them. There are electrical engineers and technicians who design electrical devices, from television sets to toasters. There are many craftspeople who build these devices. You may know an electrician who installs wiring in new houses or rewires old ones. There are also electricians and electronic technicians who service electrical devices.

The radio and television industries employ many electricians. Communications companies, like telephone companies, employ thousands of men and women to work with electrical devices.

THE NATURE OF ELECTRICITY

To understand electricity you need to know something about atoms, because atoms are the source of electricity.

All matter is made of tiny particles called *atoms*. Though they are much too small to see, atoms themselves are made up of still

88-1. *Cockpit of a DC-10 airplane. Notice all the electrical instruments and switches used to control and measure various functions of the plane.*

McDonnell Douglas Corp.

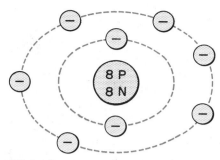

88-2. An atom of oxygen. The nucleus has eight protons and eight neutrons. There are eight electrons around the outside of the nucleus.

smaller particles called *protons*, *neutrons*, and *electrons*.

The center of the atom is called the *nucleus*. It contains the protons and the neutrons. The electrons are in orbit around the nucleus.

Figure 88-2 is a simplified diagram of an oxygen atom. In the nucleus of this atom there are eight protons and eight neutrons. Eight electrons revolve around the nucleus.

Protons and electrons have electrical charges. Protons have a positive charge; electrons, negative. Neutrons do not have any electrical charge.

Normally the positive and negative forces of an atom equal each other. In other words, there are as many protons as electrons, leaving the atom neutral. However, in some substances the outer electrons are only loosely held. These are called free electrons. When these free electrons are removed, the atom becomes positively charged. Electrons from other atoms are attracted to it, and these atoms in turn attract other electrons. *Electricity*, then, is the flow of free electrons through a material.

Some substances, such as copper and silver, allow the movement of large numbers of free electrons. These substances are called *conductors*. Other materials, such as rubber and glass, have tightly held electrons and do not conduct electricity. These materials are called *insulators*. Figure 88-3 lists some typical conductors and insulators.

To produce electricity, some form of energy must start the movement of the electrons. There are six basic sources of energy: friction, pressure, heat, magnetism, light, and chemical action. You will learn more about these sources of electricity in Chapter 91.

FOR YOUR SAFETY . . .

Electricity is powerful and potentially dangerous. Always observe these safety rules when working with electricity:

● Water and electricity are a very dangerous combination. Never handle electrical equipment with wet hands or when standing on a wet or damp floor.

88-3.
Typical Conductors and Insulators

Conductors	Insulators
Silver	Dry air
Copper	Glass
Aluminum	Mica
Brass	Polystyrene plastic
Zinc	Rubber
Iron	Asbestos

● Before you start working, make sure you know exactly what to do. If you are not sure, get help from your instructor.

● Before you work on any electrical equipment, remove the plug from the wall outlet. If you must make adjustments to equipment while it is connected, work with one hand in your pocket or behind your back. This will decrease the chances of receiving an electrical shock.

● When you are finished, have your instructor check your work before you connect it to a power source.

● Replace worn or broken electrical equipment promptly.

● Whenever possible, use electric cords with three-prong plugs. The third prong is a safety ground. *Never remove this third prong.*

QUESTIONS AND ACTIVITIES

1. What is the center of an atom called? Name the other parts of the atom; tell where each is located and what type of charge each has. Then sketch an atom of oxygen and label each part.

2. What are free electrons? What happens to an atom when free electrons are removed? What does the flow of these free electrons through a material bring about or produce? Name the six things that cause these electrons to move.

3. Define conductors. Give three examples of conductors.

4. Define insulators. Give three examples of insulators.

5. What are four safety rules to remember when working with electricity?

Electricity and magnetism are related in interesting ways. A magnet can be used to produce an electric current, and electricity can also produce magnetism. You will learn more about this in later chapters. The rest of this chapter will help you understand magnetism, as an introduction to electricity.

Magnetism has been known since ancient times. Shepherds noticed that certain small pieces of stone stuck to the iron tips of their staffs. These stones were really iron ore. The ancient Chinese discovered that a small piece of this stone on a string would always point in a northerly direction. The Greeks called these stones *magnetite*. Sailors in ancient times called them *lodestones* and used them to aid in the navigation of their ships. These were the first natural magnets.

Basically a *magnet* is a material that attracts iron or steel. A magnet may be made by stroking a piece of steel or iron with a lodestone or another magnet. A piece of soft iron may be magnetized very easily this way, but it loses its magnetism soon. Hard steel is more difficult to magnetize, but it keeps its magnetism. Therefore a magnetized piece of hard steel is called a *permanent magnet*. Permanent magnets may be made in many shapes. Fig. 89-1. Some are straight bars, called bar magnets; others are bent like horseshoes and are therefore called horseshoe magnets.

The greatest force of a magnet occurs at the ends. These concentrations of magnetic force are called the *magnetic poles*. Each magnet has a north pole and a south pole. If you hang a bar magnet on a piece of string so that it can swing freely and there is no other metal or magnet near it, one end will point north and the other south. This happens because the earth acts like a giant magnet with poles close to the north and south geographic poles. The end of a magnet which points north is marked with an N. The other end is marked with an S.

A *compass* is a small magnet balanced on a point so that it is free to turn. The magnetic effect of the earth, mentioned above, causes the end of the compass needle marked N to point north. Fig. 89-2.

The earth's magnetic poles are not in exactly the same place as its geographic poles—the points

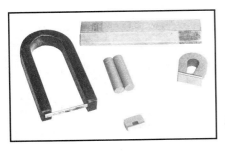

89-1. *Permanent magnets are made in many sizes and shapes.*

89-2. *A compass needle is really a magnet.*

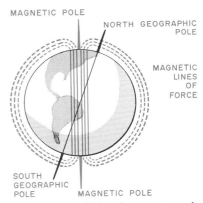

89-3. *The earth is a large magnet. A compass needle points to the earth's magnetic poles.*

around which the earth turns. Fig. 89-3. The geographic poles and the magnetic poles are about 1400 miles apart.

A compass needle points to the magnetic pole and not to the geographic pole. Navigators must make corrections for this in order to get where they plan to go. This difference between the true geographic pole and the magnetic pole is called the *angle of declination*.

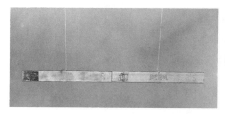

89-5. *These iron filings are arranged along the magnetic lines of force.*

89-4. *Unlike poles of magnets attract each other.*

LAWS OF MAGNETISM

As explained, magnets will attract iron or steel. Magnets can also repel (cause to move away). For example, if two bar magnets are suspended by strings, the north pole of one will repel the north pole of the other and attract the south pole. Fig. 89-4. This can be stated as a law:

Unlike poles attract each other; like poles repel each other.

Lines of force flow out of the north pole of a magnet and back into the south pole. You cannot see these lines, but you can prove that they exist. Place a sheet of paper over a magnet; then sprinkle iron filings on the paper. The filings will arrange themselves as shown in Fig. 89-5. These filings show the magnet's lines of force.

These lines make up the *magnetic field of force*. Where the lines of force flow, the magnet exerts its force, or its ability to attract or repel other particles. The flow or movement of these lines is called *flux*. Flux refers to the flow of magnetism—the stronger the flux, the stronger the magnet.

Iron, steel, and nickel can be magnetized. However, some materials cannot be magnetized. Paper, wood, glass, and copper are examples of nonmagnetic materials.

Magnetism will pass through almost all materials. For instance, air, glass, and paper allow magnetism to pass through.

89-6. *Before a bar is magnetized, the molecules are in all directions.*

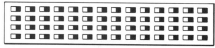

89-7. *After a bar is magnetized, the molecules are lined up in the same direction.*

89-8. *If a magnet is broken, we have two magnets. Each time a magnet is broken, the pieces that are left have a north and a south pole.*

They are called magnetically transparent.

What causes an iron or steel bar to become magnetized? No one knows exactly, but many scientists believe that the molecules in the bar act as very tiny magnets. Before the bar is magnetized, the molecules are in all directions. Fig. 89-6. The process of magnetizing makes the north and south poles of the molecules line up. After the bar is magnetized, its molecules are arranged as in Fig. 89-7.

If a magnetized bar is broken, each piece becomes a magnet with a north and a south pole. Fig. 89-8. A bar may be magnetized by rubbing it with another magnet. Commercially, magnets are made by placing the material to be magnetized in a very strong magnetic field.

CARE OF MAGNETS

Magnets are quite sturdy, but there are some things which should be done to make them last longer. When storing a horseshoe magnet, place a piece of metal, called a keeper, across the poles. Bar magnets should be stored in pairs with north and south poles together. Magnets should not be handled roughly or dropped, as this can reduce their magnetism. Also, too much heat will destroy a magnet.

QUESTIONS AND ACTIVITIES

1. Define magnet. Give an example of a naturally occurring magnet.

2. Define magnetic poles. What are these poles called?

3. Describe a compass and tell what makes it work.

4. What is the law of magnetism concerning magnetic poles?

5. What happens to the molecules of an iron bar when it becomes magnetized?

6. If a magnetized bar is broken, what happens to the pieces of that bar?

7. How should magnets be stored and cared for?

CHAPTER 90
Forms of Electricity

It is not easy to define electricity. It is a type of force, or energy, not a substance you can handle easily, like water or earth. Still, despite the difficulties and dangers of working with electricity, people have learned how to produce it, transport it, and control it.

There are two forms of electricity: static and current. Static electricity is not very useful and may even cause great trouble when large amounts of it are discharged, as in lightning. However, current electricity is valuable because it can be controlled and used to do work.

STATIC ELECTRICITY

The word *static* means at rest. Static electricity is at rest, or is stationary. This form of electricity is generated by friction. You may have combed your hair on a dry day and had the comb crackle or your hair cling to the comb. This is static electricity. So is the little shock you may get if you walk on a carpet on a dry day

90-1. *Pith balls with like charges repel each other.*

and then touch a water faucet. The static charge that has built up in you is suddenly released when you come close to the metal faucet. The friction of car tires on the road may build up static electricity in the car. At many toll booths there are wires which stick up from the pavement to discharge the electricity so that the toll taker will not get a shock. Paper going through a modern high-speed printing press

also builds up static electricity. Many times this causes the sheets of paper to stick together and jam up the press.

As you can see, static electricity is a nuisance. Yet it can be used to demonstrate one of the basic laws of electricity. This law says that *like charges repel each other and unlike charges attract each other.* This can be shown by using two pith balls as shown in Fig. 90-1 & 90-2. When the balls

90-2. *Pith balls with unlike charges attract each other.*

331

are charged alike, they repel each other; they do not touch. When they have unlike charges, they attract each other. Notice how similar this is to the law of magnetism discussed in the previous chapter.

CURRENT ELECTRICITY

To be useful, electricity must be thoroughly controlled. Static electricity can scarcely be controlled at all, but current electricity can be made to work for people.

Current electricity is a flow of electrons. The first person to produce current electricity was an Italian scientist named Alessandro Volta. He constructed what has since become known as a voltaic pile. It consisted of layers of copper, pasteboard moistened in salt water, and zinc. When wires were connected to this pile, an electric current flowed through it. Volta discovered that when two different metals were placed in a chemical which acted on them (such as salt water), an electrical current was produced.

There are two kinds of electric current: direct and alternating.

Direct Current

Direct current (DC) flows continuously and in one direction only. Storage batteries, dry cells, and direct-current generators produce direct current. It is used for telephones, automotive circuits, and wherever batteries are the source of power.

Alternating Current

Alternating current (AC) flows first in one direction, then reverses and flows in the opposite direction. Alternating current can be produced at high voltages and stepped up or down with transformers.

Most electric current is the alternating type. The current reverses direction many times per second. Two changes in direction constitute one cycle. Most electricity used in homes is 60-cycle AC. It changes direction 120 times per second.

VOLTAGE AND AMPERAGE

There are some terms which you must know in order to understand why electricity behaves the way it does. Two of these terms are *voltage* and *amperage*.

A force, or pressure, is needed to make electricity flow through a wire, just as pressure forces water through a hose. This pressure is measured in *volts*. For example, a single flashlight battery has a pressure, or voltage, of 1.5 volts.

The amount of electricity flowing through a wire depends on the number of electrons flowing through the wire. Of course you cannot count the electrons themselves, just as you cannot count the molecules of water passing through a pipe. But you can measure the flow of current accurately in terms of amperes. An *ampere* is the amount of current that one volt will cause to flow through one ohm of resistance. (Resistance and ohms will be explained in later chapters.) The important thing to remember is that amperes are a standard measurement of electrical current flow. The higher the number of amperes, the more current that is flowing.

In the SI metric system the volt and ampere are also used.

QUESTIONS AND ACTIVITIES

1. How is static electricity generated? Give two examples of static electricity.

2. What basic law of electricity does static electricity demonstrate?

3. Describe the two kinds of current electricity and give an example of where each kind can be found.

4. Explain what volts measure.

5. Explain what amperes measure.

CHAPTER 91

Sources of Electricity

As mentioned in Chapter 88, some form of energy is needed to start the movement of electrons and produce electricity. There are six basic sources of energy: friction, pressure, heat, magnetism, light, and chemical action. This chapter tells how they are used to produce electricity.

FRICTION

When two substances are rubbed together, friction causes electrons from one substance to be transferred to the other. This unbalances the atoms and an electric charge is built up. When the atoms try to balance themselves again, sparks jump from one substance to the other. As you learned in the last chapter, friction is a source of static electricity.

CHEMICAL ACTION

One of the most common sources of electricity is a battery made up of two or more cells. Remember that Alessandro Volta made a cell from layers of metal and a chemical (salt water) which acted on them. A simple cell may be made by wiring a grapefruit as shown in Fig. 91-1. Make two small cuts in the skin of the grapefruit. In one, place a penny or a small piece of copper; in the other, place a nickel or a small piece of zinc. Connect the two metals to a sensitive meter and you will see that a voltage is being generated.

A stronger cell may be made by placing a strip of copper and a strip of zinc in a glass of water. Then add a small amount of vinegar or other acid. Again, the electricity that is produced will register on a meter. Fig. 91-2.

Dry Cells

The cell just described is called a *voltaic cell*. It is one way of converting chemical energy into electrical energy. Such a cell cannot be recharged because the zinc strip will be eaten away in the chemical action that produces electricity. This type of cell is called a *wet cell* because the acid is in liquid form. Such cells are not in common use because they are not very handy. If tipped, the liquid will run out.

The *dry cell* is more convenient. Such cells are commonly used in flashlights. They have no liquid to spill. A cross section of a dry cell is shown in Fig. 91-3. This dry cell has four main parts. These are the zinc container, the blotting paper liner, the carbon rod center, and the chemical mixture which is around the carbon rod. The chemical mixture is made of powdered carbon, manganese dioxide, and sal ammoniac (ammonium chloride). As the cell is discharged (used up), water is formed inside it. Sometimes this makes the cell expand.

Newer types of dry cells on the market today have certain advantages over the zinc-carbon cell just described. The *heavy-*

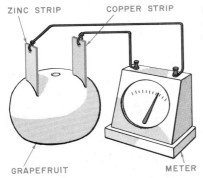

91-1. The acid in the grapefruit acts on the copper and zinc strips and makes electricity. This is a simple cell.

91-2. A better simple cell can be made with a copper and a zinc strip and an acid such as vinegar.

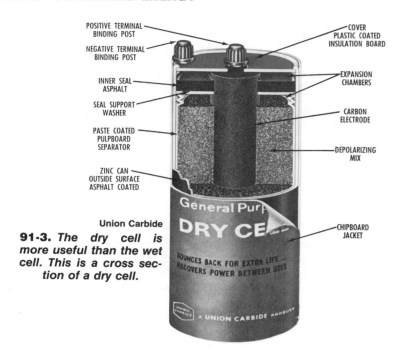

POSITIVE TERMINAL
BINDING POST

NEGATIVE TERMINAL
BINDING POST

INNER SEAL
ASPHALT

SEAL SUPPORT
WASHER

PASTE COATED
PULPBOARD
SEPARATOR

ZINC CAN
OUTSIDE SURFACE
ASPHALT COATED

COVER
PLASTIC COATED
INSULATION BOARD

EXPANSION
CHAMBERS

CARBON
ELECTRODE

DEPOLARIZING
MIX

CHIPBOARD
JACKET

Union Carbide

91-3. The dry cell is more useful than the wet cell. This is a cross section of a dry cell.

duty (zinc chloride) cell works better in cold weather. It is also less likely to corrode or leak, and it lasts longer than the zinc-carbon cell. The *alkaline* dry cell lasts about ten times longer than the zinc-carbon cell and also works well in cold weather. It is a good choice for continuous heavy-duty use, as in a cassette player. The *mercury* dry cell delivers almost constant voltage throughout its lifetime and can be stored for long periods without deteriorating. This type is used in hearing aids. The *nickel-cadmium* cell, unlike the other dry cells, is rechargeable. It provides steady power and works well at low temperatures. However, such cells are expensive, and they must be frequently recharged.

You will notice two terminals on the cell in Fig. 91-3. One is attached to the zinc container and is called the negative pole. It is indicated by a minus sign (−). The other is attached to the carbon rod and is called the positive pole. It is indicated by a plus sign (+).

You probably have heard people call a single flashlight cell a battery. Technically this is wrong. A battery is made up of two or more connected cells. An example is shown in Fig. 91-4. After the cells are connected to form the battery, they may be enclosed in one case. An automobile battery is an example.

It is important to understand how and why to connect cells. Cells are connected either in series or in parallel. If you had a small toy motor that required nine volts and you wanted to use dry cells to operate this motor, you would connect the cells in series. Fig. 91-5. Each cell produces 1½ volts; so six cells in series will supply nine volts for the motor. By connecting more cells in series, higher voltages may be obtained.

For many uses you need the voltage of only one cell, but you may need it for long periods of

91-4. A battery is really many single cells connected together.

time. All the current of one cell would be used in a short time. To make a battery that will last longer, connect several cells in parallel as shown in Fig. 91-6. The total voltage is the same as the voltage of one cell, but the life is increased because current is drawn from more cells.

Storage Cells

The dry cell is a *primary cell* because it cannot, ordinarily, be recharged. Rechargeable cells are called *secondary,* or *storage, cells.* The most familiar type of secondary cell is the lead-acid cell used in automobile batteries.

The storage battery is a reservoir for electricity. A storage battery does not make electricity. It stores electrical power it receives and makes it available as needed. In an automobile, the electrical power comes from the alternator. Running the engine activates the alternator, which in turn charges the battery. Fig. 91-7.

The storage battery is made up of two different kinds of lead plates in an *electrolyte.* The electrolyte helps current to flow. The electrolyte in the storage battery is a mixture of distilled water and sulfuric acid. (In dry cells the

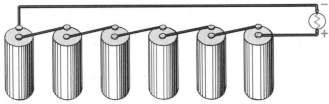

91-5. *Single cells connected in series.*

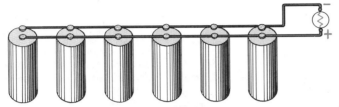

91-6. *Single cells connected in parallel.*

electrolyte is in paste form.) As the battery is charged or discharged, there is a change in the composition of the plates and the electrolyte.

Each cell of a storage battery produces about two volts. To make the twelve-volt battery used in most cars today, six cells are connected in series. Each of the cells has its own little compartment. At the bottom of each compartment is a space called a *sediment chamber.* Here particles from the plates and other material collect. This prevents short circuits.

Car batteries lose water through their vents. On conventional batteries, the water level can be checked by removing the filler cap ("gang vent plug" in Fig. 91-7). If the water level is low, add distilled water. Maintenance-free batteries have built-in indicators to show whether the water level is correct, but you cannot add water to these.

The condition of a battery may be tested by the use of a load test or a hydrometer. For the *load test* the battery is discharged at a faster than normal rate and the drop in battery voltage is noted.

A twelve-volt battery should not drop below nine volts under the load test.

A *hydrometer* is a device that tests the battery by testing the condition of the electrolyte. A fully charged battery will read 1.280 on the hydrometer. A fully discharged battery will read 1.110.

The terminals of the storage battery in a car may become corroded after a period of time. If this happens, plug the vent holes with toothpicks and pour a solution of baking soda and water on the terminals. Allow it to remain for a few minutes. Then it should be washed off and the terminals wiped clean. Remember to unplug the vent holes.

FOR YOUR SAFETY . . .
● When working with storage batteries, be careful not to spill the electrolyte. It will eat holes in your clothes and burn your skin.
● During the charging of storage batteries, highly explosive hydrogen gas sometimes forms. **Do not light matches near charging batteries**.
● Remove jewelry to avoid making sparks.

ORGANIC STORAGE BATTERIES
Currently under development are new types of storage batteries called organic batteries. These

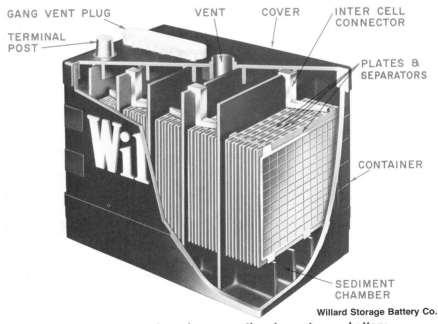

GANG VENT PLUG
VENT
COVER
INTER CELL CONNECTOR
TERMINAL POST
PLATES & SEPARATORS
CONTAINER
SEDIMENT CHAMBER

Willard Storage Battery Co.

91-7. *A cutaway view of a conventional car storage battery.*

335

use specially treated plastic electrodes (plastics are organic compounds). Organic storage batteries are lightweight. They can store three times as much energy as the conventional lead-acid battery, and their energy can be drawn off ten times faster. These advantages make them good candidates for powering electric cars.

Fuel Cells

One of the newer methods of producing electricity is the fuel cell. A *fuel cell* converts the energy of a chemical reaction to electricity. It uses low-cost fuel and an oxidant. The most common type, and the one used in the space program, is the hydrogen-oxygen cell. In this cell, electricity and water are generated from a controlled reaction of oxygen and hydrogen and an acid electrolyte. Figure 91-8 shows the action of a fuel cell.

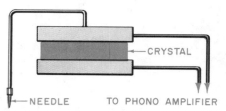

91-9. *Heating a thermocouple will make electricity.*

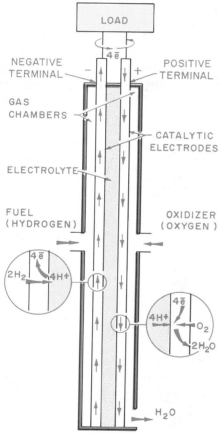

General Electric Corporation

91-8. *A typical fuel cell. Hydrogen reacts at the negative terminal to give up an electron (e⁻) to the load while releasing hydrogen ions (H⁺) in the solution. At the positive terminal, these hydrogen ions combine with oxygen and the electrons from the load circuit to produce water.*

HEAT

When two wires of different metals are twisted together and heated as shown in Fig. 91-9, an electric current will flow. Heat causes the electrons to move. *Thermocouples* operate on this principle. These are commercial devices, made of two different metals, which are used to indicate and control heat in ovens and furnaces. Thermocouples are

91-10. *Grooves in a phonograph record cause pressure on the needle. This pressure is carried to the crystal, which produces electricity.*

connected to meters which record the amount of current flowing through the wires—the more heat, the more current. A thermocouple combined with such a meter is called a *pyrometer*. Pyrometers are used for checking the proper pouring temperatures of molten metal.

Thermocouples do not furnish large amounts of electrical current. For this reason they cannot be used to provide electrical power.

PRESSURE

When pressure is applied to certain materials, they will produce a slight electric current. One substance that reacts this way is a type of crystal known as Rochelle salts. As more pressure is applied to the crystal, more electricity is produced.

One of the most familiar uses of electricity produced this way is in a record player. Fig. 91-10. A needle is attached to the crystal. When the needle slides in the record groove, it vibrates. The vibrations of the needle duplicate

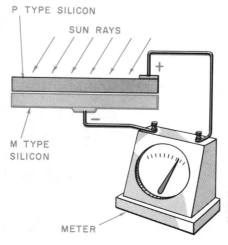

91-11. *Drawing of a solar cell. Electrons travel from one type of silicon to the other when the rays of the sun strike the cell.*

the original vibrations that were pressed into the record. The needle vibrations cause pressure on the crystal so that it produces electricity in the same pattern as the original. The amount of electricity is very weak. It is made stronger by an amplifier.

Crystals are also used in microphones. The principle of generating electricity by pressure is the same for the microphone as for the record player. The only difference is that the pressure is applied to the crystal by a voice speaking into the microphone.

LIGHT

Some materials will generate small amounts of electricity when light strikes them. This is called the *photoelectric effect*. Some of these materials are sodium, selenium, and potassium. These materials are used to make what are known as photoelectric, or solar, cells. Fig. 91-11. By controlling the amount of light that strikes one of these cells, the amount of electricity can be controlled.

Most photoelectric or solar cells consist of a metal plate coated with some light-sensitive substance. These light-sensitive cells are used for operating light meters for photography, burglar alarms, and other control devices. One of the important uses for solar cells is as a source of power for satellites. Because each cell generates only a small amount of electricity, it takes many cells to power a satellite.

MAGNETISM

The most important source of electricity is the generator. A *generator* is a device that changes mechanical energy to electrical energy. All generators work on the principle of cutting magnetic lines of force with coils of wire. Fig. 91-12. The generator is a cheap and easy way of producing electricity. It is the way most electricity is produced for your home and for industry. Automobiles also have a form of generator to keep the battery charged. It is called an alternator because it produces alternating current. In Chapter 93 you will learn more about electrical generators.

91-12. *Electricity is produced by a coil cutting magnetic lines of force. As the magnet is moved in and out of the coil, a current is generated.*

QUESTIONS AND ACTIVITIES

1. Which of the six sources of energy produces the electricity in batteries? What is the difference between a wet cell and a dry cell?

2. How is a nickel cadmium cell different from other dry cells? What are some advantages of this type of cell? What are some of the disadvantages?

3. What are the four main parts of a zinc-carbon dry cell? To what part of the dry cell is the positive pole attached? To what part is the negative pole attached?

4. How are the cells of a battery connected to make a battery that will perform for a long period of time? How are the cells connected to get higher voltage from the battery? Draw a simple diagram to demonstrate each of these types of connection.

5. Very simply, what is a storage battery and how does it work? Give an example of this type of battery.

6. Describe how to clean corroded terminals on a car battery.

7. What are two safety rules to remember when working with storage batteries?

8. What are thermocouples? What principle of electricity is involved in the operation of thermocouples?

9. Explain how a phonograph crystal produces electricity. Which of the six basic sources of energy is used?

10. What is another name for photoelectric cells? What do most of these cells consist of? What are three things these cells can be used for?

11. Define generator. On what principle do all generators work?

From Chapter 88 you will remember that electrons usually are in orbit around the nucleus of an atom. The atoms that make up certain metals have electrons which are loosely bound to the nucleus. This means the electrons are free to move from one atom to another if pushed with a small amount of force. The flow of current is really the movement of free electrons. Materials which allow the motion of large numbers of free electrons are called conductors. A good conductor is said to have *low resistance* to current flow. Copper wire is a good conductor.

The opposite of a conductor is an insulator. Insulators have very few free electrons; so little or no current will flow through them. The best conductors are used as wires to carry electrical energy; the poorest are used as insulators, to prevent current flow.

SIMPLE ELECTRICAL CIRCUIT

Most useful electrical energy is transmitted through closed electrical circuits. There are four parts to a closed circuit. They are:

● Source of electricity, such as a dry cell or battery.

● A way of turning the current on and off. One way to turn the current on and off is with a switch. Another is with a push button, as for doorbells.

● A way for the current to flow through the circuit. This is provided by wires. Electrical wires are usually of copper, but aluminum is sometimes used.

Wire sizes are indicated by gauge number—the larger the number, the smaller the wire. Electricians usually choose the smallest wire that will safely carry the current and not get hot and burn the insulation. If the wire gets too hot, it might start a fire. The smaller the wire, the more resistance it has to the flow of electricity. Resistance also increases as the length or temperature of a wire increases. And of course the kind of wire also determines resistance. Because copper has low resistance, it is often used for electrical wiring.

Joining wires is called *splicing*.

92-1. Three kinds of wire splices. The Western Union splice (left), the tap splice (center), and the pigtail splice (right) are shown. The pigtail splice is the one most often used in house wiring.

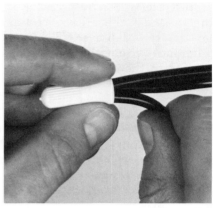

3M

92-2. Wire nuts (solderless connectors) can be used to fasten the ends of wires together.

The most common splice is the Western Union splice. Others are the tap splice and pigtail splice. Fig. 92-1. Spliced wires may be soldered, taped, or fastened together with wire nuts. Fig. 92-2.

● The fourth part of a complete circuit is the load, or device, that is to use the current. In Fig. 92-3, the load is a small light bulb.

There are many kinds of circuits. There are open circuits, closed circuits, short circuits, parallel circuits, and series circuits.

A circuit is *open* when there is a break in the wire or when a switch is not closed. A *closed* circuit is a complete circuit. For example, when a push button is pushed, this closes the circuit. A *short* circuit allows the current to flow where it is not meant to flow. Short circuits can cause injury and damage.

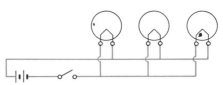

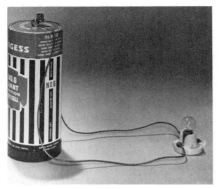

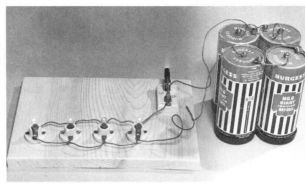

92-5. *Diagram and photograph of lights wired in parallel.*

92-3. *These illustrations show a schematic drawing and a photograph of a simple circuit.*

When lights are connected in *series*, the current must flow through each light to complete the circuit. If one light burns out, the circuit is open. Figure 92-4 shows lights connected in series. Why do you suppose this type of circuit is not used for house wiring?

The parallel circuit is used in connecting lights and other electrical appliances. In *parallel circuits* each load has its own path for current flow. If one light or load burns out the circuit re-

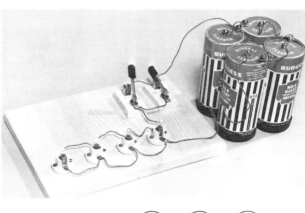

92-6. *Diagram and photograph of lights and switches wired in parallel.*

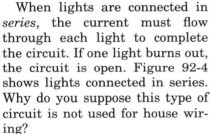

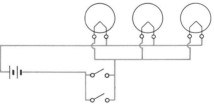

92-4. *Diagram and photograph of lights wired in series.*

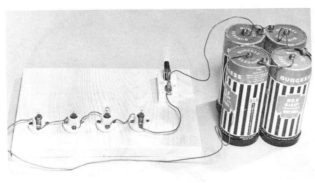

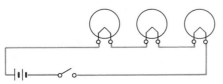

mains closed and the other loads on the circuit will continue to operate. Fig. 92-5. Two switches could also be connected in parallel to turn on the lights from two places. Each switch has a separate branch connecting it to the main path to the lights, so when either switch is turned on the circuit to the lights is closed. Fig. 92-6.

Plans for circuits are drawn,

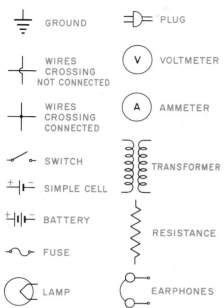

92-7. *Common symbols used on electrical drawings.*

92-8. *Lamp socket with parts separated and cord attached.*

just as plans for a building or a machine are. It would take too much time and space to draw pictures of the parts of a circuit; so symbols are used instead. Some of the more common electrical symbols are shown in Fig. 92-7.

CORDS, PLUGS, AND LAMP SOCKETS

Most electrical appliances have an attached cord so that they can be plugged into an electrical outlet. These cords should be inspected frequently and replaced as needed. Especially common problems are broken or frayed wires, and those with cracked insulation.

Wiring a lamp is one of the most common jobs in electricity. Sometimes you may need to put a new cord on the lamp. At other times you may need only to replace the plug.

To replace a lamp cord, first unplug the lamp. Obtain a sufficient length of new lamp cord. Attach the socket as shown in Fig. 92-8. Remove ½″ of insulation from the wire ends. Remove

the cap from the socket. (This is usually done by pressing as indicated on the socket.) Slip the cap over the wire. Slip the casing from the socket. Fasten the wires under the terminal screws on the socket. The wires should go on the screws in a clockwise direction. Put the casing over the socket. Snap the socket into the cap.

Plugs are of various types. With some, the wire is slipped into the plug and held by a clamp. On others the wire is soldered to the prongs of the plug. The most common type uses screws to fasten the wire to the prongs. Fig. 92-9. Fastening a cord to a plug or socket is very easy. However, be careful not to have any of the strands of wire touching, or a short circuit will result.

Some plugs have three prongs instead of two. These *grounded plugs* are especially common on appliances that may be used in moist places or near water. One end of the wire is connected to the case of the appliance; the other end is connected to the third prong. This grounds the case so that if a short circuit occurs, it will not shock the person

using the appliance. Special adapter plugs are available for connecting grounded appliances to two-pronged outlets. **CAUTION:** *The grounding wire on the adapter should be connected to a good ground.* Usually it is connected to the screw that holds the outlet cover in place.

Always be sure to read the instruction book which comes with the appliance. Never attempt any repairs unless you are qualified.

OHM'S LAW

In Chapter 90 you learned about voltage (the pressure causing a current to flow) and amperage (the amount of electric current). In this chapter you learned about electrical resistance. One of the basic laws of electricity is *Ohm's law.* This law is named after George Simon Ohm, who developed it. Ohm's law is a mathematical way of finding the

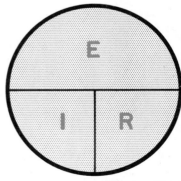

92-9. *Wiring a plug.*

92-10. *Ohm's law diagram. The letter E stands for electromotive force (voltage); I stands for intensity of electron flow (amperage); R stands for resistance, which is measured in ohms.*

resistance, the amperage, or the voltage of a circuit. It says that total amperage (I) multiplied by total resistance (R) is equal to the voltage (E).

$$E = I \times R$$

Figure 92-10 makes it easy to remember how to use this formula. The horizontal line means divide; the vertical line means multiply. If you know the amperage (I) and voltage (E) of a circuit and want to find out the resistance (R), cover the R with your finger. What is left? E divided by I. Substitute the numbers for the symbols, divide, and you have the answer.

Now try an actual example.

How much voltage would it take to push 2 amperes through a wire that has a resistance of 55 ohms? Cover the E in the circle and you see that you need to multiply I × R, or 2 × 55, for an answer of 110 volts. Engineers who design electrical equipment and appliances use Ohm's law in deciding what materials to use.

QUESTIONS AND ACTIVITIES

1. What does it mean when we say a material has low resistance to electrical current? What unit of measurement indicates resistance?

2. What are the four parts of a closed electrical circuit? Give an example of each part.

3. Define splicing. What are the three most common types of wire splices? Name and sketch the type of splice most often used in house wiring.

4. Name and describe the five types of circuits described in this chapter.

5. Describe a grounded plug and tell what the purpose of grounding an appliance is. What is an important safety rule to remember when using special adapter plugs with grounded appliances?

6. Draw the diagram of Ohm's law. Tell what quantity each letter stands for and what unit of measurement is used. Tell what the horizontal and vertical lines on the diagram indicate. If there are 220 volts and 44 ohms, what is the amperage?

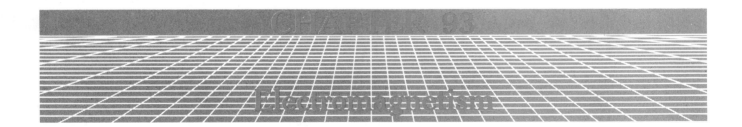

CHAPTER 93
Electromagnetism

In Chapter 89 you learned about magnetism and permanent magnets. Now you are going to study the effects of a current in a wire. Around every electrical conductor where current is flowing there are magnetic lines of force. You can see this by doing the following experiment. Through a piece of cardboard, pass a wire. Connect the wire so that it will carry a current. Fig. 93-1. Place three compasses on the cardboard and you will see that the needles point in the direction of the magnetic lines of force. If the current is reversed, the compass needles will point in the opposite direction. The direction of the magnetic field depends on the direction of the current flow.

When a wire carrying a current is wound into a coil, it is

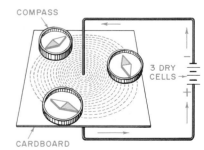

93-1. *Lines of force around a wire. Notice how the compass needles point in different directions.*

93-2. *A solenoid coil. Note that it has a north and a south pole.*

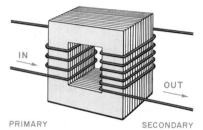

93-3. *A solenoid. The iron bar is pulled into the coil when the current is on.*

93-4. *A transformer. The current enters the primary coil and leaves the secondary coil. A transformer either raises or lowers the voltage.*

called a *solenoid*. The magnetic fields around each loop of wire join together. A solenoid will have a magnetic north pole at one end and a magnetic south pole at the other. Fig. 93-2. The strength of the magnetic field depends on the number of turns of wire in the coil and the current flowing through the coil.

A typical solenoid is a coil of wire wound on a hollow tube. A piece of soft iron fits into this tube and can be moved in and out. When the coil is connected to a current, the solenoid magnetically pulls the iron core into its center. The solenoid is useful for opening or closing valves and switches, among many other applications. Door chimes use one or more solenoids to strike metal strips and produce musical notes. Automatic washers also use solenoids for turning the water on and off. Fig. 93-3.

The solenoid coil uses air as the only conductor of the magnetic field. Other materials will conduct magnetic lines of force better than air. These materials are said to have better permeability. Soft iron is one such material. When electric current is passed through a coil of wire wound on an iron core, we have what is called an *electromagnet*. An electromagnet may be made by winding many turns of No. 24

wire around a nail, then connecting the wire to a dry cell.

With the electromagnet, iron substances may be picked up. You will notice that when the dry cell is disconnected there is no longer any magnetism; whatever was stuck to the magnet drops off. This explains how large electromagnets are used in junkyards for picking up and loading scrap iron. Electromagnets are also used in relays, in voltage regulators for cars, in doorbells and buzzers, and in magnetic switches.

Still another use of electromagnetism is in the transformer. A *transformer* is a device for transferring electrical energy from one circuit to another. A transformer consists of two coils of wire

wound around an iron core. A transformer may increase (step up) the voltage or decrease it (step it down). Fig. 93-4.

The winding in which the current enters the transformer is called the *primary coil*. The winding in which the current leaves the transformer is called the *secondary coil*.

GENERATORS

As mentioned earlier, electric current will produce a magnetic field. Michael Faraday, an English scientist of the 19th century, proved that this works both ways—a magnetic field can be used to produce electricity. His experiments led to the development of the *generator*, a device for changing mechanical energy into electrical energy.

To produce electricity with a generator there must be a magnetic field and a coil of wire (called an *armature*) which moves through the field. As the coil revolves, it takes on an electrical charge.

Because the coil turns, it is necessary to connect wires that are not turning to those that are. To do this a small ring, divided into two sections, is fastened to the ends of the coil. This ring is

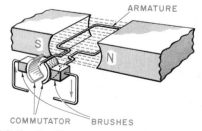

93-5. *A direct current generator. Broken lines show the magnetic field. The armature turning in the magnetic field generates electricity which is removed by the commutator and the brushes. The colored arrows show which way the current flows. The black ones identify parts.*

called the *commutator*. As the armature revolves, it causes the two sections of the attached commutator to rub against the fixed *brushes* which transfer the voltage from the revolving coil to the stationary wires. Such a generator produces direct current. Fig. 93-5.

Generators that produce alternating current have two or more *slip rings* instead of a commutator. These rings are not divided into sections as the commutator is.

Simple generators can be improved by using electromagnets instead of permanent magnets. These electromagnets are called *field coils*.

MOTORS

The operation of the electric motor is the reverse of the generator. The *motor* changes electrical energy to mechanical energy. Besides their use in almost every industry, electric motors have many household uses. They are used in refrigerators, vacuum cleaners, fans, and many other appliances. In school shops motors are used on drills, saws, lathes, jointers, and many other machines.

The operation of an electric

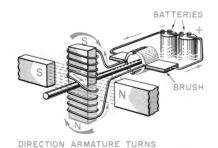

93-6. A motor. The attraction and repulsion of the magnets and the armature make the motor turn. The large, colored arrows show the movement of the armature. The smaller ones show current flow. The black arrows identify parts.

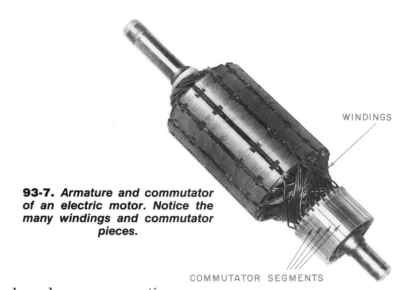

93-7. Armature and commutator of an electric motor. Notice the many windings and commutator pieces.

motor depends on magnetic fields. It particularly depends upon two principles you have already learned. One is that like poles repel each other and unlike poles attract each other. The second is that a magnetic field exists between the two poles of a magnet.

An armature is placed between the north and south poles of a permanent magnet (called a *field*). Like the armature of the generator, the armature of the motor rotates and at its ends it has pieces of metal called the commutator. Brushes contact the commutator sections and provide a way for the current to get to the armature coil. Current from a power source, such as a battery, passes through the armature coil, then returns to the power source through the other commutator piece and brush. Fig. 93-6.

Here is how one turn of the armature takes place: The poles of the field attract the poles of the armature, causing the armature to turn. As the armature turns, the brushes slide off the commutator and the circuit is broken. The armature continues to turn, and the current is reversed in the armature. This causes the poles of the armature to change so that its north pole is next to the north pole of the field. Like poles repel; so the armature continues to turn.

This process of attraction and repulsion continues until the power is disconnected. The arrangement of the commutator and brushes is very important because the direction of the current must be reversed at just the proper time.

Large electric motors have many coils on both the armature and the field. Each coil on the armature is connected to a commutator segment. Fig. 93-7.

There are many ways of winding and connecting the coils which control the speed of the motor. Motor speed is measured in revolutions per minute (RPM). Refrigerator motors run at about 1725 RPM; some others run as fast as 7000 RPM.

So far you have seen how the motor operates on direct current. Many motors will work on both direct current (DC) and alternating current (AC). One kind of motor, the *induction motor,* is designed to operate only on AC. This motor has no armature coil, no commutator, and no brushes. The rotating part, or rotor, is made of laminated pieces of metal. The change in direction of the alternating current substitutes for the commutator and brushes.

QUESTIONS AND ACTIVITIES

1. Define solenoid. Describe a typical solenoid and how it works. Name at least two things that use solenoids.

2. What is an electromagnet? Is an electromagnet a permanent magnet? Explain.

3. What is a transformer used for, and what does it consist of?

4. What is an armature and what does it do in a generator? Describe the commutator and its function on a generator. What do the brushes of a generator do?

5. What is the main difference between a generator and an electric motor?

6. What two principles does the operation of an electric motor depend upon?

CHAPTER 94
Converting Electricity to Heat and Light

HEAT

Electricity can be used to produce heat. This makes it useful for many purposes in the home as well as in industry. The iron, the toaster, the waffle iron, and the coffee maker are just a few common appliances that depend on electric heat. How many more can you name?

How does electricity produce heat? To understand this you have to remember a few facts from previous chapters. Electrical current is really the movement of electrons through a conductor. This movement causes heat.

Electrons move easily through some materials, and cause very little heat. Such materials are called good conductors. However, when electrons pass through a poor conductor, they meet more resistance. It takes more pressure (voltage) to push them through. As a rule, the more resistance the electrons meet, the more heat they will build up.

There are special materials which can be used to control the resistance to the movement of electrons. *Nichrome wire* is such a material. By using the correct size and amount of this wire, the desired degree of heat can be produced with a certain amount of electricity. This is the kind of arrangement used to provide heat in the electrical appliances mentioned a few paragraphs earlier. The device that actually produces the heat is called a resistance-heating element.

Large appliances that use a great amount of current need special circuits that have their own outlets. An electric range, for example, can only be plugged into a 240-volt circuit. Fig. 94-1. Smaller appliances are usually portable so that they can be plugged into electrical outlets wherever they are needed. Whether large or small, these appliances can be dangerous. Both electricity and heat, if used improperly, can cause personal injury and property damage. Before using any heating appliance, be certain to read the instruction book.

Industrial Uses of Electrical Heat

Industry also uses electricity for making heat. Some special steels are heated in electric furnaces that use large carbon rods and an arc of electricity. These furnaces use great amounts of electricity and produce very high temperatures.

Another important use of electrical heat in industry is for *arc welding*. Fig. 94-2. The welding

94-1. *A range outlet.*

rod is held close to the piece to be welded. Current passing through the rod arcs across the air space to the workpiece. The heat of this arc melts the rod and the piece being welded. Arc welding is used for such purposes as joining sections of pipelines and fastening the steel members of tall buildings.

Arc welding produces harmful rays which will burn much like the sun. Never look directly at the arc or you will damage your eyes. Welders wear helmets with special glass to protect their eyes.

Another electric welding method is *high-frequency resistance welding*. This method makes use of the fact that extremely high frequency current flowing through a conductor remains at or near the surface rather than penetrating the entire conductor. Low-amperage alternating current with frequencies as high as 450 000 cycles per second is sent through the metal workpieces to be joined. The surfaces of the metal become hot. They are then pressed together. High-frequency resistance welding is used for pipe, tubing, and structural shapes. It is a high-speed process done by machines.

LIGHT

Probably the most common use of electricity is to provide lighting. Most people are so used to electric light that they never stop to wonder how the light bulb works. Fig. 94-3.

As you just learned, electricity produces heat when it flows through a wire with resistance. When the wire gets hot enough, it glows, first a dull red, then brighter and brighter. However, most types of wire would get too hot and melt instead of giving a bright and lasting glow.

The first practical electrical lighting system was developed in

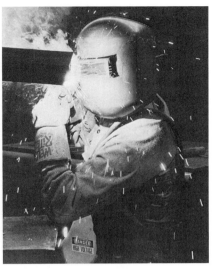

(Peoria, IL) *Observer*
94-2. Arc welding.

the late 1800s by Thomas Edison. He placed a wire in a glass bulb, removed the air from the bulb, and passed an electric current through the wire until it glowed. (Wires used this way are called *filaments*.) Edison's first bulb produced light, but also got very hot and did not last long. He used filaments of carbon and later of bamboo. Now filaments for light bulbs are made from *tungsten*—a metal that has high resistance and a high melting point. The

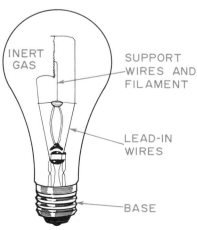

94-3. The parts of a light bulb.

INERT GAS

SUPPORT WIRES AND FILAMENT

LEAD-IN WIRES

BASE

tungsten filament gives a white light with much less heat than the Edison light bulb. Today's bulbs are filled with inert gas.

Light bulbs come in various wattages. The higher the wattage, the brighter the light.

Lighting Circuits

Most of the lights in your house are connected in a parallel circuit. This method is best because if one light goes out, they do not all go out. Today's Christmas tree lights are connected this way. Fig. 94-4.

Some lights are connected in series as in Fig. 94-5. Christmas tree lights used to be connected this way. When one bulb burns out, all the others go out because the circuit is broken. With this type of connection there must be just the right number of lights in the circuit. If there are too many lights, they will be dim; if too few, they will be too bright and perhaps burn out too fast.

Other Lighting Methods

So far we have been talking about common household light

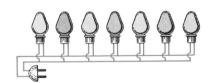

94-4. Christmas tree lights connected in parallel.

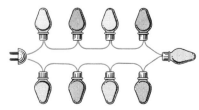

94-5. Christmas tree lights connected in series. What will happen if one light burns out?

bulbs, also known as incandescent lights. There are other ways to change electricity to light. The arc light, for instance, used to be fairly common as a street light and is still used in commercial motion picture projectors. Its light is produced by electricity passing between two carbon electrodes. This causes a great deal of heat as well as light.

Neon and some other gases will glow when an electric current passes through them. The gas is in a glass tube that has a metal electrode attached to each end. An electric current is made to flow through the gas between the electrodes. Different gases glow in different colors. The glass tubes may be bent into many shapes, making them useful for signs.

Fluorescent lights work on a principle similar to neon tubes. A coating on the inside of the tube glows when the current flows between the electrodes. These lights are often used to light stores, offices, and schools because they use less electricity than incandescent lights of equal illuminating power.

Smaller tubes filled with a gas called xenon are also used for lighting. These require a high voltage for short periods of time and give off a very bright light. These tubes are used in photographers' flashguns. Because they give off such an intense light, they are used at some airports. In fog and haze, pilots can see them better than other types of lights.

CAUTION: *The coating on the inside of fluorescent tubes is dangerous; so when handling, be careful not to break the glass tube.*

QUESTIONS AND ACTIVITIES

1. How does electricity produce heat?
2. How does resistance to electrical flow affect heat build-up?
3. What kind of wire is used in electrical appliances such as irons and toasters? Why?
4. Describe arc welding.
5. Describe high-frequency resistance welding.
6. What is the filament in a light bulb? What are today's filaments made of and what properties make this material a good choice?
7. Other than the incandescent light, what are three other types of electrical lights and what is each used for?

CHAPTER 95
House Wiring

Electricity for your home comes from a power station. Fig. 95-1. These stations are often many miles away; so the voltage is increased, or "stepped up," by transformers to many thousand volts. Then it is carried over wires to a substation, where the voltage is reduced, or "stepped down." Before it can be used in your home, it must be still further reduced, to 110 or 220 volts. This is usually done by a transformer close to your home. Fig. 95-2.

In some areas the electrical distribution system is underground. In these areas the wiring and the step-down transformers are both underground, and no wires can be seen overhead.

The voltage that comes to a home is dangerous and can cause severe shock or even death under certain conditions; so take no chances with the electricity in your home. A person with limited

Southern California Edison

95-1. *Typical electric power generating station. This plant operates on natural gas to avoid polluting the air.*

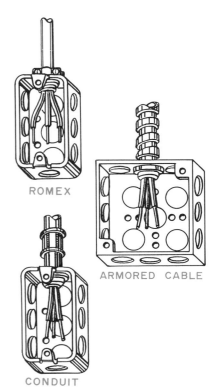

95-4. *Three kinds of cable. Connections between pieces of cable are made in metal junction boxes like the ones shown here.*

experience should not attempt to do house wiring. However, everyone should know some basic facts about house wiring and be able to make simple repairs and replacements.

Electricity is brought into the house by three wires to the watthour meter. Fig. 95-3. This meter is usually attached to a service panel which contains circuit breakers to protect the wiring in the house. The service panel also has one main circuit breaker to shut off all power to the house if that is necessary. Some older service panels have fuses. When a circuit becomes overloaded, the circuit breaker trips open or the

fuse blows. Remember, before resetting the circuit breaker or replacing a fuse, correct the cause of the overload. *Never* replace a fuse with one rated for a higher current.

From the service panel the wiring goes throughout the house. There are three common types of electrical cable used in homes. Where local regulations permit, the most common type is nonmetallic sheathed cable (Romex). Other kinds are flexible armored cable (BX) and metal conduit (rigid or flexible) through which wires are placed. Many localities require metal conduit in places where the wire might get damaged, such as in the garage. Fig. 95-4.

Many portable appliances, such as a toaster, are plugged

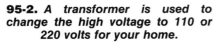

95-2. *A transformer is used to change the high voltage to 110 or 220 volts for your home.*

95-3. *A watt-hour meter measures the amount of electricity that is used in a home, industrial plant, or other building.*

347

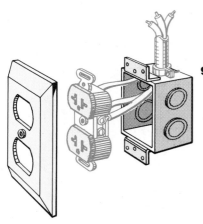

95-5. Duplex wall outlet.

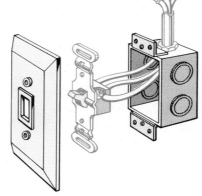

95-6. Typical wall switch.

into wall outlets. Figure 95-5 shows a common duplex outlet and how it would be connected. Outlets are usually wired with No. 12 wire. Several outlets may be connected in parallel.

The lights in a home are usually turned off and on with a wall switch like the one shown in Fig. 95-6. Since lighting circuits do not require as much current as outlet circuits, they are usually wired with No. 14 wire.

When architects draw plans for a building, they indicate where all the lights and outlets will be located. They also show on the floor plan where the service panel and other electrical devices will be located. Typical electrical symbols are shown in Fig. 95-7.

READING METERS

The power company must have some way of measuring the amount of current you use. This is done with a meter. Fig. 95-3. As the current goes through the meter, the electricity runs a motor that turns the hands on the meter dials. The faster the current flows, the faster the meter runs.

Notice the dials on the meter in Fig. 95-8. The first dial gives thousands of kilowatt hours, the second gives hundreds, the third tens, and the last units.

The hands on two of the dials move clockwise; on the other two, counterclockwise. Always read the number which the hand has just passed.

Notice the position of the

hands in Fig. 95-8. The thousands hand has just passed 8; so the reading for this hand is 8000. The hundreds hand has passed 5 and is about halfway to 6; so this would be 500. The tens hand is almost to 4, but has not reached it yet; so this would be 30. The units hand has passed 2. The reading for this meter would be 8532.

Meters are not set back to zero each month. The reading for the preceding month is subtracted from the reading for the current month. This gives the number of kilowatt hours of electricity used during the current month.

What does the term *kilowatt hours* really mean? Watts are a measurement of electrical power,

95-8. Dials on a watt-hour meter.

SYMBOLS			
⊢⌿	ELECTRIC SWITCH	◎	RECESSED CEILING FIXTURE
⊢⌿³	3 - WAY SWITCH	⊕	CEILING MOUNTED PENDANT FIXTURE
⊢⌿ᴰ	DIMMER SWITCH	✦	CUSTOM EXTERIOR HANGING FIXTURE
◀	TELEPHONE OUTLET	◇	SURFACE MOUNTED CEILING LIGHT
⊢⊖	110 V CONVENIENCE OUTLET	⊢◇	WALL BRACKET LIGHT
⊢⊖	1/2 HOT CONVENIENCE OUTLET	⬦	CUSTOM HANGING FIXTURE
⊢⊖⁻ᵂᴾ	WATERPROOF CONVENIENCE OUTLET	✳	CUSTOM CHANDELIER
⊢⊖²²⁰	220 V CONVENIENCE OUTLET	⊗	HEAT, LIGHT AND EXHAUST FAN
⊜	110 V CONVENIENCE OUTLET (FLOOR MTD)	△ FLOOD	150 V FLOOD LIGHT
⊢☓⁻ᴴᴮ	HOSE BIB		2 TUBE FLUORESCENT LIGHT
⊢☓⁻ᶠᴳ	FUEL GAS		
⊢╫	T V ANTENNA		

95-7. Symbols used by architects when drawing house plans.

95-A. *Fuses.*

95-B. *Circuit breaker.*

Fuses and Circuit Breakers

Fuses guard against one of the chief dangers of using electricity. Like many appliances, they work on the principle that electricity makes heat.

Fuses are safety devices to prevent electrical equipment and circuits from being burned out by an overload. An *overload* means that the circuit is carrying too much current. The circuit is arranged so that all the current must pass through the fuse. A special conductor in the fuse will carry only a certain amount of current. This conductor grows hot and melts when too much current passes through it. A melted fuse (usually called a "blown" fuse) breaks the circuit and prevents the overload from flowing into the rest of the circuit. You can see that the fuse operates on the same principle as electrical heating devices.

The two types of fuses usually used in the home are the cartridge fuse and the plug fuse. The cartridge fuse is commonly used for 220-volt circuits that carry up to 30 amperes. The plug fuse is usually used for circuits that do not carry over 25 amperes.

For motor circuits, such as on a furnace or air conditioner circuit, you may find a special kind of fuse called a *fusetron*. This type of fuse permits a large current flow for a short period of time. This allows the motor to get started, but prevents the heavy load from lasting too long.

Today many homes are equipped with circuit breakers instead of fuses. These devices, which look like switches, will open the circuit like a fuse if the current flow gets too heavy. Unlike fuses, circuit breakers do not need to be replaced every time there is an overload. The circuit breaker is just reset.

A blown fuse or a tripped circuit breaker means trouble. It means there is either a short circuit or an overload. In either case the trouble should be corrected before replacing the fuse or resetting the circuit breaker. Never replace a fuse with one of higher current rating. That would defeat its purpose and allow an overload to develop in the house wiring. The overload would heat up the wires and possibly cause a fire. Never place a piece of metal, such as a penny, under a fuse. Always turn off the main power supply to the house before replacing a fuse.

determined by multiplying voltage times amperage.

$$P = E \times I$$

In this equation, P is power (in watts), E is electromotive force (in volts), and I is intensity of current (in amperes).

A watt hour would be the amount of work done by one watt in one hour. It takes 60 watts to light a 60-watt bulb for one hour. This would amount to 60 watt hours. A kilowatt is simply a thousand watts. It is easier to measure large amounts of electrical power in kilowatts than in watts, just as it is easier to count large sums of money in dollars rather than in pennies. A kilowatt hour is the amount of work done by a thousand watts in an hour. Here is a simple example. If it took two kilowatts to operate all the appliances in a home for an hour, and if the appliances were used for four hours, the total electricity used would be eight kilowatt hours.

QUESTIONS AND ACTIVITIES

1. What are three common kinds of electrical cable used in homes?

2. What is the purpose of a meter? How does it work? What do each of the dials indicate? How do you read the numbers on the dials? How many kilowatt hours were used on the meter in Fig. 95-3?

3. Define kilowatt hours.

4. What is the purpose of a fuse? Explain how a fuse is "blown."

5. What are two types of fuses commonly used in the home and what size circuit is each used for?

6. What are circuit breakers? How are they different from fuses?

7. What should be done before replacing fuses or resetting circuit breakers?

8. What are three safety tips to remember when replacing a fuse?

CHAPTER 96

Telecommunication

Telecommunication is communication at a distance. This chapter discusses the ways people communicate over distances: by telegraph, telephone, radio, and television.

For centuries most information traveled only as fast as people could carry it. The use of carrier pigeons and smoke signals helped a little. However, it was only when people learned to use electricity that fast, effective communication across long distances became possible.

TELEGRAPH

The first method of using electric current for communication was the telegraph. It was developed about 1837 by Samuel F. B. Morse. In 1844 he sent and received messages between Washington, D.C., and Baltimore, Maryland. The distance was about 40 miles.

His telegraph had an electromagnet with an arm that clicked when a current flowed through the circuit. A spring pulled the arm from the magnet when the circuit was broken. The device to open and close the circuit was called a *key* and was operated like a push button.

Figure 96-1 shows the circuit of a simple telegraph. Notice that one wire is connected to the symbol for ground. This system needs only one wire because the earth serves as another "wire" and completes the circuit.

When the key is pressed down, current flows in the circuit. The electromagnet at the other end makes a click. The clicks can be varied to make a code—the Morse Code of dots and dashes. Dots are made by short contact of the key and dashes by longer contact. The International Morse Code is shown in Fig. 96-2.

Today in commercial telegraph systems thousands of messages may be sent over one line in one day. To help speed up these messages, devices other than the hand key are used. Messages may be sent on a machine that looks like a typewriter. It is called a *teletype*. At the receiving end the message may be printed automatically on sheets of paper or a paper tape.

TELEPHONE

The second method of using electric current for communication was the telephone. It was invented by Alexander Graham Bell. In 1876 he said the first words over a telephone to an assistant in another room. Today we can talk by telephone to almost any place in the world.

Of course, the sound waves do not travel from speaker to listener. Instead, the sound is

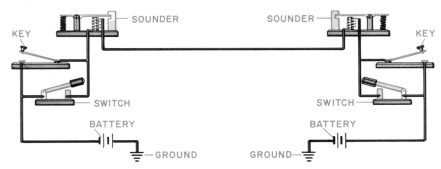

96-1. *Drawing of a simple telegraph circuit.*

INTERNATIONAL MORSE CODE

A •—	N —•	I •————
B —•••	O ———	2 ••———
C —•—•	P •——•	3 •••——
D —••	Q ——•—	4 ••••—
E •	R •—•	5 •••••
F ••—•	S •••	6 —••••
G ——•	T —	7 ——•••
H ••••	U ••—	8 ———••
I ••	V •••—	9 ————•
J •———	W •——	0 —————
K —•—	X —••—	
L •—••	Y —•——	QUESTION
M ——	Z ——••	••——••

PERIOD •—•—•—   COMMA ——••——

96-2. *International Morse Code.*

talk into the transmitter and listen with the receiver.

Inside the transmitter is a thin metal disk called a diaphragm. Fig. 96-3. Attached to the back of this diaphragm is a little cup containing small grains of carbon. Carbon is a conductor of electricity. The amount of current that can flow through the carbon grains depends on how tightly the grains are packed together. The tighter they are

packed, the more current will flow.

The receiver also has a thin metal diaphragm inside. It is held in place by a small permanent magnet. Under the diaphragm is a small electromagnet. Fig. 96-4.

Here is how the telephone works. When you talk, you make sound waves in the air. These strike the diaphragm of the transmitter, causing it to vibrate. The vibration causes changes in pressure on the carbon grains. More pressure packs the grains tightly together; less pressure gives the grains more room. An electric current is passed through the carbon. As you talk, each change in pressure on the carbon grains causes a similar change in the flow of current. In this way the sound waves of your voice are changed into electric currents that are sent over the telephone lines.

changed to electricity which travels over the telephone wires. The telephone circuit has four parts: a source of electricity, a conductor, a transmitter, and a receiver.

The source of electricity is the central office, which also has switching mechanisms for connecting the caller's telephone to the call recipient's telephone.

The conductor, usually wire, connects telephones in various places. In more and more cities, the wires are being replaced by optical fibers. (See Chapter 6 for an explanation of fiber optics.)

The transmitter is used for sending messages and the receiver is used for receiving them. When you use a telephone, you

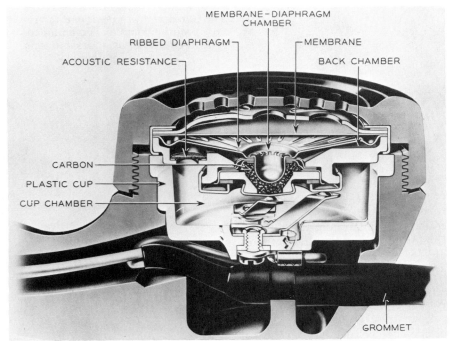

AT&T

96-3. *Cross section of a telephone transmitter.*

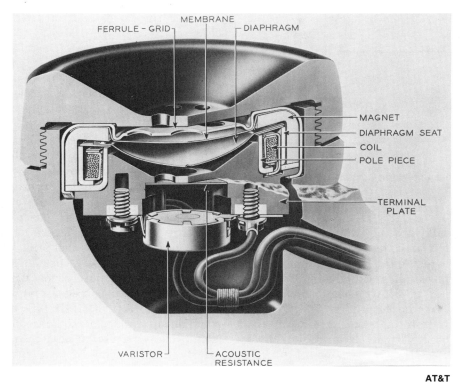

FERRULE – GRID · MEMBRANE · DIAPHRAGM

MAGNET
DIAPHRAGM SEAT
COIL
POLE PIECE

TERMINAL PLATE

VARISTOR · ACOUSTIC RESISTANCE

AT&T

96-4. *Cross section of a telephone receiver. Note the coil of the electromagnet.*

GTE

96-5. *Modern telephones like this GTE Ultrastar™ are designed for today's more complex communications needs, such as accessing long distance services.*

complicated switches pick out the correct line and connect your telephone to it. This completes the circuit so that your conversation can take place. Today it is possible not only to dial practically every place in the United States but also to talk with persons at 98 percent of the telephones in the world. Fig. 96-5.

New Uses for the Telephone System

Today the telephone system is being used for much more than conversation. It can be linked with various other devices to expand communications capabilities. For example, a system called a *facsimile transceiver* combines a telephone with an office copier. Fig. 96-6. A document, such as a report or a memo, is loaded into the machine. The machine converts the images on the document into electric pulses called bits. These are transmitted over telephone lines to an identical machine at another location. The receiving machine converts the bits back to words, numbers, and lines and produces a paper copy of the document.

When the electric current reaches the receiver, it passes through the electromagnet's coils. The electromagnet pulls on the diaphragm. As the current varies, so does the pull on the diaphragm. The movement of the diaphragm makes the sound waves which you hear.

Today many millions of telephones are connected together in one gigantic network. When you dial a telephone number, very

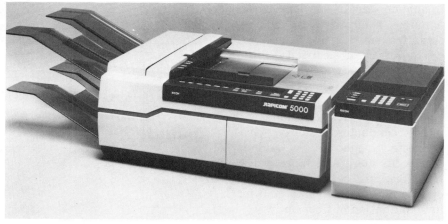

Ricoh Corporation

96-6. *This facsimile transceiver can store documents in its memory for automatic transmission at predetermined times to specified locations. The telephone keypad is built in.*

Computers in different locations can be linked by means of the telephone network. A device called a *modem* enables computer signals to be transmitted and received over telephone lines. Fig. 96-7. This technology has many applications. A stockbroker in New York can send financial reports to a client in Dallas. A publisher in Los Angeles can transmit a manuscript to a typesetter in Chicago. A hobbyist in Milwaukee can participate in computer games with people in Seattle and Orlando.

The telephone network makes videotex systems possible. *Videotex systems* link a specially equipped home television set or terminal to a computer data base by means of the telephone lines. (A data base is a collection of information arranged for rapid search and retrieval.) Users select the information they want to receive from a menu (list) on the TV screen. The appropriate words and pictures are then transmitted from the data base to the home screen. Users of videotex systems have access to a wide range of data, including news, weather, sports, travel, shopping, and entertainment. One applica-

Courtesy of Radio Shack, a Division of Tandy Corporation

96-7a. *On this modem, connection is made by dialing the desired number and placing the telephone handset on the acoustic coupler.*

Hayes Microcomputer Products, Inc.
5923 Peachtree Industrial Blvd.
Norcross, GA 30092

96-7b. *This type of modem plugs directly into a telephone outlet.*

tion of videotex is an electronic directory service which takes the place of a printed telephone directory.

RADIO

In Chapter 6 you learned about light waves. Visible light is a type of electromagnetic radiation. Radio waves† are another type. The difference between radio waves and visible light is that radio waves are longer and have a lower frequency. Fig. 96-8.

The *length* of a wave is determined by the distance from crest to crest (or trough to trough). Fig. 96-9. The *frequency* is determined by the number of crests that pass a given point in one second. The *velocity* (speed) is the length multiplied by the frequency. The velocity of all electromagnetic waves is the same: 300 000 000 metres (186 000 miles) per second. The speed re-

†In this sense, the term "radio waves" refers to a certain portion of the electromagnetic spectrum. This portion includes the frequencies used for television, radiotelephone, and radiotelegraph transmissions as well as those used for radio transmissions.

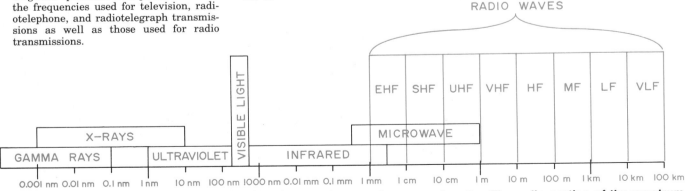

96-8. *The electromagnetic spectrum. The scale at the bottom shows wavelengths. The radio portion of the spectrum is shown in color. Radio waves range from extremely high frequencies (EHF) to very low frequencies (VLF).*

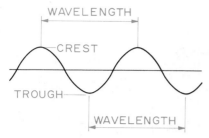

WAVELENGTH

CREST

TROUGH

WAVELENGTH

96-9. Wavelength is determined by the distance from crest to crest or trough to trough.

mains constant because, as length increases, frequency decreases.

Wavelength is measured in metres or fractions of a metre. The shortest waves, gamma rays, have wavelengths of less than 0.001 nanometre. This is smaller than an atom. (One nanometre is one-billionth of a metre. That is 0.000 000 039 inch.) At the other end of the spectrum are the radio waves, which can be as long as 100 000 metres.

Frequency is measured in hertz. One hertz is equal to one complete wave cycle (crest to crest) per second. The radio portion of the electromagnetic spectrum is generally considered to range from 45 gigahertz (45 billion hertz) down to 10 kilohertz (10 thousand hertz).

Besides length and frequency, waves have amplitude. The *amplitude* is one-half of the vertical distance between the wave's crest and its trough.

Electromagnetic waves can be used to transmit information. This is done by modulating (varying) one of three characteristics: amplitude, frequency, or duration (time it lasts). Here is how it works in broadcast radio:

At the studio, sound, such as music or an announcer's voice, causes the air to vibrate. This vibration causes a plate in the microphone to vibrate. The vibrating plate produces an electric signal. The frequency and amplitude of this signal correspond to those of the sound. Therefore, it is called the *audio signal,* although it cannot, of course, be heard. The audio signal is very weak, so it must be amplified (strengthened).

At the same time, a transmitter produces a continuous signal, called a *carrier wave.* The carrier wave must also be strengthened. The audio signal and the carrier wave are then combined to form the modulated radio waves that are sent out by the transmitter's antenna.

In amplitude modulation (AM radio) the audio signal varies the amplitude of the carrier wave. In frequency modulation (FM radio) the audio signal varies the frequency of the carrier wave. In another system of modulation, the carrier wave is switched on and off in pulses. The duration of each pulse is determined by the audio signal.

The radio waves travel through the atmosphere in all directions. When they reach a receiver that is tuned to their frequency, they are demodulated. The audio signal is fed into the loudspeaker and converted back into sound.

TELEVISION

It is not likely that you will work on a television set in class. However, you will probably be interested in learning something about how television works.

The television camera shoots a scene similar to the way a movie camera does. The camera sends electrical impulses of what it sees to the transmitting station. The transmitting station sends out the sight and sound (in the form of radio waves) to receivers that are within range.

The receiver converts the radio waves to electrical signals and separates the picture and sound signals. It sends them to the pic-

Paul M. Schrock

96-10. Antennas such as this are used to send signals up to orbiting satellites.

ture tube and the loudspeaker. The loudspeaker converts the electrical signals back into sound. The picture tube (cathode ray tube, or CRT) in a TV receiver has a screen, which is the glass seen by the viewer. This screen is coated with a compound which glows like a fluorescent light when hit by electrons. Each little area glows in brightness according to the amount of light in that part of the picture scanned by the camera. In this way, an image is produced on the screen.

For color television, the camera picks up the scene in color and sends impulses to the transmitter just as in black and white television. The transmitter sends out the signal, which is picked up by the set and changed back to a color scene in the picture tube.

Cable television works the same way, except the signals are transmitted over a coaxial cable rather than through the air. A *coaxial cable* is a tube of electrically conducting material which surrounds an insulated central conductor wire. A cluster of such lines combined into one cable is also called a coaxial cable. Coaxial cables can be used in telegraph and telephone systems as well as in cable television systems.

TELECOMMUNICATIONS SATELLITES

You have learned that information can be transmitted over wires or optical fibers, as in telephone systems, and through the air, as in radio broadcasts. These methods have their limitations. It is expensive to run cable to remote and sparsely populated areas, and geographical features such as mountains can block radio waves. Furthermore, radio waves decrease in intensity the farther they travel. Satellites provide a solution to these problems.

A telecommunications satellite is placed in geosynchronous orbit around the earth. This means it stays over the same part of the earth and is easy to locate. The transmitting station beams signals up to the satellite. Fig. 96-10. The satellite beams them back down to receiving stations.

In commercial broadcasting the receiving stations then transmit to their surrounding area. It is also possible to send signals directly to individual users rather than to a commercial station. Teleconferencing is an example. A *teleconference* is a type of closed-circuit television. A program is transmitted to a satellite, which relays it to a small receiving antenna. From there it is sent to one or more locations.

Teleconferences provide a way to convey information to large numbers of people in widely scattered areas. For example, a computer manufacturer might hold a teleconference to introduce dealers to the newest line of products. The high cost of travel and accommodations can make teleconferencing an economical alternative to seminars and conventions.

One problem with using satellites has been that, once in orbit, they could not be repaired if something went wrong. The Space Shuttle now has the capability of retrieving satellites from orbit for servicing. Fig. 96-11.

NASA

96-11. *Artist's conception of a satellite repair mission. One astronaut has brought the disabled satellite to the Shuttle. The Shuttle's Remote Manipulator Arm is pulling the satellite towards the cargo bay, where a second astronaut will repair it.*

QUESTIONS AND ACTIVITIES

1. Define telecommunication.

2. What was the first method of using electric current for communication? What modern communications system has evolved from this early method?

3. What are the four parts of a telephone circuit?

4. Name three uses for the telephone other than conversation. Describe each of these uses.

5. What is the difference between radio waves and visible light?

6. What determines the length of a wave? What unit of measurement is used to indicate wave length?

7. What determines the frequency of a wave? What unit of measurement is used to indicate frequency?

8. What is the amplitude of a wave?

9. What three things can be done to electromagnetic waves in order to use them to transmit information?

10. What two things combine to form the radio waves that are sent out by a transmitter's antenna? Define each of these two things.

11. What do the letters AM and FM stand for? What is the difference between AM and FM radio?

12. Define coaxial cable. What are three telecommunications systems that use coaxial cable?

13. What is teleconferencing and how does it work? What is the main advantage of this type of communications system?

PART 5

Power and Energy

For the cave dwellers the only power available was their own muscles. They traveled, hunted, built shelters, and raised food, all with muscle power. After a time they learned to tame some of the animals they had hunted and put the animals' muscles to work for them.

After many years people learned to use tools to increase their power. The first such tool may have been a tree branch which was used as a lever. Such a lever is an example of what is now called a *simple machine*. The other simple machines include the wedge, the wheel and axle, the pulley, the screw, and the inclined plane. Fig. 97-1. The machines we use today are combinations of these simple machines.

Only within the last 150 years have we learned how to add other than muscle power to machines. One of the first power sources was the steam engine. James Watt is given credit for producing the first really workable steam engine about 1800. After that steam was an important source of power for many years. Today steam is still used to power the turbines for producing electricity.

The next great step in the development of power sources came with the invention of the gasoline engine. Dr. Nicholas Otto developed a working engine in 1876. Today millions of cars, trucks, and buses—as well as pumps, generators, and power tools—use the gasoline engine as a power source.

The steam engine and the gasoline engine are just two of the power sources we use to help us do work. Everyone is familiar with the jet engine on airplanes. Fig. 97-2. Electrical power is used for many jobs which muscle power would not be strong enough to do. Since World War II the use of the atom as a power source has become important. Nuclear reactors generate heat which produces steam to run electrical generators.

With today's emphasis on conserving our natural resources and environment, there are many people investigating the use of solar power. The sun's rays are used in certain sections of the country for heating and cooling homes and heating water. Using wind as a source of power for generating electricity is also being researched. The heat within the earth is another possible energy source. All of the

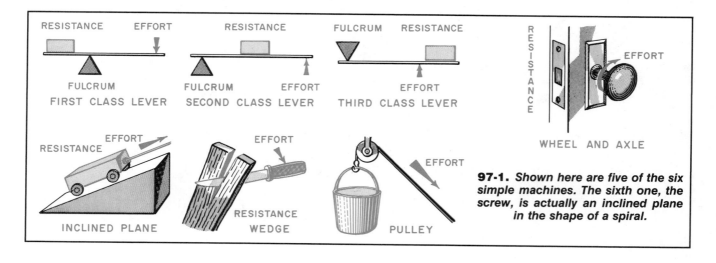

97-1. Shown here are five of the six simple machines. The sixth one, the screw, is actually an inclined plane in the shape of a spiral.

McDonnell Douglas Corporation
97-2. *A DC-10 passenger jet. This airplane can carry up to 380 people.*

97-3.
Metric Units Related to Power and Energy

Quantity	Unit of Measurement	Symbol
Speed	metre per second kilometre per hour	m/s km/h
Power	watt kilowatt	W kW
Energy	kilowatt-hour	kW · h
Electric potential difference	volt	V
Electric current	ampere	A
Electric resistance	ohm	Ω
Frequency	hertz	Hz
Pressure	pascal	Pa

power sources and machines were developed to help people do work more easily, better, or faster.

POWER AND ENERGY MEASUREMENTS

Presently both the customary units and the metric are being used by people who work with the various power generating devices. Eventually the metric units will become more commonplace. You should know and understand at least the few units given in Fig. 97-3.

QUESTIONS AND ACTIVITIES
1. Name the six simple machines.
2. What two important sources of power were developed and used in the 1800s?

3. Name three sources of power that are being investigated to help conserve our natural resources and environment.

CHAPTER 98

Power and Energy

Energy is the capacity to do work. There are six forms of energy: mechanical, thermal (heat), light, chemical, electrical, and nuclear. One form can be changed into another. For example, in Chapter 91 you learned that chemical energy can be converted to electrical energy in an electrical cell.

To make energy work for us, we must be able to control it. *Power* is a source or method of supplying controlled energy. It is a specific amount of energy applied for a specific time. The only energy the first humans could control was their own. They used their muscles to hunt food, build shelters, and do the other labor required to stay alive. Over time, people learned to use the energy in nature to help them. This chapter describes the major forms of power and energy used in today's technology.

ENGINES

Much of the power we use is supplied by engines that convert heat energy into mechanical energy. This is done by burning fuel to expand gases and increase their pressure. These pressurized gases are then used to produce motion.

Steam Engines

The steam engine is an *external combustion* engine. This means the pressurized gases which cause motion are produced outside the engine itself. Heat is applied to a *boiler,* a closed container holding water. Any energy source can be used to heat the water: coal, oil, gas, solar energy, etc. The heat causes the water to change to high-pressure steam. The steam travels through pipes to the engine, where it drives pistons or turbines. Pistons move up and down; turbines rotate.

The first industrial revolution started when steam power was used to pump water from the coal mines of England. James Watt improved the steam engine which had been invented by Thomas Newcomen.

Not only did the invention of the steam engine change factories, but it also changed transportation. It was the steam engine that George Stephenson used in the locomotive which he invented, and Robert Fulton used it in his steamboat. Until the end of World War II many steam engines were used on the railroads.

Then the diesel locomotive replaced the steam engine. Today, however, research is being done to develop an improved steam engine for locomotives. Fueled by coal, this engine would be cheaper to operate than a diesel engine. To make it more efficient, microprocessors will monitor and control much of the operation.

Steam is still an important source of power. Steam turbines are used for generating much of the electricity used in the United States. Fig. 98-1. The steam turbine is also used for power on large ships.

Gasoline Engines

Probably no invention changed transportation as much as the gasoline engine. It is used to power automobiles, motorcycles, and outboard motors. Gasoline engines are *internal combustion* engines. The fuel is burned inside the engine. The German inventors Carl Benz and Gottlieb Daimler, working independently, developed the first successful gasoline-engine automobiles in 1885 and 1886.

Gasoline engines are either four-stroke cycle or two-stroke cycle engines. The operation of the four-stroke gasoline engine is quite simple. Fig. 98-2. During the *intake stroke* the piston moves down. This makes a suction which draws the fuel and air into the cylinder. When the piston reaches the bottom of the stroke, the intake valve closes and makes the chamber airtight.

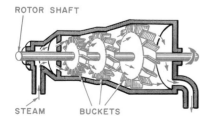

ROTOR SHAFT

STEAM   BUCKETS

98-1. *The steam turbine changes steam power to electrical power.*

PISTON ENGINE

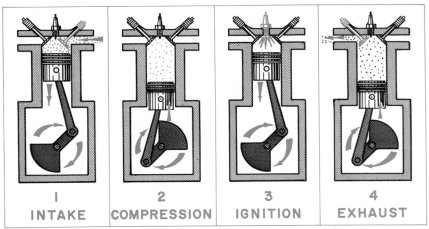

| 1 INTAKE | 2 COMPRESSION | 3 IGNITION | 4 EXHAUST |

98-2. *A four-stroke cycle piston engine.*

During the *compression stroke* the piston moves up and compresses the fuel and air. When the piston gets to the top of the stroke, a spark from the spark plug ignites the mixture of fuel and air. The fuel burns very rapidly, almost like an explosion, and produces great force. The force of this explosion pushes the piston down. This downward movement is called the *power stroke*. During this time both the intake and exhaust valves are closed. As the piston gets to the bottom of the power stroke, the exhaust valve opens. The burned gases pass out of the engine through the muffler into the air. During the *exhaust stroke* the piston goes back to the top of the cylinder, forcing the gases out.

When the piston reaches the top of the cylinder during the ex-haust stroke, the cycle begins again. When many cylinders are operated together, as in an automobile engine, each cylinder is fired in order to provide smooth operation and more power.

These four processes—intake, compression, power, and exhaust —also take place in the two-stroke engine. However, there are only two strokes per cycle:

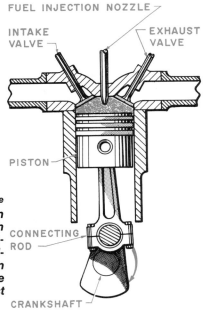

SpaN Magazine

98-3. *The diesel compresses air in the cylinder to a temperature high enough to ignite the fuel spontaneously when it is injected. Expanding hot gases drive the piston down to power the crankshaft, and the rising piston then forces exhaust gases out.*

compression and power. Chapter 99 discusses both types of engines in more detail.

Gasoline engines are made in many sizes. You know how big most automobile engines are. Engines of other sizes provide power for lawn mowers, chain saws, farm tractors, boats, airplanes, and many other pieces of machinery.

Diesel Engines

Another type of internal combustion engine important today is the diesel engine. This was invented by Rudolph Diesel. He was trying to make an engine that was more efficient than the steam engine. He found that if he compressed air enough, it became so hot that it would ignite fuel. Today's diesel engines work on this principle. To prevent the fuel from exploding too soon, the fuel is shot into the cylinder after the air is compressed. Since the compressed air is hot, as soon as the fuel is shot in (or injected) it burns. No spark is needed as in gasoline engines. Fig. 98-3.

At first diesel engines were very large and could only be used where they could be kept in one place. As the engines were improved, the railroads started using them. As the diesel engine was made still smaller, the number of its uses increased. Among other uses, these engines now supply power for trucks, buses, tractors, and boats. Today many small cars and trucks are equipped with fuel-efficient four-cylinder diesel engines.

Jet Engines

About three hundred years ago Sir Isaac Newton, the great British physicist, discovered a law of nature that explains why jet engines work. This principle is known as Newton's third law of motion. It says: For every action there is an equal and opposite reaction.

Gas rushes out of the rear of the engine. That is the action. The engine (and the airplane that contains it) surges forward. That is the reaction. The action and the reaction are opposite, obviously, because the gas and the engine move in opposite directions. They are also equal, as the law states, but this is harder to understand because other forces, such as gravity, are also acting on the airplane.

The important idea to remember is that since they are equal, the speed of the airplane depends on the speed of the gas leaving the rear. This is called the exhaust velocity and is related to the heat of the burning fuel.

Airplanes powered by jet engines are familiar to everyone today. Giant airliners carrying hundreds of people and tons of freight fly from coast to coast in about five hours. Jet airplanes are relatively new; the first jet was flown by the Germans in 1939. The first jet flight in the United States came during World War II, about 1944. However, the principle of the jet engine has been known for many years. In the seventeenth century Sir Isaac Newton designed a carriage which was to be propelled forward by shooting a jet of steam out of the back.

Several types of jet engines are in use today. Three will be de-

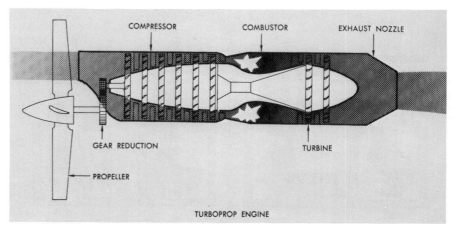

98-5. *A propeller is attached to the turboprop engine.*

scribed here: the turbojet, turbofan, and turboprop.

The *turbojet* engine shoots a jet of hot gases out the back of the engine at high speed. The plane responds by moving forward in much the same way that an inflated balloon shoots forward when you let go of the stem. In the turbojet engine a large amount of air is drawn in the front, compressed, and forced into a combustion chamber. Fuel is shot in and the hot gases try to escape. The only way they can escape is out the back of the engine, and on their way they strike a turbine wheel which drives the compressor, compressing more air. This action continues as long as fuel is injected into the combustion chamber. Fig. 98-4. These engines develop great amounts of power for their

weight. They also allow planes to fly higher and faster than the propeller-driven airplanes. Turbojet engines are used mainly on military aircraft.

The *turbofan* engine is similar to the turbojet. The main difference is a large fan which is mounted on the front of the engine. The fan is powered by turbines at the rear of the engine. This fan forces air into and around the engine, producing more thrust. At low speeds the turbofan is more powerful and more efficient than the turbojet. Many commercial aircraft use the turbofan engine.

For certain uses, such as for short runway takeoffs and flying below 500 MPH, the propeller has advantages. The *turboprop* engine combines the jet engine and the propeller. Several turbine wheels are fastened to a shaft. The power of the jet engine is used to turn the shaft, which has a propeller fastened to its end. This jet-propeller combination furnishes about twice as much power as a piston engine of the same weight. Fig. 98-5.

The prop-jet is one example of how the jet engine can be used to turn a shaft. This principle can be used for purposes other than

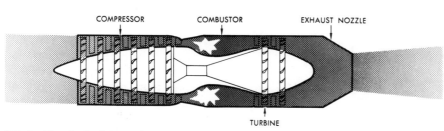

98-4. *The turbojet engine. Hot gases rushing out the exhaust nozzle force the airplane forward.*

driving a propeller. *Gas turbine* engines are now being tested by several automobile manufacturers for use in cars and trucks. Gas turbines differ from those used in airplanes. Instead of fastening extra turbine wheels to the shaft which drives the compressor, a second turbine is fastened to a separate shaft. It is not connected to the first turbine, but is driven by the same gases. This shaft can drive the wheels of a car or locomotive or drive an electric generator.

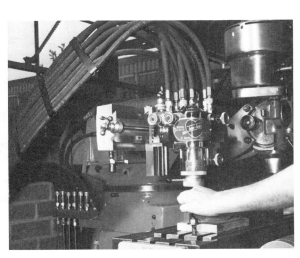

98-7. *Fluid power is used to operate this machinery. The power is transmitted through these hoses by a liquid under pressure.*

Rocket Engines

Rocket engines work on the same principle as jet engines. The difference is that rocket engines carry their own supply of oxygen, while jet engines use the oxygen from the air. Rockets are

NASA

98-6. *To escape the earth's gravitational pull requires a speed of seven miles per second. Only rocket engines provide the power to reach such speeds.*

the only engines which can provide enough power to propel a vehicle into outer space. Fig. 98-6.

There are two types of rocket engines. *Liquid-propellant* rockets use fuel and oxygen in liquid form. Common fuels are kerosene, alcohol, and liquid hydrogen. The fuel is mixed with the oxygen and ignited in the rocket's combustion chamber. *Solid-propellant* rockets use fuel and oxygen that have been premixed and produced in solid form. Solid-propellant rockets are simpler than liquid-propellant rockets, but their power output cannot be regulated. The Space Shuttle uses liquid oxygen/liquid hydrogen main engines and solid-propellant booster rockets.

FLUID POWER

Fluid power is becoming very important in industry. Fluid power is really not a source of power like the gasoline engine. Instead it is a way of transmitting power and applying it to do work. Pressure on a fluid or gas transfers the power from a source to a machine or other device that uses the power.

There are two types of fluid power systems. *Hydraulic* systems use a liquid, usually oil, to transmit power. *Pneumatic* systems use a gas, usually air. Today fluid power is used for operating car lifts, power shovels, aircraft wheels, and many other things. Figure 98-7 shows fluid power equipment for operating a piece of machinery.

ELECTRICAL POWER

Electrical power is used extensively in homes, businesses, and industry. Chapters 88-96 discuss the forms, sources, and uses of electrical power.

ENERGY SOURCES

To run engines or generate electricity, we need energy. Energy can be divided into two categories: potential and kinetic. *Potential energy* can be thought of as stored energy. A chunk of coal, for example, has potential energy. The energy is not actually doing anything, but it is ready for use. *Kinetic energy* is energy in motion. Burning the coal will release energy in the form of heat and light. The heat can be used to turn water into steam. The steam can be used to run a turbine generator to produce electricity.

Our technology requires vast amounts of potential energy

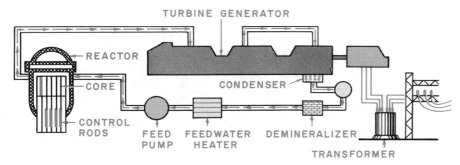

BOILING - WATER REACTOR

Iowa Electric Light and Power Co.

98-8. *A boiling-water nuclear reactor for generating electricity.*

which we can convert to kinetic energy for use in our factories, businesses, vehicles, and homes. The following pages discuss some of the sources for this energy.

Fossil Fuels

Fossil fuels come from plants and animals that lived millions of years ago in warm, swampy areas. As the plants and animals died, their bodies sank to the bottom of the swamp. Movements of the earth's crust eventually pushed these remains underground. There the heat and pressure gradually changed them into coal, petroleum (oil), and natural gas.

The energy of fossil fuels is released by burning. The heat can be used directly (as in a natural gas furnace) or it can be converted to other forms of energy. Most of the electrical power plants in this country use coal heat to drive the steam turbines that generate electricity.

For years, fossil fuels were a cheap and readily available source of energy. This is no longer true. Although the United States has large coal reserves, mining the coal and controlling the pollution that burning coal creates are expensive. The prices of oil and natural gas have risen sharply over the past decade.

Furthermore, fossil fuels are not renewable energy sources, unless one can wait millions of years. If we are to meet our energy needs for the near future, we must develop additional sources.

Nuclear Power

In nuclear reactions matter is changed into energy. The change takes place at the atomic level. There are two types of nuclear reactions. In *fission,* atoms are split apart. In *fusion,* they are combined. Both types of reactions yield enormous amounts of energy in relation to the amount of matter involved.

Fission

Nuclear power plants in operation today use nuclear fission as their energy source. Uranium atoms are split in a reactor. This process produces energy, mostly in the form of heat. A liquid passing through the reactor in pipes absorbs the heat. The liquid is next channeled to a boiler. There the heat is used to turn water into steam. The steam runs a turbine to generate electricity. Fig. 98-8.

Nuclear power plants are regulated and monitored to insure safe operation. However, there have been problems at some plants. In addition, the disposal of radioactive wastes is a problem. When it can no longer be used to generate steam, much of the material is remanufactured

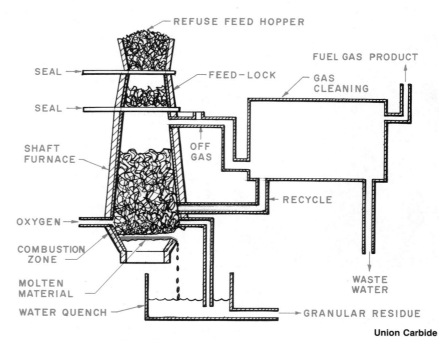

Union Carbide

98-9. *Producing fuel gas from waste.*

and used over again in other plants. Some material is waste and must be disposed of where it is of no danger to the public. This waste is usually buried underground in atomic disposal areas where it must remain for many years.

FUSION

In fusion two hydrogen nuclei are combined to form one helium nucleus. The process releases large amounts of heat and light. Fusion is the reaction which powers the sun.

The difficulty with building a fusion reactor on earth is that tremendous pressure and heat are needed to force the hydrogen nuclei together. Fusion has been achieved in experiments, but large-scale commercial reactors are still some years away.

Synthetic Fuels

Synthetic fuels, or *synfuels,* are liquid or gaseous fuels which are usually derived from existing solid fuels. Synfuels can be produced from coal, heavy crude oils, tar sands, and oil shale. Coal is the most frequently used raw material for synfuels. It can be made into various types of ready-to-use liquid and gaseous fuels. Heavy crude oils, tar sands, and oil shale yield a synthetic crude oil that must be refined before use.

Note that all these sources of synfuels are nonrenewable. Synfuels can also be made from renewable sources. For example, starchy grains, such as corn, can be used to make ethyl alcohol, or ethanol. Ethanol can be mixed with gasoline, a petroleum product, to make gasohol.

Biomass

Biomass refers to plant and animal waste, algae, and garbage. When these organic wastes are allowed to decompose, or rot, carbon dioxide and methane gas are produced. Methane gas can be burned in boilers to produce steam for running electrical generators.

A more efficient method of producing gas from solid organic waste involves the use of heat. When organic materials are heated in the absence of oxygen, they break down into several gases, liquids, and tars. Wastes are placed into a chamber, or reactor, and hot inert gas is used to heat the wastes. (An inert gas is one that will not react chemically with the wastes.) This process is called *pyrolysis*. This process can be run continuously. The gas that is produced is taken off, cleaned, and used for heating homes, cooking, or for generating electricity. The solid material that is left is used for landfill. Fig. 98-9.

Water

People have long used water to make work easier. A waterwheel placed in a moving stream will change some of the energy of the stream to other types of energy. This energy may be used for grinding grain, sawing wood, or other useful purposes. In the early days of the United States water furnished most of the mechanical power for factories.

People kept looking for improvements. A waterfall, obviously, was a source of more energy than a flowing stream. However, the early waterwheels were not very efficient and could not be used with large waterfalls. Eventually a better waterwheel, called the Pelton wheel, was invented to answer this need. The Pelton wheel was widely used in the western United States.

Today waterwheels or turbines do not supply power directly for running machinery. Instead they change water power to electrical power that can be sent over long distances and used to run machinery. Fig. 98-10. The production of electricity by water power is called *hydroelectricity.* About 14 percent of the electricity in the United States is generated by falling water.

Water is a potential energy source in another way. Water molecules are made of two gases: oxygen and hydrogen. Hydrogen is the cleanest fuel; when it burns in air, the waste product is water. Hydrogen is also plentiful; two-thirds of the earth's surface is covered with water. If a way can be found to economically split water into its component gases, we will have an abundant source of clean energy. Up to now the difficulty has been that it takes as much or more energy to split water molecules as is gained from burning the hydrogen that is yielded. Recent research on using sunlight to provide the energy for splitting water molecules has brought encouraging results.

Geothermal Energy

Geothermal power means using natural steam from within the earth to run electrical generators. In some places the magma,

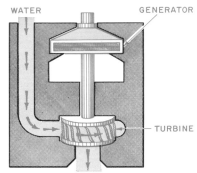

98-10. The water turbine. This is used for changing water power to electrical power.

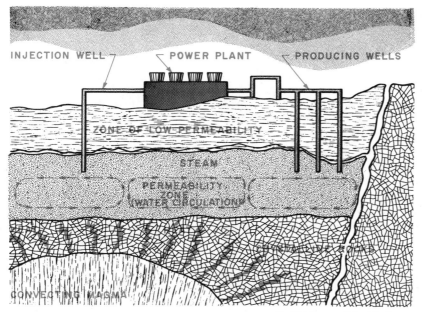

INJECTION WELL — POWER PLANT — PRODUCING WELLS

ZONE OF LOW PERMEABILITY

STEAM

PERMEABILITY ZONE (WATER CIRCULATION)

CONVECTING MAGMA

Pacific Gas and Electric Company

98-11. *A geothermal power plant uses steam from the earth to produce electricity.*

or molten rock, which is normally deep within the earth is close to the surface. Steam is formed from water which comes in contact with the magma. Wells are drilled which tap this steam. Geologists say that these steam wells may last for a thousand years.

Areas in which geothermal power is available include the western United States, Iceland, and New Zealand, as well as other areas. Fig. 98-11.

Steam from the wells is carried by pipes to turbine generators. From this point on, an electrical generating station run by geothermal steam is the same as a plant using steam from any other source, such as coal.

Wind Energy

For centuries wind has furnished free power to the world. Windmills have long been used by the Dutch to pump water from their land. In the early days of this country, windmills were used to pump water for use on farms.

Today, windmills can be used to generate electricity. Surveys have shown that there is tremendous potential in wind power, but that it is not a very reliable source of energy. The development of high efficiency windmills that will work with little wind will make the use of wind power practical.

In the 1940s there was a unit in Vermont which generated power to be fed into the regular electrical system of a town. When

the wind velocity was between 20 and 70 miles per hour, the wind generators cut into the line. Damage to the blades eventually caused this unit to be shut down and dismantled. Today Southern California Edison is using a new wind turbine generator which measures 191 feet at the highest point of rotation. This is located at a new Wind Energy Center near Palm Springs, California. Fig. 98-12.

Solar Energy

Of all the sources of power available today, solar power— power from the sun—offers the best alternative to the sources in common use. The radiation received from the sun over the total surface of the earth is many times larger than the world's energy needs. The problem is how to harness this energy efficiently and economically. Another problem is its irregularity due to clouds and the changing seasons. When these problems are taken

Southern California Edison

98-12. *A wind turbine generator. Wind power as an energy source is most promising in areas where winds are strong and steady, such as on the Great Plains, in mountain passes, and along coastal areas.*

NASA

98-13a. *This is an experimental house built by NASA to make use of the technology developed by the space program. Note the solar collectors on the roof.*

The heat of the sun can also be used to produce electricity. An array of mirrors is used to focus the sun's rays onto a boiler. The boiler produces steam to run a turbine generator. Fig. 98-14.

SOLAR PONDS

Solar ponds are shallow bodies of salt water. The water near the bottom is saltier than that near the top. Sunlight passes through the upper layer and becomes trapped as heat in the lower layer. The hot water is pumped to a turbine house where its heat vaporizes a special liquefied gas. The vapor drives a turbine generator. The vapor is changed back to a liquid by cooler water from the pond's surface. The water is then returned to the pond to keep the cycle going.

care of, solar energy can be used for a variety of purposes.

HOME HEATING

A solar heating system in a home needs an energy collection system, a heat storage system, and a pump to distribute the heat. The energy collectors are usually flat panels located on the roof of a house. The energy is used to heat water in large tanks or to heat rocks. The water or rocks will hold the heat for quite a while. When it is needed, the heat is distributed by water which is pumped through the system or by air which is moved through ducts by fans. Fig. 98-13.

While the sun shines, heat is collected by the rooftop panels and stored. During the night and on cloudy days, the stored heat is used to warm the home.

ELECTRICITY

Solar cells convert sunlight into electricity. (See Chapter 91.)

98-13b. *Heating a home with solar energy. When the sun shines on the solar collectors on the roof, warm air is circulated through the house. The cool air is sent up to the roof to be warmed by the sun.*

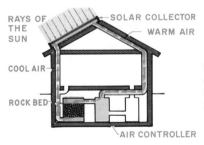

98-13c. *Once the home is at the desired temperature, the warm air from the rooftop can be diverted to the rock bed. Here the heat can be stored for future use.*

98-13d. *During the night and on cloudy days, the stored heat is used to warm the house. As you can see in this drawing and in the two before it, there is an auxiliary heater. This is a regular gas or electric heater. It supplements the solar heat and can take over completely when solar heat is not available.*

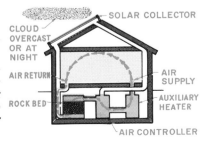

98-14. *A field of giant mirrors (heliostats) reflects the sun's rays to Solar One's boiler (receiver) atop a 300-foot tower. The heat produces steam to operate a turbine generator on the ground below. This solar-powered electric generating facility could provide enough power to meet the needs of about 6000 homes.*

QUESTIONS AND ACTIVITIES

1. Define energy. What are the six forms of energy?

2. Define power. Briefly explain how engines supply power.

3. What does the term "external combustion" mean? Briefly describe how external combustion engines work.

4. What does "internal combustion" mean? What are three things powered by internal combustion engines?

5. What are the four processes that take place in a four-stroke cycle engine?

6. Briefly describe how diesel engines operate. Name three things that are powered by diesel engines.

7. Describe the law of nature that explains why jet engines work. What determines the speed of a jet airplane? Name three types of jet engines.

8. What is the difference between rocket engines and jet engines? Name two types of rocket engines.

9. Define fluid power. Name two types of fluid power systems.

10. Define potential energy and give an example. Define kinetic energy and give an example.

11. Where do fossil fuels come from? Name three fossil fuels. Why must we develop sources of energy other than fossil fuels?

12. What happens in nuclear reactions? Name and define the two types of nuclear reactions. Which type is used by nuclear power plants?

13. Define synfuels. What are three sources of synfuels?

14. Define hydroelectricity.

15. Briefly describe three ways solar energy can be harnessed and used as a source of power.

CHAPTER 99

Small Gasoline Engines

Almost everywhere you look today there is some machine that makes use of a small gasoline engine. From lawn mowers to motorcycles, from cement mixers to portable electrical generators, the small engine has many uses. Fig. 99-1.

The small gasoline engine operates on the same principles as the larger automobile engines. A fuel/air mixture is taken into a cylinder, compressed, and ignited to produce the force to drive the piston down, creating power. This explosive up and down power of the piston is converted into rotary power by using a connecting rod to connect the piston to the crankshaft. The crankshaft drives the functional part of the machine, such as the blade of a lawn mower. Most small engines in use are either two-stroke cycle or four-stroke cycle engines.

FOUR-STROKE CYCLE ENGINE

A cutaway view of a four-stroke cycle engine is shown in Fig. 99-2. The cylinder block is a casting, usually made from aluminum for lightness. On top of the cylinder block is the *cylinder head,* which is the top of the combustion chamber in which the fuel is ignited. The head has a threaded hole into which a spark plug is placed. The *spark plug* produces a spark in the cylinder to ignite the fuel.

Inside the cylinder block is the piston. The *piston* moves up and down in the cylinder and is attached to the connecting rod, which, in turn, is attached to the crankshaft. The piston has several grooves cut around it into which rings are placed. These *piston rings* help to seal the piston to the cylinder wall and to

99-1. *Typical small gasoline engine.*

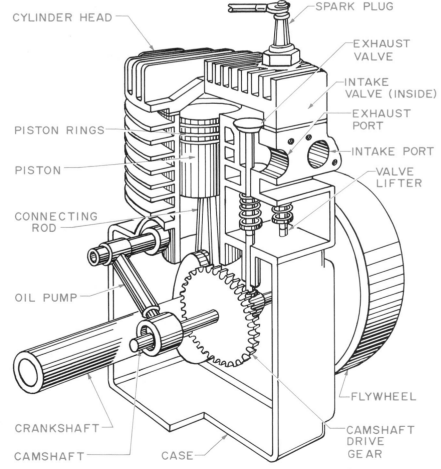

99-2. *Cutaway view of a four-stroke cycle engine.*

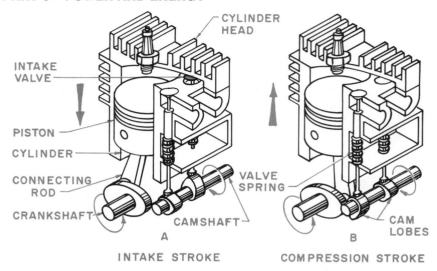

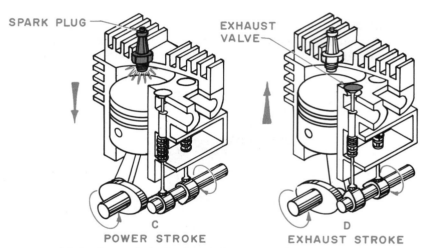

99-3. *Operation of a four-stroke cycle gasoline engine.*

keep gases and oil from escaping, ensuring high pressure within the cylinder.

The *crankshaft* has an offset so that as the piston moves up and down it causes the crankshaft to turn. The crankshaft is in the *crankcase,* which also houses the lubricating oil. All the moving parts are lubricated from the crankcase. The *flywheel* (in the case of the small engine shown here) is the part that is set into motion by the action of pulling a starter rope. It is attached to the crankshaft and creates the momentum to keep the crankshaft running smoothly.

The *camshaft* operates from a timing gear from the crankshaft. On the camshaft are two cams that open the valves as they turn. A heavy spring closes the valves. Fig. 99-3. There are two valves, intake and exhaust. The *intake valve* opens to allow the fuel/air mixture into the cylinder. The *exhaust valve* opens to allow the burned gases to escape.

In order to produce the power needed to turn the crankshaft, all engines go through four steps. The operation of a small four-stroke cycle engine is shown in Fig. 99-3. During the intake stroke the piston moves down,

creating a suction which opens the intake valve and draws fuel and air into the cylinder. When the piston reaches the bottom of this stroke, the intake valve closes and makes the cylinder airtight. Fig. 99-3(A). During the compression stroke, Fig. 99-3(B), the piston moves up and compresses the fuel/air mixture, raising its temperature. When the piston gets to the top of this stroke, a spark from the spark plug ignites the hot fuel/air mixture. The fuel burns rapidly, producing a great force similar to an explosion. This force pushes the piston down. This downward movement is the power stroke. Fig. 99-3(C). During this time both the intake and exhaust valves are closed. As the piston reaches the bottom of the power stroke, the exhaust valve opens. During the exhaust stroke, Fig. 99-3(D), the piston goes back to the top of the cylinder, forcing the burned gases from the fuel/air mixture out the engine through the exhaust valve. When the piston reaches the top of the cylinder during the exhaust stroke, the cycle begins again.

TWO-STROKE CYCLE ENGINE

The two-stroke engine operates without the camshaft and the cam-driven valves that are needed for the four-stroke engine. The two-stroke engine uses a reed valve to control the entry of the fuel mixture into the crankcase. The *reed valve* is a flexible piece of spring steel which is drawn open as the piston moves toward the spark plug. The piston acts as a "valve" to open and close the *transfer port* (which admits fuel to the cylinder) and the *exhaust port* (through which waste gases are expelled). The crankcase houses the connecting rod and crankshaft as it does with the four-stroke engine. In both engines two strokes of the piston turn the crankshaft one revolution. A four-stroke engine spins the crankshaft twice before completing one power cycle. A two-stroke engine turns the crankshaft only one revolution in each power cycle.

The power cycle of a two-stroke engine includes the same four steps or processes as that of the four-stroke engine. However, these steps are completed in only two strokes of the piston; two steps are completed simultaneously during each stroke. First the fuel mixture is drawn from the carburetor into the crankcase through the reed valve. Fig. 99-4(A). The fuel mixture flows through the transfer port into the cylinder, where the piston compresses the mixture as it moves toward the spark plug. Fig. 99-4(B). These two steps combine to make the compression stroke. When the piston is at the top of the cylinder, the mixture is ignited by the spark plug. Fig. 99-4(C). The force of the exploding fuel pushes the piston back, uncovering the exhaust and transfer ports and allowing the exhaust gases to escape as fresh fuel from the crankcase enters the cylinder. Fig. 99-4(D). These two steps make up the power stroke.

Two-stroke cycle engines are small engines built for brief but demanding use. The quick acceleration that is the result of one full turn of the crankshaft during each power cycle makes them good for driving chain saws and small weed-trimming tools. However, these fast revolutions also cause the engine to become hot in a short period of use. For jobs that require lower-speed operations over extended periods of time, like lawn mowing and snow blowing, the larger, cooler-running four-stroke engines are often used.

OPERATION OF GASOLINE ENGINES

There are two main systems necessary for the operation of any gasoline engine. These are the fuel system and the ignition system.

Fuel System

The fuel system must provide the air/fuel mixture for combustion. The basic parts of the fuel system include a gasoline tank to store the fuel, a fuel line to carry

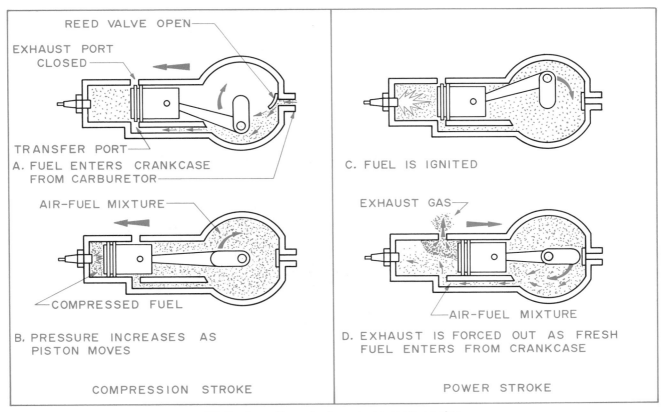

REED VALVE OPEN

EXHAUST PORT CLOSED

TRANSFER PORT

A. FUEL ENTERS CRANKCASE FROM CARBURETOR

AIR-FUEL MIXTURE

COMPRESSED FUEL

B. PRESSURE INCREASES AS PISTON MOVES

COMPRESSION STROKE

C. FUEL IS IGNITED

EXHAUST GAS

AIR-FUEL MIXTURE

D. EXHAUST IS FORCED OUT AS FRESH FUEL ENTERS FROM CRANKCASE

POWER STROKE

99-4. *Operation of a two-stroke cycle engine.*

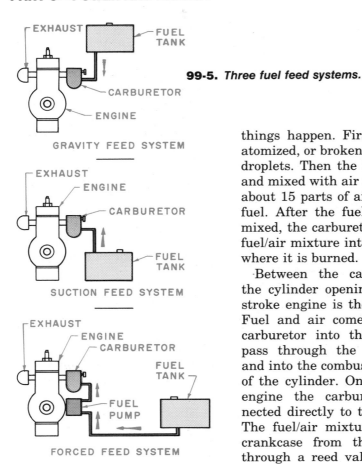

99-5. *Three fuel feed systems.*

things happen. First the fuel is atomized, or broken up into small droplets. Then the fuel is moved and mixed with air in the ratio of about 15 parts of air to 1 part of fuel. After the fuel and air are mixed, the carburetor directs the fuel/air mixture into the cylinder where it is burned.

Between the carburetor and the cylinder openings of a four-stroke engine is the intake port. Fuel and air come through the carburetor into this port, then pass through the intake valve and into the combustion chamber of the cylinder. On a two-stroke engine the carburetor is connected directly to the crankcase. The fuel/air mixture enters the crankcase from the carburetor through a reed valve. From the crankcase, the mixture enters the combustion chamber through

transfer ports that open when the piston is at the bottom of the stroke. Fig. 99-4(D).

Four-stroke engines use plain gasoline. Two-stroke engines use a mixture of gasoline and oil, usually 25 parts gasoline to 1 part oil. However, you should check the owner's manual to be sure you are using the correct ingredients in the right proportions.

CARBURETOR

The essential parts of the carburetor are shown in Fig. 99-6. Fuel enters the carburetor through a fuel inlet and flows into a storage area called a *float chamber.* This chamber has a float in it made of hollow brass. When fuel comes into the chamber, the float rises and closes the inlet. As fuel leaves the chamber, the float drops down. The inlet opens again, allowing more fuel to enter the float chamber.

Next to the float chamber is the *air horn,* which is the carburetor proper. The air horn is a hollow metal tube connected to

the fuel to the carburetor, and the carburetor itself. The fuel must get from the fuel tank to the carburetor before it can be burned. This is done by one of three methods. Fig. 99-5. In the *gravity system* the fuel tank is placed higher than the carburetor and the fuel flows by gravity. In the *suction system* the fuel tank is usually placed below the carburetor and the fuel is sucked from the tank to the carburetor by the intake stroke of the piston. The *forced feed system* uses a fuel pump that draws the fuel from the tank and pumps it to the carburetor. This type of fuel feed system is used on automobiles. Small gasoline engines use either gravity or suction feed.

After the fuel arrives at the carburetor, several important

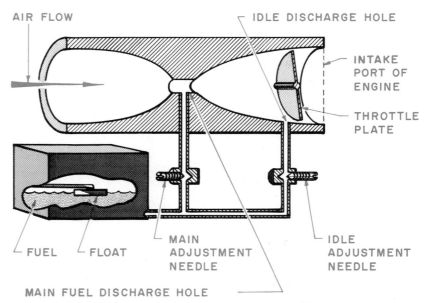

99-6. *Parts of a carburetor on a small gasoline engine.*

the air cleaner at one end and to the engine at the other end. (The air cleaner is important because it filters out dust and dirt from the incoming air.)

Air coming through the air horn mixes with fuel from the float chamber to create the proper fuel/air mixture for combustion. This may be done in various ways. Some air horns have a narrower diameter near the middle. This narrow area is called a *venturi*. As air comes out of the venturi, its speed increases and its pressure decreases. The decreased pressure around the main discharge hole causes the fuel to spray out in a fine mist.

The same result can be obtained in some small engines without a venturi. A tube runs from the float chamber to the air horn. When the piston in the cylinder moves down, it causes a partial vacuum in the cylinder. Outside air pressure forces the air through the carburetor (air horn) to create the fuel/air mist.

Between the venturi and the lower part of the carburetor is the *throttle valve* (butterfly). When an engine is idling, the throttle valve is closed, and only a small amount of air flows past it. In fact, the air flow is so small that it cannot pick up enough fuel to keep the engine running. To supply fuel when an engine is idling or running at low speeds, another fuel passage is used. This passage is a tube which connects with the carburetor below the throttle valve. Fig. 99-6. The amount of fuel that passes through this tube can be controlled by the idle adjustment needle. At higher speeds, the throttle valve is opened, blocking this passage so it does not interfere with the flow of air and fuel.

A cold engine cannot be started on the same fuel mixture it uses for running. It must have a rich mixture; that is, one that has more than the normal amount of fuel in the air. To create a rich mixture, a *choke valve* is used. The choke valve is located in the upper part of the air horn. By closing the choke, air intake is reduced and a vacuum is formed in the carburetor. The vacuum increases the flow of fuel into the carburetor, creating the rich mixture needed for easy starting.

Once the engine is running, the choke valve should be opened to allow more air to enter. If the choke valve on a small engine remains closed too long, raw fuel will enter the combustion chamber. The raw gasoline can damage the cylinder walls.

Ignition System

An internal combustion engine must have a way of igniting the air/fuel mixture to make it burn. Most small engines have an ignition system called a magneto system. Magnetism is used to help generate the electricity that creates a spark to ignite the fuel.

The magneto ignition system consists of several important parts. Fig. 99-7. The *armature* (sometimes called the iron core) consists of several thin strips of soft iron laminated together. The armature makes a path for the magnetism to travel and holds the coil. The *coil* acts as a step-up transformer (changes low voltage to high voltage) and consists of a primary winding and a secondary winding. The *primary winding* is fairly heavy gauge copper wire which is wrapped around part of the armature. One end of the primary winding is attached to the armature and the other end is attached to the breaker points. The *breaker points* act as a switch for the electrical current. They are opened and closed by a cam which is operated by the crankshaft on two-stroke engines or the camshaft on four-stroke engines. The *secondary winding* of the coil is much lighter gauge copper wire and has many more windings than the primary coil. One end of the secondary winding is attached to the armature. From there the wire is wrapped around the primary coil about 20 000 windings (compared to about

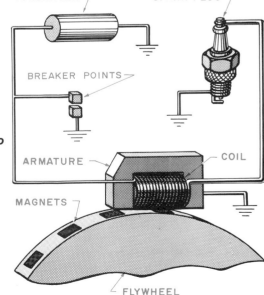

99-7. *A typical magneto ignition system.*

200 windings of the primary coil) and then it leads to the spark plug. The *spark plug* produces the spark that ignites the fuel.

The *secondary circuit* of the magneto system consists of the spark plug and the secondary winding of the coil. The *primary circuit* consists of the primary winding of the coil, the breaker points, and the condenser. Fig. 99-8. The *condenser* acts as a small electrical storage tank which absorbs any remaining surge of electron flow when the breaker points open, cutting the primary circuit. If the condenser did not absorb this remaining current, the current would arc across the gap between the open breaker points, burning the points and weakening the magnetic flow which reverses the current to the secondary coil.

The driving force of the magneto system is the magnets that are attached to the edge of the flywheel. As the flywheel revolves and these magnets near the armature, the magnetic field passes through the armature and builds up low voltage electricity in the primary winding. This current turns the primary winding into an electromagnet and surrounds the secondary winding with a magnetic field. At this time the breaker points are closed and the current passes freely through the primary circuit. Just as the current reaches its maximum strength, a cam opens the breaker points and the primary circuit is quickly switched off. The current swiftly reverses to the secondary winding, creating a surge of power. The fine windings of the secondary coil outnumber the windings of the primary coil by 100 to 1, and the voltage is multiplied accordingly. (If the primary winding has 50 volts, the secondary winding will have 5000 volts.)

99-8. *Schematic of the primary and secondary circuits of the magneto ignition system.*

The current surges through the fine strands of the secondary winding into the spark plug. The current goes through the spark plug and jumps the gap between the terminals of the plug to reach the grounded end of this secondary circuit. Fig. 99-8. The spark created when the current jumps this gap ignites the air/fuel mixture.

Cooling System

During operation, all engines must be kept from overheating. Some small engines, such as outboard motors, are water cooled. Other small engines, such as those in motorcycles and lawn mowers, are air cooled. In Fig. 99-2, you will notice there are fins extending from both the cylinder head and the part of the cylinder block the piston moves in. These fins help keep the engine cool by providing a larger surface for the heat to escape and for the cooler air to move over. The flywheel is often finned, too, so that while it revolves it can act as a fan to help cool the engine.

SMALL ENGINE MAINTENANCE

Although small engines are built to provide trouble-free operation, they still require periodic maintenance and repair. The best way to see that an engine is maintained properly is to refer frequently to the owner's instruction manual provided by the manufacturer and to carefully follow its procedures and specifications. The following paragraphs describe general procedures for engine maintenance.

FOR YOUR SAFETY . . .

• Work on an engine and add fuel or oil only when the engine is stopped and has cooled at least thirty minutes.

• When working on an engine, disconnect the spark plug to avoid accidental starting.

• Do not operate or refuel an engine without adequate ventilation.

• Keep all flames and heat away from gasoline. *Never smoke when refueling.*

• Do not store large quantities of fuel.

Fuel

Use the proper fuel for the type of engine you are using. Two-stroke engines use a mixture of gasoline and oil. Check your owner's manual for the proper mixture. Mix only one gallon of this fuel at a time because gasoline evaporates much faster than oil, and the ratio of gas to oil could be changed by evaporation. Do not use a gasoline and oil mixture in four-stroke engines.

Regular grade leaded gasoline is used in these engines unless the manufacturer recommends another type. Before starting the engine, make sure the vents in the gas cap are not clogged.

Two-stroke and four-stroke engines also use fuel filters to help keep impurities out of the engine. Check your owner's manual for the location of this filter and for instructions on cleaning and changing it.

Oil

All engines need oil between their moving parts. Oil allows the parts to move more easily because it prevents friction. Of course when the parts do not rub against each other, they do not wear out as fast either. Oil also flushes dirt off the engine parts and helps cool the engine by carrying some of the heat away from hot engine parts. Substances that reduce friction, heat, and wear in this manner are called *lubricants*. Two-stroke engines are lubricated by the oil in the fuel mixture and need no additional lubrication. Four-stroke engines are lubricated with oil poured directly into the crankcase through a filler plug hole. The level of oil should be checked every time before the engine is used. Check the oil by removing the filler

99-9. *Checking oil level in a four-stroke cycle engine.*

plug. Fig. 99-9. If the oil is at the top of the filler plug hole the level is high enough. If it is not, add enough oil to reach this level. Some engines have a dipstick to help you check the oil level. Remove the dipstick, wipe it dry, re-insert it as far as it will go, then withdraw it again and check the indicated level.

Oil needs to be drained and changed periodically, usually about every 25 hours of normal operation or 12 hours of operation in dusty conditions. Check your owner's manual for a timetable for oil changes and the type, weight, and quantity of oil to use.

Air Cleaners

Air entering the carburetor first passes through the air cleaner. The cleaner removes dirt from the air to prevent clogging the carburetor or damaging the engine. Periodically the air cleaner must be removed and cleaned. One type of air cleaner, the dry element, is shown in Fig. 99-10. The air cleaner is removed from the engine. Then the element is removed from the air cleaner. Tap the element sharply on a flat surface to dislodge the dirt. This element may also be cleaned by using the pressure of compressed air to blow out the dirt. If the filter is badly clogged, replace it with a new one.

Some small engines have oiled-foam air cleaners. These are made of polyurethane foam which must be kept moist with small amounts of oil. The oily pockets of the foam trap the dirt

99-10. *Removing the dry element from an air cleaner.*

from the air. These filters are cleaned by rinsing them thoroughly in kerosene, squeezing them dry, rinsing in water, and drying. Then drip engine oil into the filter and squeeze it to distribute the oil evenly. The oiled-foam filter is now ready for reuse.

Oil-bath air cleaners are sometimes used on engines that are used under very dusty conditions. This cleaner is like a can with air passages in it and a pool of oil at its bottom. It is mounted on top of the carburetor, as are the other types of air cleaners. Incoming air is directed into the bottom of the can where the pool of oil removes the larger dust particles. The air is then pushed up through an oiled mesh or foam filter which further filters impurities. The clean air then goes down the center of the can into the carburetor.

Spark Plugs

The spark plug supplies the spark needed to ignite the fuel. It too needs periodic cleaning for best operation. The burning fuel leaves a deposit on the spark plug which must be removed. The deposit may be removed by scraping or with a wire brush. If the electrodes are badly burned or the insulator cracked, the plug should be replaced. If you are

going to reuse the plug, set the gap to the manufacturer's specifications by using a feeler gauge. Fig. 99-11. Check the spark plug wire to be sure that it is not cracked and that the terminal going to the spark plug is clean.

Starting Problems

Probably the most common complaint with a small engine is failure to start. If an engine fails to start, there are certain things to check for:

1. Check to see that there is fuel in the tank.

2. Check for spark. Remove the spark plug wire from the spark plug. Wearing a glove to protect yourself from shock, hold it close to the frame of the engine and turn the flywheel. There should be a spark. If not, check the breaker points, the coil, and the condenser.

0.030" FEELER GAUGE

99-11. *Adjusting a spark plug with a feeler gauge.*

3. Check to see if there is compression. Remove the spark plug and place your finger over, but not in, the spark-plug hole in the engine. Pull the starter cord. If there is proper compression the air pressure in the cylinder should draw your finger toward the hole, then blow it away. Lack of compression can be caused by a leaking head gasket, worn valves, or a faulty piston or piston rings. A compression tester is helpful in diagnosing the cause.

4. Check to see that there is gas flowing to the cylinder. Remove the spark plug and crank the engine a few times. There should be gas in the cylinder by now. If there is not, check to see if the fuel line is plugged or if the fuel valve is turned off.

5. Check to see that the engine is not flooded. This can usually be determined by a strong smell of raw gasoline. If the engine is flooded, open the choke and turn off the fuel. Crank the engine until it starts; then open the fuel valve.

Remember: Be sure to follow the manufacturer's directions for the operation and repair of any engine.

QUESTIONS AND ACTIVITIES

1. Briefly describe the principles on which a small engine runs.

2. Where are the moving parts of an engine lubricated?

3. What is the main difference between a two-stroke cycle engine and a four-stroke cycle engine? What type of fuel does each use?

4. Describe the function of each of the following: spark plug, camshaft, piston rings, fins on cylinder head, reed valve.

5. Name and describe the three fuel feed systems.

6. What is the function of the carburetor?

7. Where is the choke valve? What is its purpose? Describe how it is used.

8. Name the parts of the ignition system and describe the function of each part.

9. What are three safety rules to remember when working on engines?

10. Why do engines need lubrication?

11. What is the purpose of an air cleaner? Name three types of air cleaners.

PART 6

Transportation

One of the amazing features of American life is our convenient and efficient transportation system. Goods as well as people are moved from place to place quickly, comfortably, and safely. This not to say that the system cannot be improved. Trains and planes sometimes run late, and baggage is often lost or delayed. But when you consider the size of this nation, with all its people, one must agree that the system is good.

We take our transportation system for granted today. An air trip from Detroit, Michigan, to Washington, D.C., takes about one hour and fifteen minutes. Think of traveling that distance by carriage, when the trip took many days! Efficient transportation has contributed greatly to our economic and industrial growth.

THE TRANSPORTATION SYSTEM

Transportation, like farming or manufacturing, is an important business. Transportation is the business that has to do with travel and communication, with the movement of people and things. It is one of the *service industries* because it serves to move people and materials from one place to another. Fig. 100-1. It does not manufacture anything. In contrast, industries which are *goods-producing* make products.

In order to provide transportation service, we need vehicles such as trucks and ships, and highways and waterways on which they can travel. We also need people to operate the system—drivers, highway engineers, designers, traffic control technicians, and many other skilled workers. Many people play a part in designing, planning, operating, and maintaining our transportation system.

Essentials of Transportation

There are two essential parts to any transportation system: the *way,* or travel space, and the *vehicle,* or carriage unit.

TRAVEL SPACE

The way for travel varies with the type of vehicle. There are highways for trucks and cars, airways for aircraft, railways for trains, and waterways for barges and ships. Some of these ways, such as highways and railways, are clearly visible to us. Airways and some waterways such as those on the open sea are not easily "seen," but they are clearly marked on maps and charts for navigators to follow. Fig. 100-2. Our government maintains over

3 million miles of highways and 25 000 miles of inland waterways and regulates 230 000 miles of airways. Railways, on the other hand, are usually privately owned and maintained.

The terminals or stations are also important parts of the travelways. Airports, bus stations, etc., are needed to provide a scheduled place for loading and unloading people and materials. And, of course, the operations of each travelway must be under the direction of some authority to guard against accidents and to build an efficient and reliable transport system. Control tower operators at airports, signal and control technicians for rail and seaways, a traffic officer or signal light—all of these serve to direct traffic.

VEHICLES

Many types of vehicles haul freight and people. Fig. 100-3. There are airplanes, automobiles,

Amtrak

100-1. Our transportation system moves both people and materials. This electric train carries passengers at speeds of 120 miles per hour.

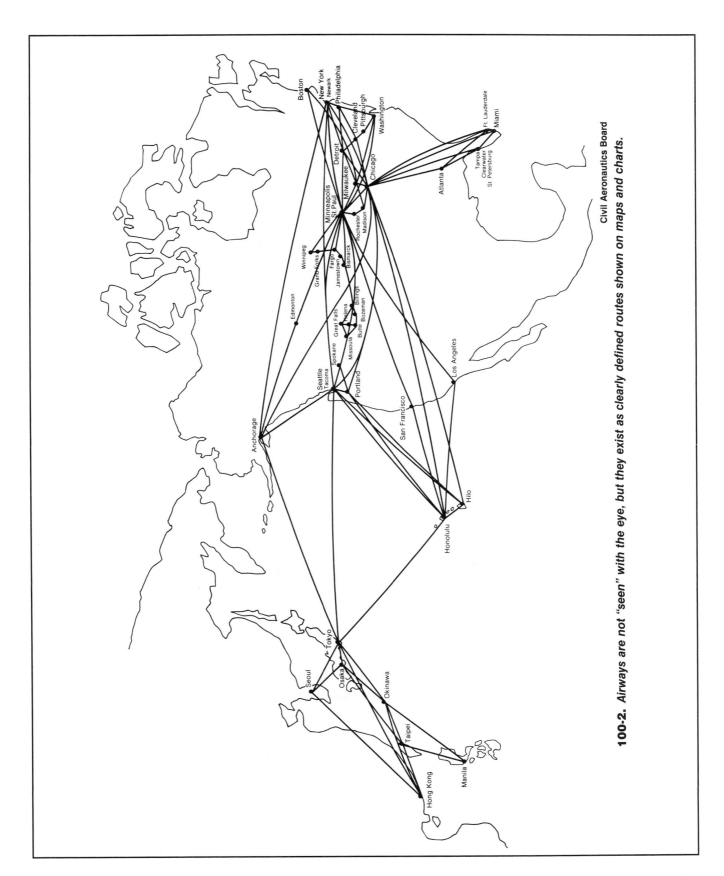

100-2. *Airways are not "seen" with the eye, but they exist as clearly defined routes shown on maps and charts.*

Civil Aeronautics Board

Map labels: Boston, New York, Newark, Philadelphia, Cleveland, Pittsburgh, Washington, Ft. Lauderdale, Miami, Detroit, Milwaukee, Chicago, Atlanta, Tampa, Clearwater, St. Petersburg, Minneapolis, St. Paul, Rochester, Madison, Winnipeg, Grand Forks, Fargo, Jamestown, Bismarck, Edmonton, Great Falls, Helena, Billings, Butte, Bozeman, Spokane, Missoula, Seattle, Tacoma, Portland, Los Angeles, San Francisco, Anchorage, Hilo, Honolulu, Tokyo, Seoul, Osaka, Okinawa, Taipei, Hong Kong, Manila

Automation in Public Transportation

Public transportation is becoming more efficient through applications of computer technology. Metrorail (or Metro) is the modern transportation system in Washington, D.C. It serves 60 stations and has 61 miles of track. Passengers can use this system conveniently and easily, without having to stand in long lines to buy tickets or to have an agent check and take the tickets. This is all done for the passenger with the Automatic Fare Collection System (AFCS).

This system is built around a farecard that has a magnetic strip like the ones on credit cards. The strip is both encoded and decoded by the three types of machines found in every station: farecard vending machines, faregates, and addfare machines. The code on the magnetic strip tells where the user entered and left the rail system. It also tells the time of entry and exit, the amount paid for the card, and the amount of credit remaining.

Metro uses this system so that the fares can be tailored to attract riders. For example, during rush hours riders can pay according to the distance they travel. During other hours, they pay a flat discount fare. Without the AFCS, riders would pay according to some version of the flat fare system. This would mean that every rider would pay the same amount regardless of distance traveled.

Kenworth Truck Co.

100-3. *Vehicles of many types are used to carry materials and people. This truck is hauling logs to a lumber mill.*

cycles, trucks, buses, ships, barges, and trains. In addition, there are other special types of vehicles which do not usually use travelways. Among these are military vehicles, off-road vehicles such as farm tractors and earthmoving machinery, and experimental vehicles and competition vehicles such as racing cars. Fig. 100-4.

Each vehicle is designed to be used in some special way. In order to perform its specific task or function, its structure must conform to this function. For example, earthmoving vehicles must have a strong frame to withstand twisting and turning while pulling or pushing heavy loads. In addition, the vehicle must have a propulsion system, or engine, which will supply enough power to move these heavy loads. Finally, the vehicle must have a proper control system to maintain its direction, speed, and performance.

The elements of vehicle design, then, are structure, propulsion, and control. These elements, along with the essentials of transportation, will be related to rail, air, sea, and automotive transportation systems in the chapters that follow. (More information on power systems is found in Chapters 97-99 of this book.)

100-4. *Earthmoving vehicles do not normally use established travelways. They are built to withstand the rigors of off-road operation.*

QUESTIONS AND ACTIVITIES

1. Name and briefly discuss the two essentials of any transportation system.

2. Name and describe the three elements of vehicle design.

3. If you have a public transportation system in your area, discuss the advantages of using it and how it might be improved.

CHAPTER 101
Rail Transport Systems

Railroads have played an important part in the development of the United States as a major industrial nation. They opened up the western part of this country. They provided fast and reasonably efficient long distance transport before the days of trucks, buses, and airplanes. Since the opening of the first American public railroad, the Baltimore and Ohio, in 1828, we have seen the railroad grow and prosper, decline, and now begin to grow again.

The railroads remain the most important hauler of freight in the nation. Today's railroads carry over 41 percent of all intercity freight. In terms of ton-miles they carry more than all the trucks, planes, and barges combined. (One ton-mile is equal to one ton of freight moved one mile.) America's railroads serve thousands of towns and cities on over 200 000 miles of track. Fig. 101-1.

One reason for their importance is the efficiency of rail operations. The metal wheel on the metal rail produces a minimum of friction and permits railroad

Amtrak

101-1. *Railroads are an important link in the American transportation system.*

length of 39 feet. Modern railways are made of welded track sections of these rails; each welded section is a mile long. The distance between the rails is called the *gauge*. Rails are spaced 4'8½" apart in the United States, Mexico, and Canada. The thousands of miles of such track serve many communities in North America.

There are also other types of railways, such as the monorail. Fig. 101-2. Monorail cars run along a track located either above or below them. Monorails are less costly to build and are popular for city-to-city travel.

RAILROAD OPERATIONS

An important part of a railway system is the control of its operations. This requires stations, freight terminals, classification yards, signals and control centers. Some of the more important operations are described here.

Railroad stations are familiar to almost everyone. Here passengers board trains which take them to their destinations. But

locomotives to pull their loads with less energy than other vehicles. A locomotive uses about one-fourth as much energy as a highway truck in moving one ton of material one mile. Because of this railroads can move three times as much freight per gallon of fuel as big trucks and 125 times as much as cargo aircraft. This is a huge fuel savings, which also reduces air pollution. For example, railroads—on a ton-mile basis—give off less than half the amount of exhaust emissions released by diesel trucks. A recent study indicates that railroads—despite carrying the lion's share of the intercity ton-mileage—are responsible for only slightly more than one percent of all air-pollutant emissions from transportation sources. Railroads are truly an important part of our nation's commerce.

THE RAILWAY

The train travels on a bed of ties and steel rails. These *ties* (crossties) are square wooden logs about six feet long and are

laid on a bed of gravel or crushed rock called *ballast*. The ties are spaced 21 inches apart, from center to center, and provide support for the steel *rails* that the wheels of the train travel on. The rails weigh from 60 to 155 pounds to the yard and have a standard

Seattle Center

101-2. *Monorail trains ride on only one rail. They are faster and less expensive to operate.*

hauling freight by rail is another matter. In the *railroad classification yard,* loaded freight cars from all over the nation are joined to others to form a train. Fig. 101-3. The classification yard is a massive complex of tracks. A few inbound tracks expand into a dozen or more parallel receiving tracks. Then these merge into a few tracks, which in most modern classification yards is called a "hump yard" because the separated cars go up, over, and down a hump in the track to the area where they are classified. Beyond the hump yard, the tracks fan out again, this time in the "bowl," where row on row of parallel tracks, in turn, join to handle the outbound traffic.

Obviously, the simplest way to run a railroad would be to make up a full train at Point A and run it straight through, without stopping, to its destination at Point B. There are such trains, called *unit trains,* and they are, indeed, an efficient way to transport goods. But every car in such a train must be headed for the same place. If railroads tried to make up every train that way, there would either be a lot of very short trains, or locomotives would spend a lot of time at shipping points waiting for a full complement of cars for a single destination. That wouldn't be very efficient. Therefore most trains are made up of cars headed for a wide variety of destinations.

That is why a classification yard is needed. The train pulls into a yard and is broken up, with the cars sorted according to final destinations. New trains pick up the sets of reclassified cars and leave the yard in different directions.

Through the years this has mostly been done manually. Workers read the numbers on in-coming cars, learned their destinations from advance reports, and then assigned the cars to classification tracks where they awaited pickup. In this system, of course, there's an obvious danger of a yard becoming bottlenecked. In fact, by its very nature, it must be a sort of bottleneck—a place into which incoming trains are squeezed, then separated and sent out in all directions. Railroaders constantly look for ways to reduce the "squeezing" time and cut the period between the funneling in of trains on one side and the release of trains on the other.

In modern yards the most important time-saving tool is the computer. It can "see" and "think" faster and "remember" more than any human being. Automatic Car Identification (ACI) scanners can be used to "read" color-coded labels on incoming cars. (The ACI system will be discussed in detail later in this chapter.) Because the scanners can read very fast, an incoming train can travel three times faster than when a person had to read and write down the car numbers. The computer's memory banks can also store the ex-act location of every car in the yard and make this information available almost instantly. As a series of cars approaches the top of the "hump"—over which they will roll freely to the proper track—scanners transmit the number of each to a screen in the hump yardmaster's office. As a car rolls down the hump, a machine automatically slows its speed to prevent it from damaging the cars it is to join. To do this, the computer must calculate the car's gross weight, the curvature of the track, the distance the car must travel, and even the prevailing weather conditions. Among other things, this means the car may be weighed while in motion. When the cars have been classified, the computer can produce lists of trains ready for departure, along with the total tonnage.

Not every railroad yard is computerized yet. Such yards are very expensive, but the ones that have been constructed are proving their worth. One automated yard in the Midwest cost $12.5 million. It has taken over the operations of four other yards in the area and relieved a burden on yet another, and handling

Association of American Railroads
101-3. *A railroad classification yard. Here freight cars are joined to become trains.*

time in that area has been cut in half.

Today's railroaders are constantly looking toward the future—a future when a growing population and economy will need much more of every product than it does now. It is a future when peak efficiency will be essential. There will be no time for unnecessary delays or avoidable errors. The computerized classification yard is a giant step toward that peak efficiency.

Still another electronic development on the railroads is CTC—Centralized Traffic Control. CTC enables an operator seated at a remote terminal to "see" and direct traffic on hundreds of miles of railroad track. From the terminal the operator pushes a button or moves a lever, actuating switches and signals miles away. Thus one train can be routed briefly onto a siding while another, coming from the opposite direction along the same track, speeds safely by. One of the values of CTC is that more trains can use fewer tracks, and they can use them safely and efficiently. Other applications for computers and computer-related

Canadian National
101-5. *Automatic Car Identification scanner unit reads the rail car label.*

systems are being developed at a rapid pace.

Automatic Car Identification

As mentioned earlier, the Automatic Car Identification system (ACI) uses computers to speed the process of car classification. This system also makes it possible to give a customer accurate information on the location of a shipment of automobiles, iron ore, or other freight. Before ACI this information was collected visually. This led to many delays and errors. Now both the customer and the manufacturer always know the whereabouts of a freight shipment.

There are three basic parts to ACI. A *label,* Fig. 101-4, is applied to both sides of a railway car. This label is made up of strips of reflective color-coded tape. This special tape reflects

light in such a way that it is 200 times brighter than paint. The labels contain optical information on car serial numbers and ownership codes. A *scanner,* which reads the labels, is mounted in freight yards. Fig. 101-5. This sensitive instrument can read each label four times when a car is moving at speeds of one hundred miles per hour. (Of course, the railroad car would not be moving that fast in the classification yard.) The third part of ACI is the *decoder.* This device converts the scanner signal into meaningful numbers suitable for a teletype printout or for input to a computer control center.

The next time you are at a rail crossing, notice the colorful labels on the freight cars. You will know that this label is a part of ACI and that it contributes to a more efficient railroad freight system.

RAILWAY VEHICLES

Locomotives, freight cars, and passenger cars are the standard railway vehicles. The steam locomotive was the first type of engine used to pull railway cars.

101-4. *This rail car identification label is made from color-coded reflective tape. A scanner reads the label and sends the information directly to a computer control center.*

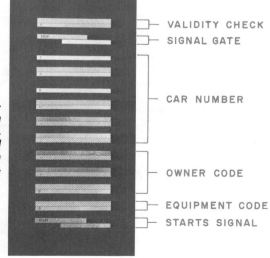

VALIDITY CHECK

SIGNAL GATE

CAR NUMBER

OWNER CODE

EQUIPMENT CODE

STARTS SIGNAL

Southern Pacific Company

101-6. *The first successful steam locomotive was built in England in 1812. Pictured here is a steam locomotive from 1866.*

Fig. 101-6. Diesel engines, introduced in 1925, were cleaner, more powerful, and more efficient. This added power enabled the railroad companies to increase the number of cars in a train and cut shipping costs. Fig. 101-7. Locomotives of more than 6000 horsepower are used today.

The locomotive of today is usually a diesel-electric. It uses an electric generator and wheel-mounted motors to transfer power from the engine to the drive wheels. It is the most familiar locomotive on today's railroads. Fig. 101-8.

Electric locomotives are also in use. They derive their power either from a third rail that is laid beside the track or from an overhead wire. The biggest advantage of the electric locomotive is simply that it is powered from a central generating station. It can therefore draw extra power as needed to start a heavy train or go up an incline. But the higher unit horsepower and the low maintenance costs are offset by

the high cost of transmission facilities along the right-of-way.

Newer gas turbine locomotives have also been developed. The train in Fig. 101-9 has five cars, with a power unit at each end. The total train length is 423 feet.

Each power car is propelled by a turbine engine similar to those developed for aircraft use and is rated at 1140 horsepower. A second turbine engine rated at 430 horsepower drives an alternator, which supplies electric power for the train's auxiliary system, lighting, bar-grill unit, and air conditioning.

The locomotive has simple control devices for safe, easy operation. Many safety features are built into its controls, such as automatic braking in case the operator has a disabling accident. The all-electric train shown in Fig. 100-1 has only one control to dial train speed from zero to a maximum of 120 MPH.

Concern over the supply and price of diesel fuel has renewed interest in steam-powered locomotives. Under development is a new type of steam-engine locomotive. This locomotive consists of a power unit (two steam engines) and a service module that carries coal and water. The basic operation is similar to that of steam locomotives from the

Santa Fe Railway

101-7. *A modern diesel locomotive. For long trains two or even three locomotives are joined together to provide the needed power.*

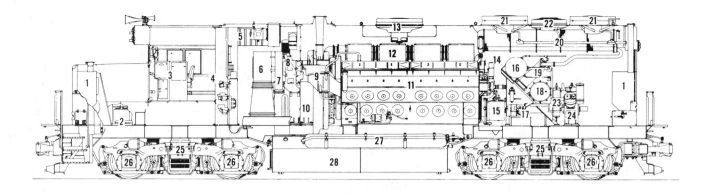

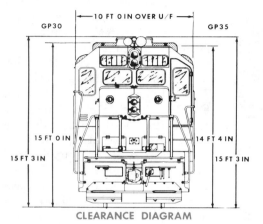

CLEARANCE DIAGRAM

LEGEND

Length 56'2"
Weight 242,000 lbs.

1. Sand Box
2. Batteries
3. Loco. Controls
4. Electrical Cabinet
5. Inertial Air Filter
6. Traction Motor Blower
7. Generator Blower
8. Aux. Generator

9. Turbocharger
10. Main Gen. & Alt.
11. Engine 16-567 D3A
12. Exhaust Manifold
13. Dyn. Brake Fan
14. Governor
15. Lube Oil Filler
16. Eng. Water Tank
17. Fuel Pump
18. Lube Oil Filter
19. Lube Oil Cooler
20. Radiator

21. 48" Fan and Motor
22. 36" Fan and Motor
23. Fuel Pressure Filter
24. Air Compressor
25. Trucks
26. Traction Motors
27. Main Air Reservoir
28. Fuel Tank—1700 gals. 2600 gal. tank available at extra cost.

101-8. *Cross section of a diesel-electric locomotive.*

1950s, but microprocessors are used to monitor and control the stoker, the combustion and exhaust air, boiler water feed, and valve settings. To prevent air pollution, ash is removed from the exhaust gases and stored until it can be unloaded.

The new locomotive can attain speeds of up to 80 MPH. It will run 500 miles between fuel stops and 1000 miles between water stops. Its tractive (pulling) power is equal to that of a 3000-horse-power diesel locomotive, and its high-speed power is even greater.

Other "rolling stock" vehicles include freight and passenger cars. The important freight cars are shown in Fig. 101-10, and

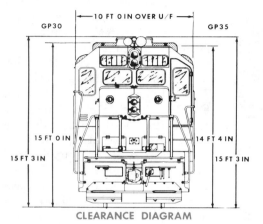

101-9. *A gas turbine locomotive, capable of speeds up to 125 MPH. It is rated at 1140 horsepower.*

standard flat

single deck livestock cars

box car

standard refrigerator car

hopper—open

standard gondola car

101-10. *Railway freight cars. Study the design of each in order to understand what it is used for.*

Santa Fe Railway

101-11. *Truck trailers ride "piggyback" on this special freight car.*

at speeds up to 250 MPH. Its propulsion system is almost noiseless and does not pollute the atmosphere. Fig. 101-14. A second experimental vehicle is the Tracked Air Cushion Research Vehicle (TACRV). Fig. 101-15. These vehicles operate over a guideway suspended by air without touching the guideway. They run at speeds up to 300 MPH. They are very efficient, using a LIM to generate 10 000 pounds of thrust.

each is designed for a special use. Some of these cars are designed to carry truck trailers "piggyback." Trucks pull these trailers to freight yards where the trailers are loaded on special railroad cars. At their destinations the trailers are unloaded, and trucks once again pull them to their customers. Fig. 101-11.

A newer type of freight car is the RoadRailer. This is a 45-foot truck trailer that has both steel and rubber-tired wheel systems. The combination makes this unit able to travel on either highways or railways. Individual units can be linked to a truck for road travel or linked together in a series behind a locomotive for rail travel. In a test run, the Road-Railer used only about half as much fuel as a conventional piggyback train. Fig. 101-12.

NEW DIRECTIONS IN RAILROAD VEHICLE TECHNOLOGY

The need for faster, safer, and more efficient rail systems has led to some interesting new vehicles. The Linear Induction Motor (LIM) test car is a lightweight, high-speed vehicle designed for intercity rail travel. Fig. 101-13. This vehicle operates

Bi-Modal Corporation

101-12. *A RoadRailer. Note the steel wheels at the rear.*

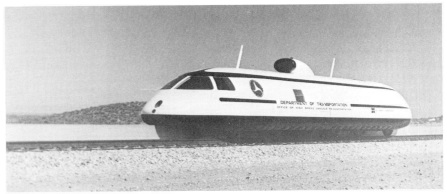

Departmet of Transportation

101-13. *The LIM test vehicle.*

LINEAR INDUCTION MOTOR TEST CAR

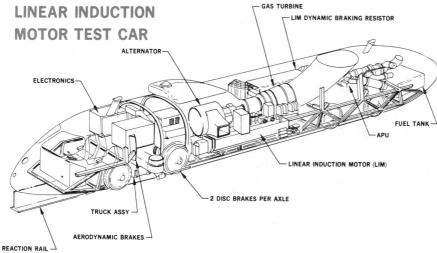

ELECTRONICS

ALTERNATOR

GAS TURBINE

LIM DYNAMIC BRAKING RESISTOR

FUEL TANK

APU

LINEAR INDUCTION MOTOR (LIM)

2 DISC BRAKES PER AXLE

TRUCK ASSY

AERODYNAMIC BRAKES

REACTION RAIL

Department of Transportation
101-14. *Structural diagram of the LIM test vehicle.*

Department of Transportation
101-15. *The Tracked Air Cushion Research Vehicle.*

QUESTIONS AND ACTIVITIES

1. Define ton-mile.

2. What are some advantages of transporting goods by rail rather than truck or aircraft?

3. Briefly describe what makes up the railway bed, or track, that trains travel over.

4. Briefly describe what is done in classification yards. Why are these yards necessary?

5. What is ACI? Describe the three basic parts of ACI and the function of each. What are some advantages of using this system?

6. Name four types of freight car and tell what you think each might be used to carry.

CHAPTER 102

Air Transport Systems

The carrying of people and materials by aircraft is perhaps the most exciting and interesting form of transportation today. Huge airplanes roaring from runways and climbing into the skies cause people to stop and stare at this marvelous invention. Fig. 102-1.

AIRWAYS

Like other forms of transportation, airplanes need paths or routes to follow. These airways are strictly controlled for air traffic so that planes will not collide. Of course, the airways are invisible except on flight maps and charts.

The useful airspace is considered to have a ceiling (top limit) of 75 000 feet. Its base is not even, of course, but follows the contours of the earth. It changes

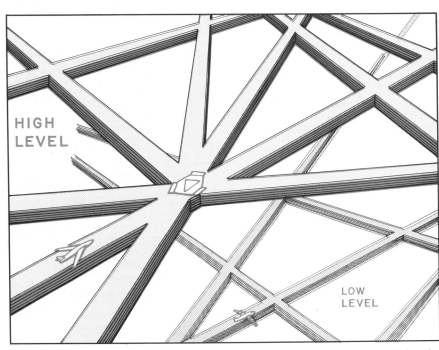

Federal Aviation Administration

102-2. *Flight levels separate airplanes' flying altitudes.*

British Airways

102-1. *Airplanes are an important form of modern transportation.*

above cities, mountains, tall towers, and other obstructions and reaches down to the ground at airports.

In the sky above the United States, there are more than 280 000 miles of federal airways. These sky highways are divided into two different systems. The low-altitude airway system generally begins at 1200 feet above the earth and goes up to, but does not include, 18 000 feet. The high-altitude airway system goes from 18 000 feet to 45 000 feet. Airspace above 45 000 feet is reserved for point-to-point flights. Fig. 102-2.

Pan American

102-3. *Mobile lounges are part of the facilities at many modern airports.*

Airplanes traveling west or south must fly at even numbered altitudes. Those flying east or north use odd numbered altitudes. Planes must also fly at least 1000 feet above cities and 500 feet above open country.

Airways are the freeways of the sky, complete with an aerial version of signs, access roads, directional guides, and even "parking" places—areas over airports known as holding points. Airplanes hold at these points, flying in an oval pattern, so that they may be spaced in an orderly fashion before moving along the airways or into airports for landing.

THE FAA

In the United States, the agency which controls the airways, aircraft, and air traffic is the Federal Aviation Administration (FAA). Its responsibilities begin at the drawing boards where aircraft are designed and at the factories where they are made. It has authority over the people who dispatch the aircraft from airports, the crews who fly the planes, the aviation mechanics who maintain them, and

other specialists (parachute riggers, flight instructors, etc.). FAA responsibilities include the airspace, the navigation aids, the airway system, the airports, and the research needed to continually improve the performance and safety of aircraft. FAA also provides inspectors to help the National Transportation Safety Board in accident investigations.

Providing communities and airport owners with planning and engineering advice is another important FAA function. The FAA also provides these services to other countries. There are over 11 000 airports on record with the FAA.

FAA operates two major airports—Washington National Airport and Dulles International Airport. Washington National is located in Arlington, Virginia, four miles from downtown Washington, D.C. Dulles, one of the world's largest jetports, is located near Chantilly, Virginia, 26 miles west of the nation's capital. Mobile lounges are used for transporting passengers to jet planes, which are parked far from the terminal building so that the noise and fumes do not cause discomfort. Fig. 102-3.

AIR TRAFFIC CONTROL

Air traffic is similar to automobile traffic; it cannot operate helter-skelter. Instead, it must operate according to established rules—Instrument Flight Rules (IFR) and Visual Flight Rules (VFR).

In general, pilots flying IFR navigate mainly by instruments. They rely on them and on the instructions they get by radio from air traffic control specialists to stay safely separated from other aircraft. They file a flight plan before taking off and are given a clearance that keeps them away from other planes flying IFR in the same area.

Pilots flying VFR, however, rely on their own sight to avoid other aircraft and must follow the idea of: "See and be seen."

IFR operations are required when weather conditions fall below the minimum for cloud ceiling heights and visibility. In order to fly IFR, a civilian pilot must pass a written and a flight test and receive an instrument rating from FAA. When flying through "positive control" airspace (generally above 24 000 feet), pilots must fly under IFR regulations.

Northwest Orient Airlines

102-4. *The DC-3 was the first of the modern airliners. It carried 28 passengers and a crew of 3.*

102-5. *The Boeing 377 Strat-o-cruiser, 1949. This airplane carried 80 passengers 3000 miles at a speed of 340 MPH. It was one of the last of the generation of propeller-driven aircraft.*

Air traffic in the United States operates under a "common system." This means that military and civilian aircraft are controlled by the same facilities. Traffic near an airport is controlled either by a military control tower or by one of the FAA airport control towers.

Traffic on the airways is controlled by Air Route Traffic Control Centers located throughout the country. Each center and tower handles traffic within its own area, using radar and communications equipment to keep aircraft moving safely. As the flight progresses, control is transferred from center to center and from center to tower.

In addition to control towers and Air Route Traffic Control Centers, there is a third air traffic facility called the Flight Service Station. It provides valuable information and other important services to civil aviators, air carriers, and military pilots. About 385 of these stations and combined station/towers are scattered around the nation, each covering an area of roughly 400 square miles. Flight Service Station specialists, all of them expert on their area's terrain, provide preflight and inflight briefings, weather information, suggested routes, altitudes, and any other information important to the flight's safety.

If an airplane is overdue at its reporting station or destination, the Flight Service Station starts a search and rescue operation. If a pilot is lost or is having some trouble, it will give instructions and directions to the nearest emergency landing field.

AIRCRAFT

The first powered, sustained airplane flight took place at Kitty Hawk, North Carolina, on December 17, 1903. There the Wright brothers flew an airplane some 120 feet, a flight which lasted 12 seconds. Since then, considerable progress has been made in the design of aircraft.

One of the most remarkable airplanes was the famous Douglas DC-3. It entered service in 1936, had a top speed of 230 MPH, and a range of about 1500 miles. (The *range* is the distance the airplane can fly without refueling.) Fig. 102-4. Following World War II, faster and larger airplanes with greater ranges were developed. Fig. 102-5. All were driven by turbine or piston-powered engines. The first jet engines were used on military aircraft. The first commercial jet airliner was a British airplane, the Hawker-Siddley Comet, which flew in 1958. It traveled at a speed of 526 MPH, carried 81 passengers, and had a range of 3225 miles. This plane ushered in a new air age. Since that time, a number of modern jetliners have been designed. Fig. 102-6.

Aircraft are designed for carrying freight and passengers on short and long hauls. While most airplanes carry both passengers and freight, there are models designed for carrying freight only. These are able to carry large pieces because they have specially made cargo doors and compartments. Fig. 102-7. Loading is done mechanically, and the cargo is placed in containers to speed up loading and unloading. Airplanes can also be converted to carry only passengers, only freight, or both. Fig. 102-8. Air freight is, of course, very fast and is growing in popularity.

There are many other types of

102-6. *This huge Boeing 747 "Jumbo Jet" is the largest commercial jet airplane.*

Boeing Company

102-7. *The nose of this huge 747 freighter lifts up to make loading more convenient.*

aircraft designed for special uses. Light airplanes are for business and recreational travel. Both light and heavy helicopters serve as a means of passenger travel, as well as for construction work. Fig. 102-9. Military aircraft are used for our nation's defense.

Aircraft Structures

Airplanes are built of lightweight, tough alloys of aluminum and titanium. Composite materials are also being used. One is a graphite-epoxy compos-

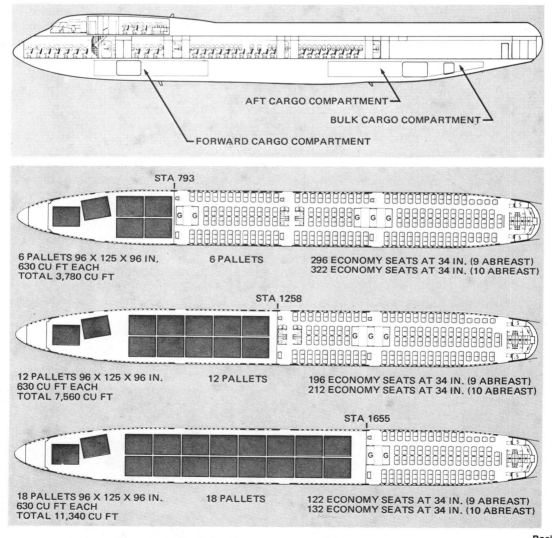

AFT CARGO COMPARTMENT

BULK CARGO COMPARTMENT

FORWARD CARGO COMPARTMENT

STA 793

6 PALLETS 96 X 125 X 96 IN.
630 CU FT EACH
TOTAL 3,780 CU FT

6 PALLETS

296 ECONOMY SEATS AT 34 IN. (9 ABREAST)
322 ECONOMY SEATS AT 34 IN. (10 ABREAST)

STA 1258

12 PALLETS 96 X 125 X 96 IN.
630 CU FT EACH
TOTAL 7,560 CU FT

12 PALLETS

196 ECONOMY SEATS AT 34 IN. (9 ABREAST)
212 ECONOMY SEATS AT 34 IN. (10 ABREAST)

STA 1655

18 PALLETS 96 X 125 X 96 IN.
630 CU FT EACH
TOTAL 11,340 CU FT

18 PALLETS

122 ECONOMY SEATS AT 34 IN. (9 ABREAST)
132 ECONOMY SEATS AT 34 IN. (10 ABREAST)

Boeing Company

102-8. *Airplanes such as this 747 can be converted to carry different amounts of cargo and numbers of passengers.*

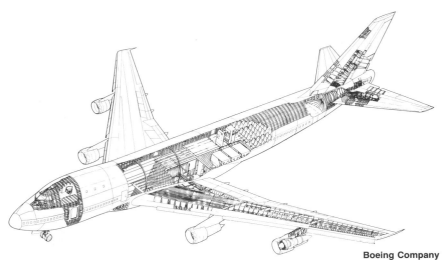

Boeing Company

102-10. *Cutaway showing the structure of a Boeing 747.*

Rotor Way, Inc.

102-9. *This lightweight helicopter is designed to carry two passengers.*

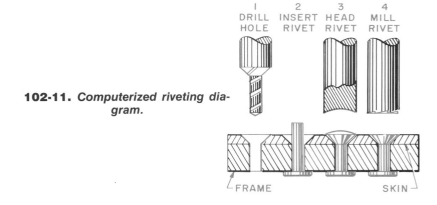

102-11. *Computerized riveting diagram.*

102-12. *Structural testing takes place in huge frames such as this.*

Boeing Company

ite as strong as titanium but much lighter. They are designed to be strong to withstand the forces of flight such as wind and pressure. Figure 102-10 shows a typical aircraft structure. Note how the frames and ribs are covered with a thin, strong metal sheathing, or skin. Riveting is the process generally used to fasten aircraft sections together. Recently, computerized riveting has been developed. Huge machines clamp the skin to an entire wing frame section. An automatic riveting machine is programmed to locate a rivet position, drill a hole, insert a rivet, head the rivet, and then mill it flush with the surface of the skin. Fig. 102-11. This is a fast, efficient way to set the 1 800 000 rivets contained in a Boeing 747.

Once built, newly designed aircraft are tested in huge structure-test frames. Fig. 102-12. Here they are twisted, pulled, and shaken to test the airworthiness of the craft.

Modern aircraft are powered by piston, turbine, and jet en-

gines. More information on these engines can be found in Chapter 98.

Much research is taking place on rocket engines and supersonic aircraft. Supersonic commercial aircraft, or SSTs, have been built by Russia and as part of a joint French-British program. In addition, research is being done by the National Aeronautics and Space Administration (NASA). Much has been learned about flight, structure, power, and electronic guidance systems through NASA research.

Aircraft are an important part of our transportation system. They join with rail, land, and water vehicles to provide this nation with a fast, efficient, and safe way to move people and materials.

QUESTIONS AND ACTIVITIES

1. Describe several responsibilities of the Federal Aviation Administration.
2. Describe IFR.
3. Describe VFR.
4. What is the function of a Flight Service Station?
5. What is meant by the "range" of an airplane?
6. What types of engines are used to power modern aircraft?

CHAPTER 103

Water Transport Systems

Carrying people and materials by water is one of the oldest forms of transportation. Even though traveling by water dates back to the ancient Egyptians, it is still an important method of moving materials between countries, as well as within countries.

A *ship* is a powered vessel designed for navigation on the sea. A ship is larger than a boat. A boat is small enough to be carried on a ship. A vessel that sails the ocean is called a ship.

A ship's size is given by tonnage. Originally the term was used to mean how much the ship could carry. Now weight is measured by *deadweight* tonnage,

Smithsonian Institution
103-1. Flying Cloud, *one of the most famous clipper ships. In 1854 this ship set a record by taking only 89 days to travel from New York to San Francisco.*

103-2. *Tanker under construction in a shipyard. Today ships are constructed of steel which is welded together.*

which describes the capacity of tankers, and *displacement* tonnage, which equals the weight of the volume of water displaced by the ship as it floats.

In ancient times ships were powered by the wind or by slaves manning oars. Sailing ships dominated the sea lanes from about 1450 until the Civil War. The greatest advantage of the sailing ship was that wind costs nothing. The one big disadvantage of the sailing ship was that it depended on the wind. For this reason people could never be sure when the ships would arrive at their destination. Sailing ships were also difficult to maneuver in battle because one was never sure of the wind.

The sailing ship reached its peak during the nineteenth century with the streamlined, fast clipper ships. One of the most famous of the clipper ships was the *Flying Cloud*. Fig. 103-1.

The first successful steamboat in the United States was the

Clermont, built by Robert Fulton in the early 1800s. This boat carried cargo and passengers on the Hudson River. Early oceangoing steamships used both steam and sail for power. Many early steamships suffered mechanical failures and had to use sails to finish their voyages.

By the late 1860s ships powered by steam were replacing the sailing ships. The last large-scale service of sailing ships came during World War I. By the 1920s there were practically no large sailing ships competing with steamships.

The first steamships were powered by reciprocating (piston) steam engines. Later, a new type of engine was developed: the

steam turbine. The turbine engine was more efficient and used less fuel. Another advance was the use of the diesel engine. Today many ships are powered by diesel engines.

Modern ocean ships are made of steel. Ships are constructed on land in shipyards. They are launched into the water after completion. Fig. 103-2.

Ships have changed greatly since the days of sail power. Today they can carry thousands of tons of cargo or millions of barrels of oil. New methods of loading and unloading have shortened the time a vessel must remain in port. Figure 103-3 shows one new type of cargo vessel. With this ship, cargo containers or vehicles can be driven onto the ship, transported, then driven off.

In this chapter, you will learn that water transportation is an important part of our total transportation system. As ships and

103-3. *Roll on/roll off cargo vessel. Vehicles are driven onto the stern of the ship and transported on deck. On reaching port, they are driven back off the stern.*

Sea-Land Service

103-4. *Modern container ship on the open sea.*

stevedores. Since the early 1950s a new type of cargo ship called the container ship has been increasingly used. Fig. 103-4.

Container ships have structures which permit the cargo containers to be stacked one on top of the other. The containers are the size and shape of semitrailers. They can therefore be transferred quickly and easily from the boat to a truck and driven to their final destination. This eliminates the handling of individual items on the dock and reduces both the cost and the time it takes to load and unload a ship.

Special cranes on the docks load and unload the containers. Fig. 103-5. An all-container ship may spend one day in port as compared with six or more days for a conventional ship handling noncontainerized cargo. Figure 103-6 shows a container port facility.

Today other types of cargo, like liquids and ore, are handled by

propulsion systems continue to improve, this form of transportation will continue to compete with air and land transportation systems for both cargo and passengers.

MATERIALS TRANSPORT

By far the greatest number of ships that are in use today carry cargo. Ships carry everything from safety pins and plastic dolls to automobiles and large machinery.

Since 1950 cargo shipping has changed greatly. Before this time all the cargo was carried inside the ship in large compartments called holds which were loaded and unloaded by workers called

Sea-Land Service

103-5. *Loading a container onto a ship. Note the wheels from which the container has been lifted.*

Port of Long Beach

103-6. *Container port facility.*

103-7. *A supertanker.*

103-9. *The* Queen Mary, *one of Enland's most luxurious passenger ships, sailed between England and New York.*

large "superships." For carrying large quantities of crude oil and gasoline, supertankers have been developed. These are so large that they cannot use port facilities. The tankers anchor outside the harbor and unload through pipes and hoses or to smaller tankers which can use the port facilities. Fig. 103-7.

On the inland waterways such as the Mississippi River much bulk cargo, such as petroleum products, coal, and ore, is still shipped by river barge. The old river steamboat has been replaced by diesel vessels and barges.

On the Mississippi, towboats push barges that are tied securely together and then fastened to the towboat. The largest tows of barges carry up to 40 000 tons of cargo. Fig. 103-8.

Another of the important inland waterways is the Great Lakes. Iron ore mined in Minnesota is shipped in large ore carriers to the smelters in Illinois and Pennsylvania. Since the St. Lawrence seaway opened, ocean-going vessels have been able to travel all the way to Chicago to unload their cargo.

PASSENGER TRANSPORT

Before the invention of the aircraft, the only way to cross the ocean was by ship. Passenger ships were an important part of shipping companies' fleets. Before World War II the passenger ships achieved a status they have not had since. Ships such as the *Queen Mary,* the *United States,* and the *Queen Elizabeth* provided luxurious ways to cross the ocean. Fig. 103-9. With the growth of commercial aviation, the numbers of passengers on ships declined. Today most of the passenger ships that are still in service are holiday cruise ships.

Although there has been a decrease in passenger ships crossing the oceanways, there is still a need for fast, economical water transport between land areas separated by small areas of water. For example, *ferries* are used to transport people (and sometimes their cars, too) across rivers, lakes, and bays. Fig. 103-10.

Union Mechling Corporation
103-8. *River tow on the Mississippi.*

S.H.M. Marine International, Inc.
103-10. *This ferry boat transports commuters across Burrard Inlet between Vancouver and North Vancouver in British Columbia.*

The Moulin à Vent, *a prototype vessel used to test the wind cylinder system of propulsion.†*

Windships

Because of the high cost of conventional fuels, shipbuilders are looking at alternative sources of energy. One of these sources has been in use since ancient times: the wind. As an energy source, the wind has several advantages over fossil fuels. It is nonpolluting, abundant, and free. Until the Civil War, most ships were powered by the wind. By the early twentieth century, however, the faster and more maneuverable steamships had largely replaced sailing vessels. Now the sailing vessel is coming back, redesigned by high technology to meet today's needs.

One of the new designs is the *Moulin à Vent,* developed by a French research team that was assembled by Captain Jacques Cousteau and funded by the French government. The *Moulin à Vent* (the name means "Windmill") is a wind-powered boat, but it has no sails. Instead, there is a hollow cylinder, 44 feet high and 5 feet in diameter. Two perforated lateral vents run lengthwise down the cylinder. A computer-controlled, movable flap can be positioned over either vent, and the cylinder itself can be turned to take advantage of the prevailing winds.

The propulsion system works as follows: At the top of the cylinder a fan rotates, pulling in air through an uncovered vent. The air intake and the position of the flap cause a deflection of the air flow behind the cylinder. The resulting differences in air pressure create lift and help propel the boat.

Although still in a developmental stage, the wind cylinders show promise as a propulsion system to be used in conjunction with conventional systems. It is estimated that a ship equipped with two or more wind cylinders could use between 30 and 40 percent less fuel than ships without the cylinders. The time may not be far off when wind power once again becomes a major source of energy for seagoing vessels.

†Photo © 1983 by The Cousteau Society, Inc., 930 West 21st Street, Norfolk, VA 23517, a nonprofit, membership-supported environmental organization.

Another type of boat used for short trips is the jet hydrofoil developed by the Boeing Company. A *hydrofoil* vessel actually rides above the water on fins called foils. Fig. 103-11. The boat is propelled by jets of water from pumps driven by gas turbines. As the speed of the boat increases, the hull of the boat lifts out of the water. The boat settles into the water at the dock to board passengers. The Boeing Jetfoil has a cruising speed of 51 MPH even over 12-foot waves. In shallow water the foils are retracted against the hull.

Another type of craft using a special propulsion system is the *surface effect ship,* often called a *hovercraft.* The hovercraft operates on a cushion of air which is created under the craft by large fans. The air is pumped down from above and comes out the sides. This craft can operate on both land and water. Figure 103-12 shows the principle of the surface effect ship. The hovercraft can be used to transport either people or cargo for short distances. It can also be used as a short-distance land-sea military craft.

Boeing Company

103-11. *Hydrofoil vessel running on its three fins, or foils.*

103-12. *Drawing of a surface effect ship showing how the stream of air moves.*

103-13. *The nuclear submarine* Bluefish. *Using nuclear power enables ships like this one to remain underwater for many weeks at a time.*

MILITARY SHIPS

Since the beginning of water transportation, ships have been used for military purposes. Some famous naval battles were fought using sailing ships. One such battle was fought in 1588 between English and Spanish forces. The defeat of the Spanish fleet (the Armada) marked the beginning of England's rise as a world power.

With the advent of steam power, navies began designing new ships. The combination of steam power and metal construction brought about new ships and new battle tactics. Today's navies have ships which travel under the sea as well as on top. Some ships today are powered by nuclear reactors. This type of power is used mostly on submarines to enable them to remain underwater for a long time. Fig. 103-13.

QUESTIONS AND ACTIVITIES

1. Define *ship*. How does a ship differ from a boat?

2. Name and describe the two types of tonnage measurements used to describe a ship's size.

3. Describe container ships and the advantages they have over earlier cargo ships.

4. What kind of ships are used to carry large quantities of crude oil across the seas? What kind of vessels are used to carry bulk cargo on inland waterways? What kind of craft can operate on both land and water?

5. Describe the propulsion systems of the hydrofoil and the hovercraft.

CHAPTER 104

Automotive Transport Systems

Probably no single invention has changed the United States as greatly as did the automobile. Think of all the ways in which cars, and other vehicles such as buses and trucks, affect your life. Perhaps you ride a bus to school. Probably your parents drive to work in cars. When you go on a vacation, you probably travel by car.

Because it is a quick, convenient form of transportation, the automobile contributed to the growth of suburbs and shopping centers, thereby changing the makeup of our cities. We have drive-in movie theaters, restau-

104-1. Duryea car.

rants, and banks. The road system, the oil industry, the steel industry—all of these have been greatly affected by the automobile.

Before the development of the automobile, travel was often slow and difficult. The roads in the United States were trails used by horseback riders to travel from one town to another. Most early settlements were on river banks or bays because the waterways were used for transportation. The first long, hard-surfaced road was completed in 1795. It was the Lancaster Turnpike in Pennsylvania. It was surfaced with broken stones and gravel and extended 62 miles. In 1830, the steam locomotive was used successfully; as a result, road building slowed. Railroads were to be the transportation of the future. Thus from 1850 to 1900, there was little change in the way that roads were built.

With the coming of the automobile, there was a demand for better roads. Farmers were asking for more roads so that they could get their products to the railroad for shipment to market.

The first concrete road was built in Detroit in 1908. By 1924 there were 31 000 miles of concrete roads in the United States. In 1925 a system of numbering highways was established. From 1924 until after World War II, little was done about building more concrete highways. Following World War II, truckers and automobile clubs realized some-

thing must be done to improve the highway system. To help solve the problem of rapid highway transportation, the National System of Interstate and Defense Highways was established. Today there are over 42 000 miles of this superhighway.

The interstate highway system connects 90 percent of the United States cities of over 50 000 population with multilane divided highways.

Today there are more than 123 million automobiles and 35 million trucks and buses in the United States. They operate on nearly 4 million miles of streets and highways. The United States is a nation on wheels, and the automotive vehicle has become a necessity.

THE AUTOMOBILE

The automobile is a self-propelled, wheeled vehicle, designed to carry passengers on highways and streets. Automotive vehicles that carry a large number of people are buses, and those that carry freight are called trucks.

The automobile was invented in Europe, but it has had its greatest development in the United States. In 1892 the Duryea brothers built the first successful gasoline-powered automobile in the United States. A Duryea car won the first American automobile race in 1895. Fig. 104-1.

By 1900 there were a number of successful cars being made in the United States. Some of these

Smithsonian Institution
104-2. *Model T Ford.*

cars were made by men whose names are still familiar: Ford, Olds, and Dodge. About this same time automobile racing and transcontinental runs became popular. The French Grand Prix was first held in 1906 and the Indianapolis 500 in 1911.

Experiments continued with types of locomotion other than the internal combustion engine. One was the steam car. The Stanley Steamer (built by the Stanley brothers) set a world speed record of 127.66 miles (205.4 kilometres) per hour in 1906. Propulsion by electricity was also tried and achieved a measure of success. Still, the internal combustion engine proved the most satisfactory. It became the standard engine for cars.

Until about 1900, automobiles were built by hand. Each varied from the other, and parts had to be hand-fitted to the machine, a slow and costly process. Henry Ford began to standardize parts. He also used the conveyor assembly line to speed up production. When the car reached the end of the line, it was driven off on its own power. With these improvements, the time needed for producing a car was greatly reduced, and the price was lowered so that the average person could afford one. The Ford Model T car was the first mass-produced automobile. Fig. 104-2.

During the early 1900s, a number of engineering improvements took place. The electric self-starter replaced the hand crank. Multiple cylinders, better tires, all-steel bodies, steel disc wheels, and closed bodies were features of the new cars.

After World War II, travel by car increased greatly. What was most needed was a dependable car that could travel at a good speed and would sell at a reasonable price. From the 1940s to the 1970s, the manufacture of cars and improvement of highways increased at a rapid pace. The car of today is a far cry from Henry Ford's Model T. Modern cars may have FM/AM radios, tape players, four-wheel disc brakes, air conditioning, and steel-belted tires. Many of the changes in the automobiles of the 1960s and 1970s were closely related to safety. Today, laws require safety glass, seat and shoulder belts, and front bumpers that can with-

stand collision at low speeds. In addition, manufacturers have developed other safety features such as the collapsible steering column, dual braking systems, inflatable air bags, and stronger passenger compartments.

In the 1980s, as car engines become more complex and pollution controls more strict, car makers are using electronics to control many engine functions. Today microprocessors monitor engine exhaust and adjust the timing to improve fuel economy, calculate how far one can drive with the amount of fuel in the tank, and indicate engine malfunctions.

TRUCKS AND BUSES

The automobile, like the railroad, was built to carry passengers. But the car companies soon discovered a demand for rapid shipment of freight. Larger cities were the first to use trucks. Bakers, dairies, and finally the post office began to use trucks. Until after World War II, the truck did not really compete with the railroad for business. Today, however, almost everything you use has been transported by a truck at some point. Manufacturers make trucks in many sizes, from small half-ton pickups to large semitrailers that can carry thousands of pounds of freight. Fig. 104-3.

While the large trucks were changing freight transportation, passenger transportation was

104-3. *Large over-the-road semitrailer trucks haul freight from one city to another.*

American Trucking Association

104-4. *Transit bus for use in city transportation systems.*

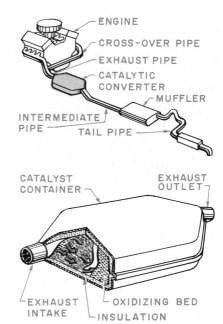

104-6. *Example of a catalytic converter.*

being affected by motor buses. Before World War I there was a slow beginning to the bus industry. During the period from 1920 to World War II, the bus industry grew, but it did not really reach great importance until after World War II. Today almost every city has buses as part of its rapid transit system. Fig. 104-4.

The railroads once carried passengers from city to city. Motor buses have taken over much of this service. Large buses travel from city to city, carrying passengers in comfort unknown to early bus travelers. Fig. 104-5.

ENVIRONMENTAL IMPACT

Automotive transport has been a mixed blessing. It has made Americans the most mobile people in the world. As the number of vehicles has increased, however, pollution has become a serious problem. As gasoline is burned in the engine, carbon monoxide, nitrous oxide, and hydrocarbons are produced. These are the main ingredients of the pollution we call smog. In 1970 Congress enacted the Clean Air Act, which limits the amount of pollutants an automobile may emit.

To comply with these rules and reduce the pollutants, automobile manufacturers have modified their engines in several ways. Fuel metering devices have been added to pace the amount of fuel fed to the engine. Since 1975, new cars have had to use unleaded gas. To avoid the wrong fuel being used, the filler pipe of the cars accepts only the fuel nozzle of unleaded gas.

The use of unleaded gas is required because of the addition of another smog reducing device. New cars have a *catalytic converter* added to the exhaust system. This converter is a device that resembles a muffler but is filled with platinum and palladium materials. Fig. 104-6. These ma-

terials reach a very high temperature (1000°F or 538°C) as the hot exhaust gases pass through them, causing a chemical reaction to occur. The carbon monoxide and hydrocarbons in the exhaust gases are converted to carbon dioxide and water.

Another solution to the problem of making a car burn fuel more efficiently and cleanly is called a *stratified charge*. A rich mixture and a lean mixture enter the combustion chamber. Fig. 104-7. The piston compresses the mixture. The rich mixture fires,

104-5. *Large overland buses such as this one can carry 50 to 60 passengers in comfort.*

Greyhound Bus Lines, Inc.

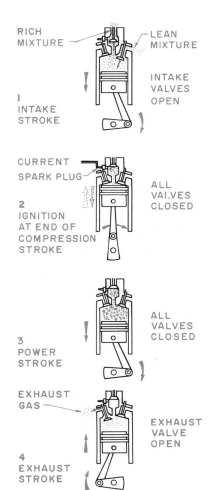

104-8. *An electric-pow-ered vehicle, one alterna-tive to the gasoline auto-mobile.*

104-7. *Stratified charge engine: (1) A rich mixture and a lean mixture enter the cylinder. (2) The rich mix-ture is ignited by the spark plug. The rich mixture then ignites the lean mixture. The power stroke and the exhaust stroke (3 & 4) are the same as on a regular engine.*

even though several companies have spent millions of dollars on development work. One of the main problems which must be overcome is the time it takes to produce the steam in a cold en-gine.

Another form of power that shows potential, and is already being used in some vehicles, is electric power. Electric-powered vehicles are equipped with bat-teries. The batteries must be re-charged periodically. Two prob-lems with electric vehicles are the weight of the batteries and the space they require. Organic batteries (see Chapter 91) may solve this problem. With further development, the electric vehicle may become practical. Fig. 104-8.

FUTURE AUTOMOBILES

The *Aero 2000* is a futuristic automobile that General Motors created with some of the most ad-vanced, computer-aided technol-ogy available today. Fig. 104-9. The car has a sleek, aerody-namic form that reduces wind re-sistance. The tires have new tread patterns that are expected to re-duce roll resistance, increase fuel economy, and improve handling and traction. This vehicle has front wheel drive and is powered by an experimental turbocharged, three-cylinder diesel engine. Fig. 104-10. It is expected to get up to twice the miles per gallon that many small cars get today.

The car weighs about 1800 lbs. It has a wheelbase of 102.7″ and

igniting the lean mixture. The leaner the mixture, the fewer pollutants.

ALTERNATE SOURCES OF MOTOR POWER

With the interest in improving the environment and reducing pollutants, several other forms of power for vehicles are being in-vestigated. One that has pro-duced great interest goes back to the early days of the automobile: steam power. So far the steam car has not proven practical,

104-9. *The Aero 2000. A single sliding door on each side gives access both to the front and rear seats.*

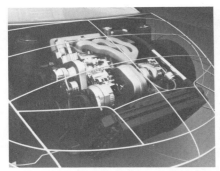

General Motors Corporation

104-10. *The* Aero 2000 *has a three-cylinder diesel engine.*

General Motors Corporation

104-11. *Interior of the* Aero 2000. *The handle on the center console replaces the steering wheel and the accelerator and brake pedals.*

General Motors Corporation

104-12. *Instrument panel of the* Aero 2000.

an overall length of 171.5″. It has room for four people.

Aero 2000 features new lightweight seats which are upholstered in breathable fabrics and an aircraft-type steering control. Fig. 104-11. It has a "head-up" instrument display that can be read while looking through the windshield, a video camera that

provides a 180-degree view of the rear, and a CRT road map to create a turn-of-the century interior. Fig. 104-12. The GM designers have proposed an advanced lap-shoulder belt that is under study.

The *Aero 2000* and other experimental automobiles lead the way to the comfortable, safe, convenient, and fuel-efficient automobiles you will soon be driving.

QUESTIONS AND ACTIVITIES

1. What did Henry Ford do to change the way cars were built? What advantages did his improvements offer?

2. How do cars cause air pollution?

3. Describe a catalytic converter, its purpose, and how it works.

4. What are two alternate sources of power that

are being investigated to power vehicles and at the same time reduce pollution? What are some disadvantages of these types of power sources?

5. What are three features of the *Aero 2000* experimental car that were designed to increase fuel economy?

PART 7

Modern Industry

CHAPTER 105

Introduction to Modern Manufacturing

In the first chapter of this book you were introduced to the world of industry and technology. Now, in this chapter, we shall look at the methods used by industry to manufacture things. Very simply, an *industry* may be defined as all the work needed to produce a certain kind of goods or services and to make those goods or services available to the people who need or want them. *Goods* are material things such as clothes and cars and computers. Fig. 105-1. *Services* are nonmaterial. For instance, the television and radio programs you enjoy are a service produced by the entertainment industry.

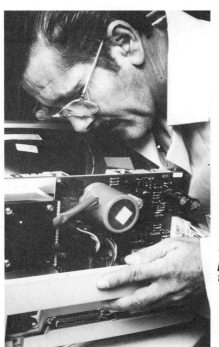

105-1. *This man works for a goods-producing company—an electronics firm. He is inspecting a computer terminal.*

Most of this book has been about the production of goods. You have learned to work with metal, plastic, and wood. In this chapter you will learn how a goods-producing industry is organized.

THE ESSENTIALS OF INDUSTRY

A manufacturing (goods-producing) industry needs three basic types of resources:

● Material resources, such as timber, petroleum, and iron ore, from which goods are made.

● Human resources. The people —everyone from computer operators to managers—whose work helps produce the goods.

● Capital resources. The factories, equipment, and money needed to turn raw materials into usable products.

Without these resources, industry could not exist. That is why they are called the *essentials of industry*.

Yet it is not enough for an industry to have the essentials for production. Industry must organize these essential resources in such a way that the mass-production of high-quality goods can take place. The elements of in-

dustry can provide such an organizational plan. Fig. 105-2.

THE ELEMENTS OF INDUSTRY

The elements of industry are seven key steps for organizing production:

● Research and Development (R&D). Inventing and designing products, processes, and materials.

● Production Tooling (PT). Designing and making special tools used to manufacture products.

● Production Planning and Control (PPC). Planning and controlling the flow of materials through the production and assembly lines.

● Quality Control (QC). Setting and maintaining standards of acceptability for products.

● Personnel Management (PM). Selecting and training workers.

● Manufacturing (MF). Changing raw materials into usable products by cutting, forming, fastening, and finishing.

● Marketing (MK). Creating a demand for and distributing finished products to the people who are to use them.

Not all industries use these same seven steps. Some give them different names or group them differently. But the logical pattern of action remains the same.

Research and Development

The new products, materials, and processes used in industry do not just happen accidentally.

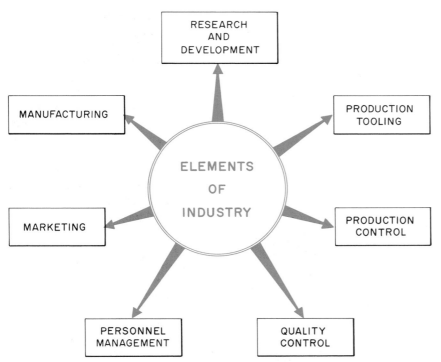

RESEARCH AND DEVELOPMENT

MANUFACTURING

PRODUCTION TOOLING

ELEMENTS OF INDUSTRY

MARKETING

PRODUCTION CONTROL

PERSONNEL MANAGEMENT

QUALITY CONTROL

105-2. *The elements of industry are a way of organizing people, materials, and machines to produce the things a modern society needs.*

They are invented and perfected by people who are trained in art, science, technology, and engineering. These are the people who work in research and development. Research and development (or R&D for short) is the planning of new products, processes, or materials and the improvement of old ones to meet the needs of people who use them. Research and development is sometimes called "the industry of discovery" because it is such a large, important part of the industrial world and requires so many people with different talents. Fig. 105-3.

Production Tooling

You may have heard the term *tooling up*. These words are often used in the manufacturing industries. For example, when a company comes out with a new model motorcycle, it must "tool up" for the product. This means that special tools must be made to stamp out the newly designed bodies. New frames, wheels, and headlights also require new tools to make them.

Production tooling (PT) is the element of industry concerned with these tools. Those responsible for production tooling obtain the tools, machines, and equipment needed to make a product. Fig. 105-4. Usually engineers design the tooling and tool and die makers produce it.

Fast, efficient mass production often requires the use of special tooling devices. Fig. 105-5. Some of these devices are described here.

A *jig* is a device which holds a workpiece firmly and guides a drill or other tool to an exact location. Look at the drilling jig shown in Fig. 105-6. Note that the steel rod workpiece is held in such a position that a hole will always be drilled exactly through the diameter of the rod. By using this jig you do not have to locate the hole with a center punch and clamp the piece in a vise before drilling. The drill bushing prevents the guide hole from enlarging. Jigs are valuable tools in production work.

A *fixture* holds workpieces during machining or assembly. Fixtures are usually attached to a specific machine. For example, there are milling fixtures, lathe fixtures, grinding fixtures, assembly fixtures, and so on. Figure 105-7 shows a fixture which clamps a workpiece in place during milling. With fixtures, workpieces can be fastened quickly and easily, and the finished pieces will always be alike.

While jigs and fixtures are two of the most important kinds of tooling used in industry, there are many others which are also necessary. Pressing punches and dies, extrusion and drawing dies,

U.S. Department of Labor
105-3. *Research and development of new products and processes is an important element of industry.*

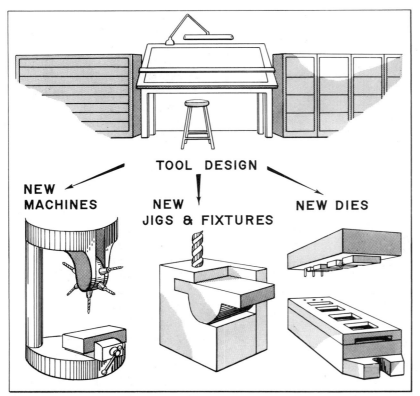

105-4. *Production tooling acquires the special tools, machines, and equipment necessary to manufacture a new product. These drawings show some of the things which have to be designed and built.*

This kind of careful planning is called production planning and control, or PPC. Its most important parts are routing, scheduling, dispatching, and plant layout.

Routing is preparing a plan of the steps required to make something. A route sheet (or plan of procedure) is usually prepared for every kind of part to be made. For example, if the product were a simple footstool with a top and four legs, one route sheet would be needed for the top and one for the legs. A bill of materials must also be made as a part of routing. This bill lists all the items needed to make the product.

Scheduling is fitting jobs into a timetable so that materials and parts enter the production line at the right place and time. A schedule, then, is a method of organizing facilities, orders, materials, and time.

One type of chart commonly used to prepare schedules is a

and casting patterns are some further examples. Fig. 105-8. The people who make these devices, the tool and die makers, are highly skilled crafters and must work and study for many years before becoming qualified.

Production Planning and Control

If you were building a table, you would not glue the legs and rails together until the dowel holes had been drilled. Neither would you paint the top if it had not been cut to shape and sanded. The same kind of clear thinking is needed when mass-producing products.

In order to make certain that a product will be made properly, the right materials must arrive at the right place, in the right amounts, and at the right time.

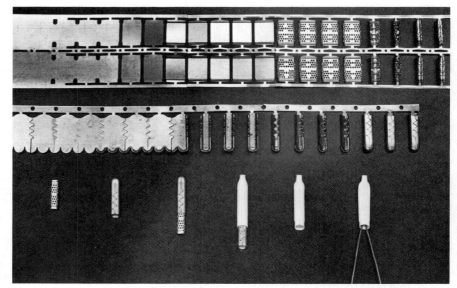

Western Electric

105-5a. *Progressive dies are used to cut and form wire connector inserts (top) and shells (middle). At bottom, left to right, are: the insert, outer shell, assembly of these two parts, application of the plastic jacket, the completed unit, and the connector crimped to join two wires. This is a good example of tooling for production.*

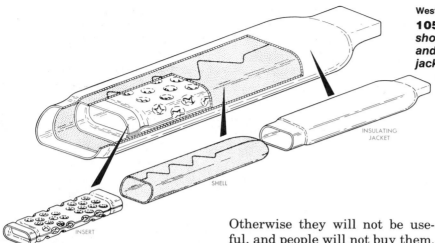

Western Electric

105-5b. *This cutaway diagram shows how the formed metal shell and insert fit into the insulating jacket to make a wire connector used in telecommunications.*

INSULATING JACKET

SHELL

INSERT

Gantt chart. One such chart is shown in the next chapter. Note that it shows the amount of work planned as well as the amount actually completed. Symbols are used to show why work was not completed on schedule. Schedules are also called operation or progress charts.

Dispatching is issuing work orders to set the production line in motion. For example, the supervisors on the assembly line are told when production should begin and how many pieces must be made each day. The supervisors make sure these work orders are followed. These orders are important, for they tell when materials should be released from storage and when production should begin.

Another important part of production planning and control is *plant layout*. The machinery and equipment must be arranged so that production can take place smoothly, without wasted time and effort. Usually this kind of layout is done by engineers and technicians who understand the production process.

Quality Control

Products made in a factory must meet certain standards.

Otherwise they will not be useful, and people will not buy them. The element of industry concerned with this part of production is quality control, or QC. Quality control can be defined as

105-6. *This drilling jig holds the workpiece so that it can be drilled.*

DRILL

BUSHING

JIG

WORKPIECE

CUTTER

WORKPIECE

CLAMP

FIXTURE

105-7. *This fixture holds the workpiece in the milling machine.*

those activities which prevent defective articles from being produced or, if they are produced, prevent them from reaching the market. In this way management tries to insure that a product will

J. A. Richards Company

105-8. *These dies are used to produce a metal part (shown by arrow).*

be acceptable to the buyer. Fig. 105-9.

There are three steps in a QC program: specifications, tooling, and inspection.

Specifications are detailed descriptions of the standards for a product. Some typical standards might include rules about size, material, function, and shape. For example, a specification might state that a cutting board must be ¾″ thick, 7″ wide, and 12″ long and be made out of basswood. It might further state that boards ⅛″ smaller or larger (± ⅛″) in any of these dimensions will be acceptable. This specifies the amount of error that will be allowed in each piece. This allowance is called the *tolerance*. It is important to state a tolerance because tool wear and operator error make it impossible for all the boards to be exactly the same size.

Tooling in QC refers to the special devices needed to measure the accuracy of parts. A "go/no-go" gauge is an example of such a tool. By slipping a part into this device the inspector can tell at a glance whether or not it is the right size. It is the responsibility of the QC staff to design

and build, or purchase, such inspection devices.

Inspection is done to make sure products meet the specifications. Inspections are done on purchased materials or parts, on goods being produced or assembled, and on finished products. In addition to special inspectors, machine operators are expected to examine their own work. They are qualified to do this because they know what the part should look like, and they can detect errors early.

Personnel Management

Personnel management (PM) is concerned with the selection, hiring, training, and supervising of workers for industry. The production line cannot operate efficiently unless the right person, properly trained, is doing the right job. Personnel management is also involved in providing good working conditions and employee benefits.

Manufacturing

Earlier in this chapter you learned that the manufacturing industries are those which produce goods for people to use. This is done by changing raw materi-

als, such as wood or metal, into usable products. You have also learned that not every department does the actual work of making the product. For example, personnel management and production control are not responsible for the actual working of the raw materials.

Those departments which actually convert the raw material into finished products make up the manufacturing (MF) element of industry. It is in these departments that the industrial processes of cutting, forming, fastening, and finishing take place. Fig. 105-10.

Cutting, forming, fastening, and finishing operations are shown in chart form in many sections of this book. For example, Chapter 34, "Metal Cutting Principles," describes various cutting operations. Study these charts to learn about the many ways in which raw materials are made into useful products.

Marketing

Marketing (MK) is the process of getting products from those who make them to those who use them. Marketing helps to deliver the right kinds of goods to us, in the right form and amount, at the right time and price. In order to do this, the people in marketing must work at market research, advertising, packaging, distributing, selling, and servicing.

Market research gathers information about products and the people who use them. Have you ever answered a questionnaire about a product or participated in a taste test? These are examples of market research. The people doing the research want to find out what people like or dislike so that their product can be made to appeal to the greatest number of people.

105-9. Quality control is a very important element of industry. This woman is inspecting a blood analyzer.

The usual purpose of *advertising* is to make the public aware of and interested in certain goods. Manufacturers advertise on television and radio, in newspapers and magazines, and on billboards and other signs. Advertising is a valuable service. For instance, through advertising we can learn about different brands of the same kind of product so that we may select the one which suits us best.

In industry, *packaging* refers to the containers used to hold products. You see and use many kinds of packages each day. The gum you chew comes in a wrapper; the shoes you buy come in a box; the soda you drink comes in a can or bottle. For some products, packaging is mainly functional. A cardboard and plastic bubble pack for nails, for example, serves to hold and display the product. For other products, however, the packaging is part of the product's image. Colognes, for example, are put in very attractive packaging to convey a sense of glamour, excitement, or romance. Some cologne manufacturers even package their product in unusually shaped bottles to attract collectors.

Distributing refers to all the ways of getting a product from the manufacturer to the place where it will be used. Trucks and trains, ships and barges, all play an important role in distribution; but perhaps the greatest advance has come through air travel. For example, one company that makes earthmoving equipment promises delivery of replacement parts anywhere in the world in 48 hours or less. Without air delivery, a road construction project in a distant place—India, for instance—could be delayed for weeks because of a defective part.

After the product reaches the store, someone has to *sell* it to

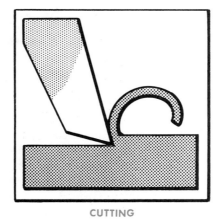
CUTTING

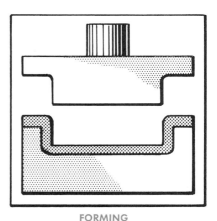
FORMING

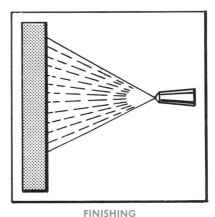
FASTENING

FINISHING

105-10. *After the designing, planning, and tooling have been done, the next step is to make the product. This is called manufacturing, and it involves four operations: cutting, forming, fastening, and finishing. Often the materials used in manufacturing have already been partially processed. Still they are considered raw materials because they are not yet a finished part of a useful product.*

the consumer (user of the product). Salespeople serve an important function by selling products. They also perform other valuable services such as keeping doctors up to date on new medicines, making manufacturers aware of new equipment, and informing teachers about new books. Selling is the final step in the making—using cycle.

To *service* a product is to maintain and repair it when needed. Servicing is becoming more and more important to the consumer. The reason for this is that so many of the things we use are be-

coming more complicated, and we cannot always repair them ourselves. Electric appliances, automobiles, computers—all need regular attention so that they will work properly. Service representatives do these jobs for us. Good service helps to sell products because people want to know that they can get items repaired without too much trouble. As more and more of our products require regular servicing, more people trained to do this work are needed. This is one reason the service industries are growing.

In this chapter you have

learned that industry must organize its resources carefully in order to manufacture its products. In the next chapter you will learn how to mass-produce a product in your school shop.

THE ROLE OF THE ENTREPRENEUR

It's part of our folklore—the hard-working young man or woman who starts a business on a shoestring budget and builds an empire. In real life, the story doesn't always end happily. Nevertheless, many people *do* succeed as entrepreneurs.

An *entrepreneur* is someone who organizes and manages a new business. In our free enterprise system, the entrepreneur has always played an important role. Many of our most successful companies were started by people who had a new idea, a little money, and a lot of ambition. Herman Hollerith, the inventor of the electrical census counting machine (see Chapter 2), founded his Tabulating Machine Company in 1896. Through various mergers, the company grew into the International Business Machines Corporation—IBM.

Today the entrepreneurial spirit is stronger than ever. Every year, about 600 000 new businesses are started. In the 1970s such businesses created 6 million new jobs.

One reason for the increase in new businesses is high technology. Semiconductors, genetic engineering, computers, lasers, robots—these technologies have given rise to industries which didn't even exist when your parents were in school. For example, the popularity of microcomputers has created a huge market for software, and many entrepreneurs have taken advantage of this opportunity. One 22-year-old started a mail-order software company in his apartment. He had one program and a $5000 loan to get started. Five years later, his company's sales were $10 million.

Another reason for growth has been the availability of money. Today, very few people can afford to start a business entirely from their own savings. They obtain loans from banks or sell shares of the company to investors. Money used to start a business is called *venture capital*. In 1969 Congress increased the maximum tax on long-term capital gains from 25 percent to 49 percent. (*Capital gain* is the profit an investor makes from the sale of stock, real estate, or other property.) The effect of this law was to dry up venture capital. People did not want to risk their money on new businesses if the price of success was a 49 percent tax. In 1978 Congress rolled back the tax to 28 percent, and venture capital became available once more.

In helping to finance a new business, the investor risks a portion of his money. The entrepreneur may risk his entire future. What sort of person will take such a risk? Although personalities vary, there are three traits most entrepreneurs have in common. These are self-confidence, imagination, and the need to be in charge. Entrepreneurs like to be their own boss, even when that means risking financial ruin.

Although they are independent, successful entrepreneurs know when and how to get help. As their company grows, they often hire experienced managers from older companies. Some entrepreneurs sell their company and then use the profits to start another business. One has even started a business to help other entrepreneurs! The business offers new companies such services as advertising and accounting in return for a share of the company's ownership.

The entrepreneurs play a vital role in American business. Their imagination and drive to succeed spur the growth of new industries. In the coming years, the entrepreneur will continue to be the pioneer who leads the way to new, unexplored regions of economic opportunity.

QUESTIONS AND ACTIVITIES

1. Define industry.

2. Name the three essentials of industry and give examples of each.

3. Name the seven key steps for organizing industrial production. Briefly describe what is done in each of these steps.

4. Define and describe tooling up.

5. What is a jig?

6. What is the function of a fixture?

7. Name and describe the four important parts of PPC.

8. Name and describe three important things that must be done to insure that a product will be acceptable to the buyer.

9. What does the term *tolerance* refer to and why is it important in industry?

10. What is a go/no-go gauge used for?

11. What is market research and why is it done?

In Chapter 105 you learned how industry works to produce large quantities of products. As you know, this is called mass production. Mass production can be done in the school shop or by student clubs on a smaller scale. You will make fewer products, but you will be using the same elements of industry that are used in large factories. Such activities will help you learn about industry and production, and they will give you an opportunity to develop your leadership skills. Fig. 106-1.

In a mass-production activity, there are many ways to practice leadership skills. For example, you may be in charge of planning the assembly line. You would need to direct the team of workers who put the product together. Or, you might be in charge of the sales program. In this job you would need to plan and direct the advertising for the product. Whatever job you have, there will be responsibilities. Showing you can handle responsibility is an important step toward becoming a leader.

The mass production experience demands much planning by all those who participate. The work begins by forming the student company. Sometimes this is done by selling shares in the company to people outside the class or club. Then the class or club divides into teams to plan each element of production. Remember that these elements are research and development

106-1. *Student organizations such as Junior Achievement can help you learn how industry works. Here a JA company and its sponsoring firm have set up booths side by side at a trade fair.*

(R&D), production tooling (PT), production planning and control (PPC), quality control (QC), personnel management (PM), manufacturing (MF), and marketing (MK). Review Chapter 105 for more information about the duties of each team. (You may be asked to participate on more than one team.)

In this chapter you will learn how to mass-produce a simple wood & rope trivet. Fig. 106-2. This is an inexpensive project, easy to manufacture yet very useful.

RESEARCH AND DEVELOPMENT (R&D)

The first step in production is choosing or designing the product. The trivet was chosen by a group of students in an industrial education class. They talked

106-2. *This wood & rope trivet makes a good mass-production project. It can be used to protect a table from hot food containers.*

about the product's function and sketched some ideas. They found that many shapes could be used, such as a circle, a rectangle, or even a fish shape. The shape shown in Fig. 106-2 was selected because it could be used for a variety of different pans or food platters.

An important part of R&D is the preparation of a final working drawing. Fig. 106-3. Materials must also be chosen. Figure 106-3 shows hemp rope. Plastic rope could also be used. In addition, you will need brass escutcheon pins, copper nails, or steel brads.

PRODUCTION TOOLING (PT)

This team is responsible for designing and making the special

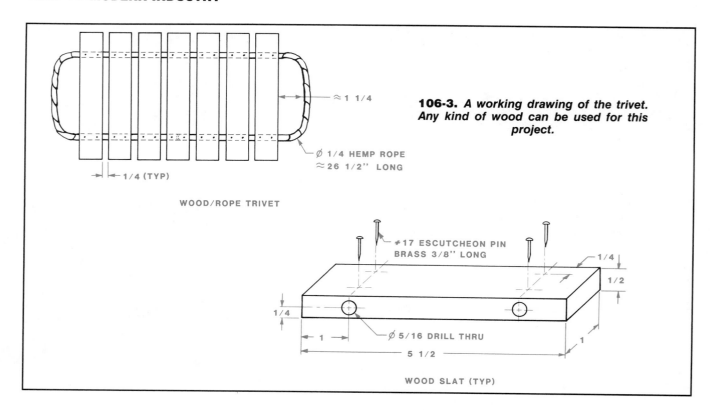

106-3. *A working drawing of the trivet. Any kind of wood can be used for this project.*

≈ 1 1/4

Ø 1/4 HEMP ROPE
≈ 26 1/2'' LONG

1/4 (TYP)

WOOD/ROPE TRIVET

#17 ESCUTCHEON PIN
BRASS 3/8'' LONG

1/4
1/2

1/4

Ø 5/16 DRILL THRU

1
1
5 1/2

WOOD SLAT (TYP)

jigs and fixtures needed to mass-produce the project. For the trivet, one jig was designed to use when drilling the rope holes. Fig. 106-4. Another one was made for the nail holes. Fig. 106-5. These are but two examples. You may choose to design a different kind. Remember to try the jigs on several different slats to make sure they work.

The assembly jig is shown in Fig. 106-6. After the rope has been strung through the slats, nail the two ends of the rope through the middle slat to secure them. Then fit the slats into the jig, even out the rope "handles," and complete the nailing.

PRODUCTION PLANNING AND CONTROL (PPC)

The task of this team is to prepare a bill of materials, route

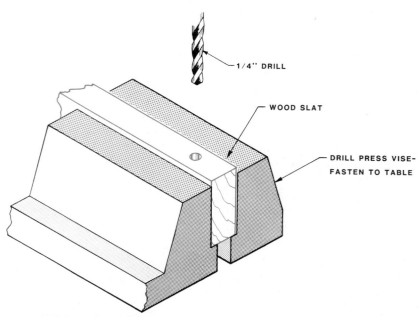

1/4'' DRILL

WOOD SLAT

DRILL PRESS VISE-
FASTEN TO TABLE

106-4. *The rope hole drill jig. Note that the slat is set even with the end of the vise. When the hole is properly located, the vice is fastened to the drill press table. Now all rope holes will be drilled in the same position in each slat.*

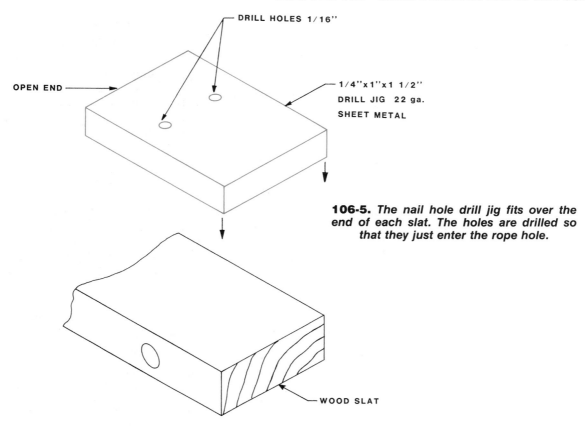

DRILL HOLES 1/16''

OPEN END

1/4''x1''x1 1/2''
DRILL JIG 22 ga.
SHEET METAL

106-5. *The nail hole drill jig fits over the end of each slat. The holes are drilled so that they just enter the rope hole.*

WOOD SLAT

106-6. *The assembly jig. Remember to apply your finish to the wood slats before using this jig.*

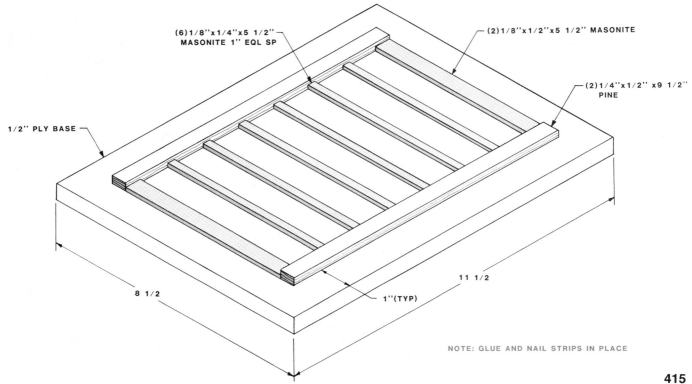

(6)1/8''x1/4''x5 1/2''
MASONITE 1'' EQL SP

(2)1/8''x1/2''x5 1/2'' MASONITE

(2)1/4''x1/2'' x9 1/2''
PINE

1/2'' PLY BASE

11 1/2

8 1/2

1''(TYP)

NOTE: GLUE AND NAIL STRIPS IN PLACE

106-7.
Bill of Materials

| No. | Size | | | Name of part | Material | Unit cost | Total cost |
	T	W	L				
7	½	1	5½	slat	walnut		
1	¼		26½	rope	hemp (or plastic)		
28	#17		⅜	escutcheon pin	brass		
As needed					oil finish		

sheet (or plan of procedure), flow chart, and a Gantt chart. Some of the information for making these is obtained from the drawings prepared by the R&D team. (Fig. 106-3.) The tools designed by the PT team are also important. The charts are shown in Figs. 106-7 through 106-10. Note that these plus the drawings show all of the information needed to make the project.

QUALITY CONTROL (QC)

The purpose of quality control is to make sure that the project is made correctly. Parts must fit together properly, and the wood must not be chipped or scratched. The slats must be straight and even. No special quality control devices or tools need to be made. Most of the inspections are visual.

PERSONNEL MANAGEMENT (PM)

An important part of industry is to train workers for jobs on the production line. By studying the production control sheets, the personnel management team can see what operations have to be done at the various work stations. Your teacher or club leader will help by demonstrating how to do these tasks. Fig. 106-11. The students on the personnel management team are helpers. Students in the class or club are assigned jobs and trained to perform them. It is very important

to stress safety. The student workers must also learn to read the drawings and to follow the procedures necessary to make each part. Their work must be checked from time to time to make sure they know their jobs. Only then can the work move smoothly.

MANUFACTURING (MF)

The manufacturing phase of the mass production activity is where it all comes together. Here all the planning and training are put into action to change raw materials into a usable product: the wood & rope trivet. The production line is set up; student workers move into their work stations. They read their plans and do their jobs. Quality control

---

Student Clubs

The development of leadership skills is important. In school, such skills can be developed in many ways. One of the easiest ways to develop leadership skills is by joining a club. In most schools, there are a variety of clubs. Each club brings together people with an interest in one activity. For example, there are stamp clubs, music clubs, and speech clubs. There are also industrial arts clubs. Many industrial arts clubs are part of a national organization. There are two national organizations of industrial arts clubs. These are the American Industrial Arts Student Association (AIASA) and the Vocational Industrial Clubs of America (VICA). These organizations have similar goals. Among their goals are the following:
- To develop contacts with people in industry.
- To develop an understanding of our technology.
- To improve consumer understanding.
- To develop productive use of leisure time.
- To plan, organize, and carry out worthwhile activities and projects.
- To recognize high standards of achievement, scholarship, leadership, and safety practices.

By taking part in the activities of a club, you can develop your leadership skills. In most clubs, you will be able to take part in the following activities. You will learn:
- To conduct a meeting using parliamentary procedure.
- To do fund raising.
- To plan an open house for the community.
- To be an officer in the organization.
- To develop technical skills.
- To participate in regional, state, and national contests.

Taking part in any group activity will help you develop your social skills. It is these skills that help you get along with others. All of the following skills are important in everyday life. They also are important in developing leadership.
- To communicate effectively, both in writing and in speaking.
- To do the best job possible.
- To adapt your behavior to a variety of situations and individuals.
- To react positively under pressure.
- To complete a job with or without supervision.
- To work safely in all situations.

Participating in the activities of a student club can help you develop your leadership and social skills while teaching you about the American system of free enterprise.

106-8.
Route Sheet

PROGRAM _Trivet_
PART NAME _Slat_
PART NUMBER 1
ISSUE DATE

FOR MODELS
MATERIAL _Walnut_ **WT./LBS.** **RGH.** **FIN.** **SHEET** 1 **OF** 1
RELEASE

LINE NO.	OPER. NO.	OPERATION DESCRIPTION	TOOL—MACHINE—EQUIPMENT DESCRIPTION	UNITS REQ'D	TOOL OR B.T. NUMBER	HOURLY CAPACITY GROSS	NET
1	1	cut width	table saw—set fence				
2	2	cut length	table saw—set stop rod on miter gauge				
3	3	drill rope holes	drill press—use jig #1, $5/16''$ bit				
4	4	drill nail holes	drill press—use jig #2, $1/16''$ bit				
5	5	sand all over	medium and fine sandpaper				
6		finish all over	oil finish, clean cloth				
7							
8							
9							
10							

PROGRAM _Trivet_
PART NAME _Rope_
PART NUMBER 2
ISSUE DATE

FOR MODELS
MATERIAL _Hemp_ **WT./LBS.** **RGH.** **FIN.** **SHEET** 1 **OF** 1
RELEASE

LINE NO.	OPER. NO.	OPERATION DESCRIPTION	TOOL—MACHINE—EQUIPMENT DESCRIPTION	UNITS REQ'D	TOOL OR B.T. NUMBER	HOURLY CAPACITY GROSS	NET
1	1	cut to length	knife, rule				
2							
3							
4							
5							

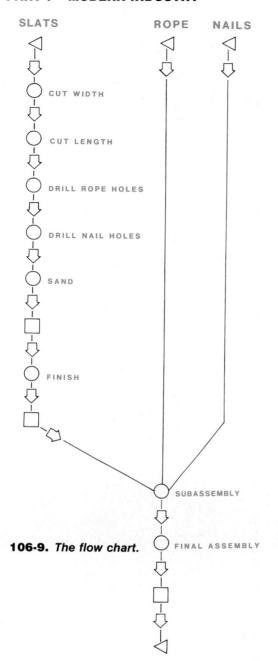

SLATS   ROPE   NAILS

CUT WIDTH

CUT LENGTH

DRILL ROPE HOLES

DRILL NAIL HOLES

SAND

FINISH

SUBASSEMBLY

106-9. *The flow chart.*

FINAL ASSEMBLY

Fred W. Gillman

106-11. *Training workers for the production line is a part of personnel management.*

JOB TRIVET	MON	TUE	WED	THU	FRI
CUT SLATS	▬▬▬				
DRILL ROPE HOLES	▬▬▬ R				
DRILL NAIL HOLES	▬▬▬ R				
CUT ROPE	▬				
SAND SLATS		▬▬ M			
FINISH SLATS		▬▬▬ M			
ASSEMBLE TRIVETS					

PLANNED WORK: ———
ACTUAL WORK: ▬▬▬

106-10. *A Gantt chart for scheduling work on the trivet. Note that on Tuesday the drill press broke down, causing delays in drilling, sanding, and finishing.*

SYMBOLS USED

A. OPERATOR ABSENT
G. GREEN (INEXPERIENCED) OPERATOR
I. POOR INSTRUCTION
L. SLOW OPERATOR
M. MATERIALS HOLDUP
R. MACHINE REPAIR
T. TOOLS LACKING
V. HOLIDAY

checks are made. The product is assembled and the final tests are made. Each trivet is checked to make sure it meets the standards of quality. The product is now ready for marketing.

MARKETING (MK)

There are several different plans you can use to market the trivet. You could make one for each student in the class. Divide the total cost of the trivets by the number made to get the cost per unit. This is the amount each person must pay for a trivet. For example, suppose you produced 20 projects. The costs might read as follows:

140 slats	=	5.00
50′ rope	=	2.50
nails	=	1.00
finish	=	1.50
Total		$10.00
$10.00 ÷ 20 = $.50		

Each student must pay $.50 for his or her trivet.

Another plan could be to advertise the projects in the school, take orders, and produce them for sale. If advertising is to be used in selling, this also becomes a cost. For example, the cost of paper and paint for posters must be figured into the total cost of the trivet. The cost per trivet could be increased to perhaps $1.00 or $1.50 and the trivets could then be sold at a reasonable profit to raise money for some school project or for an organization such as the Red Cross, United Fund, or a children's hospital.

QUESTIONS AND ACTIVITIES

1. Make a questionnaire to use in market research of your mass-production product. Poll people in your school or area of distribution to determine such things as the demand for the product, size, shape, color, acceptable price, etc.

2. Discuss the best ways and places to advertise your finished product and decide the most efficient and economical way to distribute it.

3. Discuss any problems you encountered in your mass production and how they might have been avoided.

Careers in Industry and Technology

One of the most important things you can get from your experience in school is to discover your job interests. Learning about your job interests is called career awareness and career exploration. What do you want to be doing in five years? Ten years? What kinds of jobs are available in industry? If you could pick any job, what would it be? What do you have to do in order to prepare for it? How do you go about getting a job? These are some of the questions which will be discussed in this chapter.

SKILLED CRAFTS

Industry depends heavily on people skilled in the crafts to produce goods. These are the people who make the drawings, patterns, models, and tools and dies needed for production. Fig. 107-1. They operate machines and use tools and measuring instruments. These skilled workers also print newspapers, repair automobiles, build cabinets, install electrical wiring, and repair and adjust machines and instruments. You can see that the crafts include many different kinds of jobs in industry. Some examples of the careers available are described here.

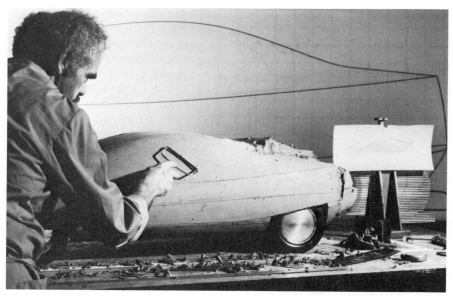

General Motors Corporation

107-1. *This modeler is sculpting a scaled-down clay model of an experimental automobile.*

Machinists are people skilled in using metalworking machines such as lathes, milling machines, and drill presses. They must know how to set the workpieces in the machine, read blueprints, and use measuring tools. Tool and die makers are skilled machinists who specialize in making devices and instruments used in production, such as jigs and fixtures. The machinists of today must work with computer controlled machines of many types.

Carpenters are among the skilled workers in the building trades. They construct, maintain, repair, and remodel residential and commercial buildings. Carpenters work mainly with wood. In house construction, for example, they build the house frame, erect the walls and roof, and install doors, windows, floors, paneling, cabinets, and molding. In commercial construction, they build the scaffolding and the forms for concrete. Carpenters must know how to follow blue-

prints and diagrams, and they must have a working knowledge of math in order to measure and lay out materials. They must be skilled in using hand and power tools such as saws, drills, and hammers. Fig. 107-2.

Automobile mechanics are valuable workers who keep automobiles, trucks, and other vehicles operating. They must know how to read service manuals, figure out what is wrong with an engine, and proceed to repair it. Mechanics use many tools, instruments, machines, and test equipment in their work. Fig. 107-3. Today especially, they have to know about emission control in engines to reduce the pollution of our environment. A closely related trade is that of the auto body technician, who straightens dents, replaces parts, and refinishes automobile equipment.

From these descriptions of jobs in the skilled crafts you can see that these jobs require special

U.S. Department of Labor

107-2. *Carpenters are skilled workers who must have a sound knowledge of construction materials and techniques.*

Bernice Q. Johnson

107-3. *Automobile mechanics must know how to diagnose and repair engine problems.*

training. You can get this training in a number of ways. The industrial shop courses you take in school will give you a good background in tool and machine skills. After you graduate from high school, you can go to a vocational or technical school for further work.

Many workers learn their skills in apprenticeship programs. Most apprenticeship programs include classroom training in blueprint reading, mathematics, and skills related to that trade, as well as on-the-job experience. You can get information on these programs from your local trade unions.

TECHNICAL CAREERS

Technicians work closely with scientists and engineers. Their main responsibility is to carry out the details of projects which are planned by a scientist or engineer. Technicians are employed in industrial laboratories and manufacturing plants, medical research centers, and experimental engineering programs. In fact, any place where there is work of a scientific or technical nature, it is almost certain that technicians also are there, working in some important job. Fig. 107-4.

Today's technicians are usually graduates of technical institutes or of junior or community colleges having two-year technology programs.

In addition to technicians with formal training, some others have become technicians because of training received during military service. Some technicians have obtained their present jobs after completing part of a regular engineering technology course at a college or university.

In summary, then, formal training at a technical institute or junior community college pro-

Gov't. of Puerto Rico
107-4. *This technician is testing an electronic measuring machine called a spectrometer.*

vides the best route for becoming a technician. Other ways are open, however, if the basic qualifications and desire are present. Some typical technician jobs are described here.

Drafting and design technicians are first of all skilled drafters. They know layout and detail work and are neat and accurate. They prepare drawings from specifications (written descriptions of details not shown on a drawing), sketches, or notes furnished by engineers or scientists. They must have some knowledge of production machinery and methods to construct layout drawings quickly and accurately. Today these technicians

are trained to work with computer-aided design systems. Fig. 107-5.

Computer-aided manufacturing (CAM) technicians operate and service robots, lasers, automatic material handling equipment, and computerized inspection systems. These skills will be in great demand in the near future.

McDonnell Douglas Automation Company, a division of McDonnell Douglas Corp.
107-5. *Drafting and design technicians use modern tools of high technology, such as computer-aided design.*

ENGINEERING AND SCIENCE

Engineers and scientists invent, develop, design, and refine ideas and things that are used in production processes. They are leaders in the production team made up of crafts workers and technicians.

How do you know if you can become a scientist or an engineer? First take a look at your school record. How are your grades in science and mathematics? Good marks are required for college admission. Prospective engineers and scientists are usually interested in these subjects and may find themselves active in scientific projects or science fairs. Good grades usually reflect the type of study habits and mental discipline required for success in both college and career.

The scientist or engineer has an inquisitive mind, wanting to know why as well as how. Such a person may like to tinker with things such as model aircraft, automobiles, or radios—discovering how they work or trying to improve them. He or she should have the ability to design and create—to transform ideas into sketches and, in turn, be able to visualize a form or device from drawings.

These are some of the signs. Don't be discouraged if they don't

IBM

107-6. *The computer is a common tool of engineers.*

all fit your situation. Some people do not display all these traits early in life while others who do possess them become successful in other fields. But these signs can be used as a rough guide.

If you think you are interested in this field, your goal in high school should be to build a firm foundation for training in college. To qualify for most professional schools you must take several basic subjects, many of which you would probably take anyway. These provide the groundwork for any field of science or engineering and represent their basic tools. They are mathematics, science, computers, technology, and communication.

THE PRODUCTION TEAM

The scientist on the production team serves as the discoverer since she or he is a person who seeks new knowledge. The scientist is an inquirer—a person trained in the laws of nature who investigates the world and its se-

crets to learn more about why and how things behave as they do. He or she is interested in both basic and applied research. Scientific study areas include physics, chemistry, mathematics, computer science, materials science, and biology.

The builder on this team is the engineer. He or she applies knowledge gained from scientific discoveries to the practical problems of life to create things that people can use. A basic knowledge of science, mathematics, and computers is combined with judgment and experience to design new products and systems. Some people call engineering the art of making science useful. Fig. 107-6.

Technicians on the team act as assistants to both scientists and engineers by helping them convert their ideas into accomplishments. They possess certain skills which enable them to perform many tasks in support of the scientist's work, such as building test equipment, conducting tests, and recording data. Fig. 107-7.

The worker skilled in a craft is an important member of the team who puts into practice the results of development and plan-

Western Electric

107-7. *This technician is inspecting electric coils used in telecommunications.*

Bernice Q. Johnson

Many people are returning to school to learn new job skills. This trend will continue.

The Future of Work

You have probably heard our era described as "the post-industrial society" or "the information age." These terms are used to describe an economy based on the production of information rather than the production of goods. Today, only about 20 percent of American workers have jobs in manufacturing. The rest work in service industries, most of them in information jobs.

However, it would be wrong to assume that manufacturing will disappear from the American economy. After all, you can't eat, wear, drive, or live in information. Computers, the symbol of the information age, are themselves manufactured objects. The post-industrial society will not make manufacturing obsolete, any more than the Industrial Revolution made farming obsolete. The new technology *will* make manufacturing more efficient, and fewer people will be needed to produce goods.

The information age is bringing enormous changes in the way we earn our living. Following are some descriptions of today's trends and predictions about how they will affect tomorrow's jobs.

● Industry in the United States is moving away from high-volume mass production. Computer-aided design and manufacturing (see Chapters 3 & 4) make it possible to produce smaller numbers of goods economically. This type of short-run manufacturing is called batch production. With smaller production runs, it is possible to manufacture goods in a wider variety.

● Computers and robots will take over much of the routine labor. Demand for assembly-line workers will continue to decrease. Demand will increase for technicians and engineers who know how to build, operate, or repair computers and robots.

● Relatively few people will be making products. Most will be creating, processing, or distributing information. They will be working for telephone and other communications companies, broadcasters, banks and other financial institutions, insurance companies, and publishers. And of course many people will be working in information jobs that are part of a goods-producing industry. Most information jobs will require some knowl-

edge of computers. In fact, computer programming is one of the fastest-growing occupations. About 50 000 programmers are needed now. By 1990, as many as 1 million jobs may be available for programmers.

● While many of our old industries are declining, new ones are springing up. Many of these are related to computers—building them, improving them, and developing software to run them. These new industries are started by entrepreneurs, people who organize and manage new businesses. Entrepreneurship is booming now because there are so many opportunities for someone with imagination, ability, and a willingness to take risks. Most of the new jobs created within recent years have been generated by new, small businesses.

● The job market will continue to change. Most people will have several careers during their lifetime. Education and job training will not stop with graduation; they will continue throughout the working years.

107-8.
Personal Inventory of Interests and School Activities

I. Work Preferences

Check your likes with a plus (+) and your dislikes with a circle (○).

1. Work indoors. ____
2. Work where I would sell things. ____
3. Work involving mathematics and science. ____
4. Work where I can be my own boss. ____
5. Work with tools and machines. ____
6. Work in drafting. ____
7. Work requiring patience and accuracy. ____
8. Work with other people. ____
9. Work which is clean and neat. ____
10. Work involving original thinking. ____
11. Work directed by someone else. ____
12. Work depending on my writing skill. ____
13. Work which is physical. ____
14. Work where I would meet people. ____

II. School Subjects

Use check marks to indicate the subjects you like, and also how well you do in them.

Subject	Like it	Below Average Grades	Average Grades	Above Average Grades
1. Science				
2. Mathematics				
3. Social studies				
4. English				
5. Drafting				
6. Shop courses				
7. Business				
8. Art				
9. Physical education				
10. Others				

III. Activities

List the clubs, teams, or other groups to which you belong.

_____

_____

_____

_____

List your hobbies.

_____

_____

_____

_____

IV. Work Experience

List the jobs you have held. Check those you liked.

_____

_____

_____

_____

If you could choose anything, what kind of job would you like to have?

_____

_____

_____

ning. Goods roll off the production line through her or his efforts. All together—the scientist, the engineer, the technician, and the skilled worker produce the goods and services which provide us with the things we must have in order to live safely, healthfully, and comfortably.

BUSINESS CAREERS

Many persons are needed in industry aside from those skilled in crafts, technical jobs, and engineering. People are needed to manage and supervise sales and marketing as well as other activities.

These people work as supervisors of departments, managers of divisions, or vice-presidents or presidents of companies. They work together in planning and managing the operation of the industry. Such operations include sales, production, services, purchasing, accounting, research and development, and training. These people are trained in business schools and colleges. Their jobs require the ability to work closely with people and to understand the finance and management of industry.

LEARNING ABOUT YOURSELF

You should begin to think seriously about your future and your career while you are in school. The classes you select in high school can and should relate to your occupational interests. One good way of beginning the search for a career is to think about your interests and school activities. You should begin to learn as much about yourself as you can. The classes you take, the spare-time jobs you have had, and your hobbies will all help in learning about yourself. Look at the form in Fig. 107-8. Answer the questions as accurately and honestly as you can. (Do not write in the book.) Think carefully about each question before you answer it. When you have decided on your answers, you may want to talk about them with your parents, teachers, and counselor. They can help you in planning a school program which will help you to prepare for your career.

STUDYING OCCUPATIONS

At this point you have learned about typical kinds of jobs in industry. You may have also learned something about yourself. The next step is to select an occupation which interests you

and to study it. You may feel that a career in drafting is one you wish to explore. The form in Fig. 107-9 will aid you in this study. There are many ways to get information on the occupation or job you have chosen to study. Talk to your teacher and to drafters in industry. Go to your counselor and get a booklet on careers in drafting. Look in magazines and in newspapers. These will have articles on drafting; your librarian can help you get these. Look at the job advertisements in newspapers. They will give you an idea of how much money a beginning drafter will make. Talk to persons who employ drafters to learn about promotions, the kinds of work drafters do, and the kind of people they want to hire. In short, learn all you can about the job that interests you. In this way you can begin to think and plan for the career you may work in when you finish your schooling.

107-9.
Occupational Study Form

1. What is the occupation which interests you? (Carpenter? Mechanic? Other?)

_____

2. How much education or training is required for this job?

_____

3. Is there now, or will there be in the future, a great demand for workers in this occupation?

_____

4. What are the promotion possibilities?

_____

5. Describe the salary, pension, vacation, and other benefits.

_____

_____

6. How many persons are now employed in this type of work?

_____

QUESTIONS AND ACTIVITIES

1. Name and describe three skilled crafts.
2. What is the main responsibility of technicians?
3. What is the function of scientists on the production team? Of engineers?
4. Copy and then fill out the two career information sheets in Figs. 107-8 and 107-9. After com-

pleting these forms, discuss them with your parents, teacher, and counselor. Find out as much as possible about any career that interests you by checking with your library, reading any related pamphlets, checking newspaper articles and ads, and talking with people who are knowledgeable and experienced in your chosen field.

Projects

These projects are practical applications of many processes described in this book. You will get practice in:
- Reading drawings.
- Measuring and marking workpieces.
- Cutting, forming, fastening, and finishing wood and metal.

Your class or student club may want to mass-produce one of these projects. They can be made as shown or changed to suit your needs.

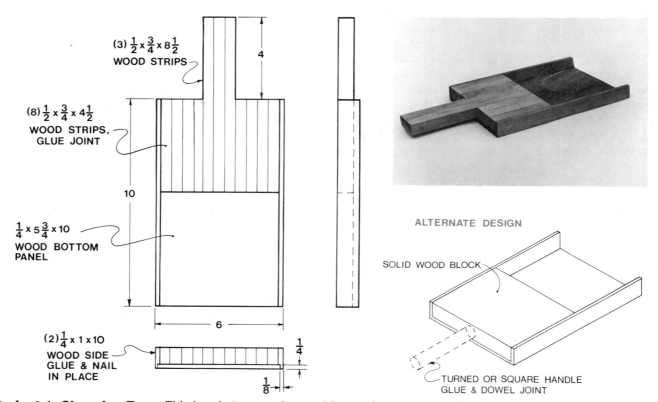

(3) $\frac{1}{2}$ x $\frac{3}{4}$ x $8\frac{1}{2}$ WOOD STRIPS

(8) $\frac{1}{2}$ x $\frac{3}{4}$ x $4\frac{1}{2}$ WOOD STRIPS, GLUE JOINT

$\frac{1}{4}$ x $5\frac{3}{4}$ x 10 WOOD BOTTOM PANEL

(2) $\frac{1}{4}$ x 1 x 10 WOOD SIDE GLUE & NAIL IN PLACE

4

10

6

$\frac{1}{4}$

$\frac{1}{8}$

ALTERNATE DESIGN

SOLID WOOD BLOCK

TURNED OR SQUARE HANDLE GLUE & DOWEL JOINT

Project 1: Chopping Tray. *This handy tray can be used for cutting up vegetables or fruits used in salads or cooking. Small pieces of scrap wood can be used. An alternate design is also shown. Finish with corn oil or light mineral oil.*

Millimetre	Inch
6	1/4
20	3/4
45	1 3/4
100	4
125	5

*Not exact conversions.

Project 2: Model Plan Holder.
This handy holder can be used to hold plans or directions for model airplanes or cars. It also can be used for holding recipes. This plan is shown with metric dimensions. If you wish to use customary inch measurements, use the replacement value chart. Finish the holder with a penetrating oil finish.

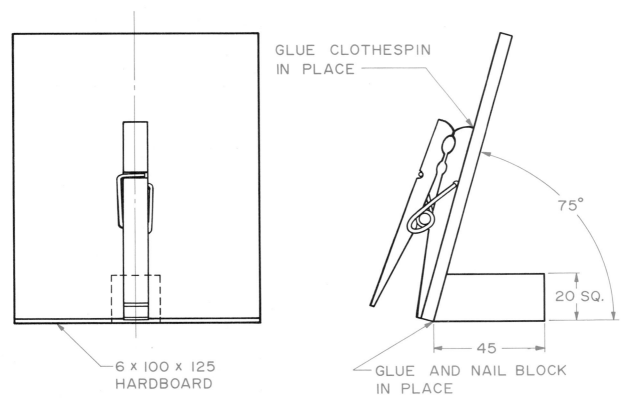

GLUE CLOTHESPIN IN PLACE

75°

20 SQ.

45

GLUE AND NAIL BLOCK IN PLACE

6 x 100 x 125 HARDBOARD

DIM. IN mm

PROJECTS

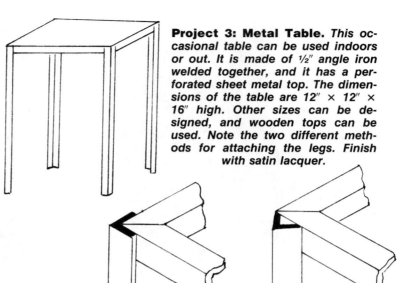

Project 3: Metal Table. *This occasional table can be used indoors or out. It is made of ½″ angle iron welded together, and it has a perforated sheet metal top. The dimensions of the table are 12″ × 12″ × 16″ high. Other sizes can be designed, and wooden tops can be used. Note the two different methods for attaching the legs. Finish with satin lacquer.*

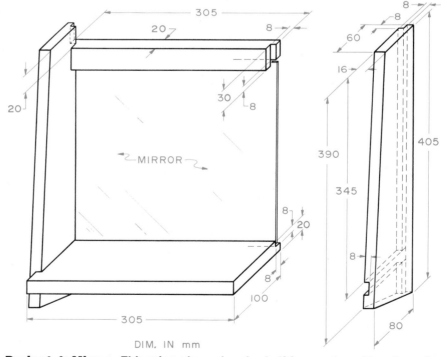

Replacement Values*

Millimetre	Inch
8	⅜
16	½
20	¾
30	1⅛
60	2½
80	3
100	4
305	12
345	13½
390	15⅜
405	16

*Not exact conversions.

DIM. IN mm

Project 4: Mirror. *This mirror has a handy shelf for comb and brush, or for a small vase of flowers. It can be fastened to the wall with screws or with a picture hanger. Finish with paint or clear varnish. The mirror should be cut slightly undersize after the mirror frame has been assembled.*

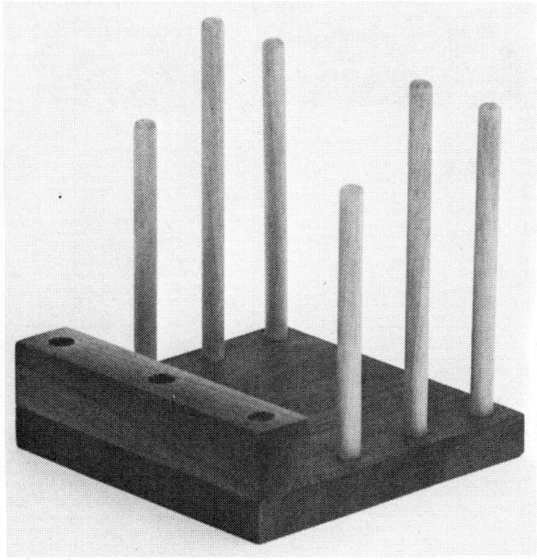

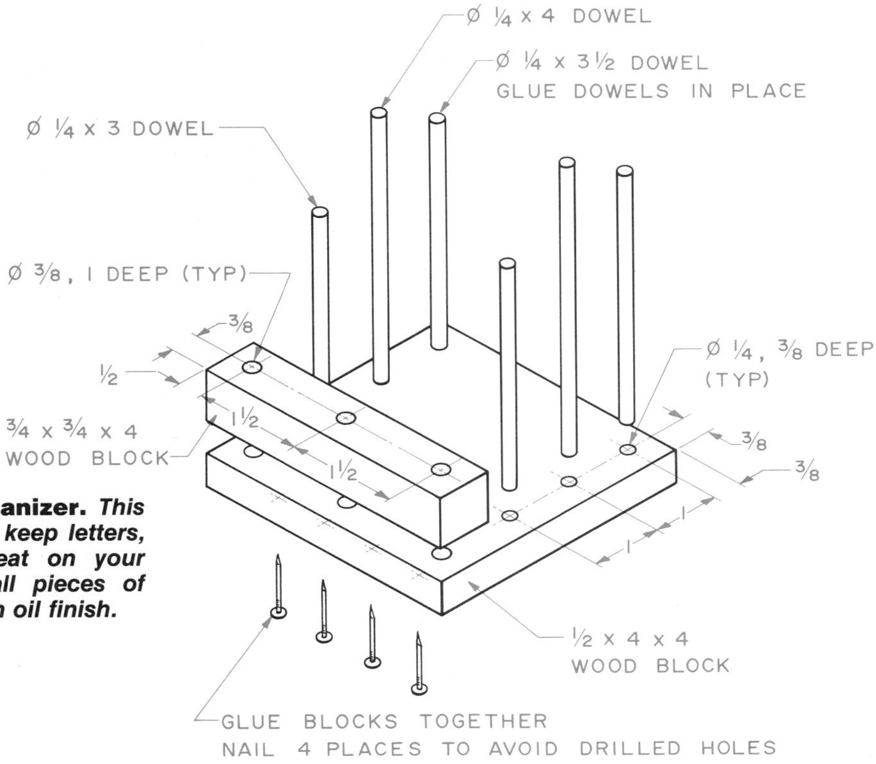
Ø ¼ x 4 DOWEL
Ø ¼ x 3½ DOWEL
GLUE DOWELS IN PLACE
Ø ¼ x 3 DOWEL
Ø ⅜, 1 DEEP (TYP)
⅜
½
1½
1½
¾ x ¾ x 4
WOOD BLOCK
Ø ¼, ⅜ DEEP
(TYP)
⅜
⅜
1
1
½ x 4 x 4
WOOD BLOCK
GLUE BLOCKS TOGETHER
NAIL 4 PLACES TO AVOID DRILLED HOLES

Project 5: Desk Organizer. *This is a convenient way to keep letters, notes, and pencils neat on your desk. Make from small pieces of scrap wood; apply an oil finish.*

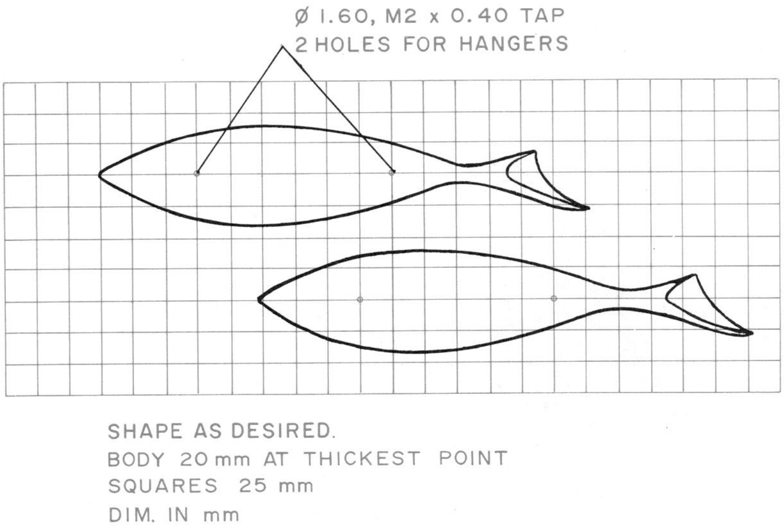
Ø 1.60, M2 x 0.40 TAP
2 HOLES FOR HANGERS
SHAPE AS DESIRED.
BODY 20 mm AT THICKEST POINT
SQUARES 25 mm
DIM. IN mm

Project 6: Cast Metal Wall Sculptures. *A variety of fish, animals, or boat shapes can be used to create cast wall decorations. These can be cast in aluminum or bronze. Roundhead machine screws can be used for hangers. Spray the sculptures with flat black lacquer, and highlight the texture by rubbing with steel wool.*

Replacement Values*

Millimetre	Inch
20	¾
25	1
Ø 1.60, M2 × 0.40 tap	No. 29 drill, 8-32 tap

*Not exact conversions.

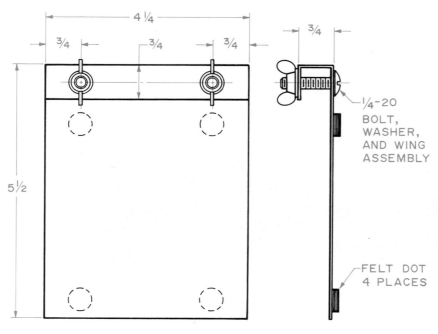

MATERIAL: 22-GAUGE SHEET STEEL

¾
4¼
¾
¾
5½
¾
¼-20 BOLT, WASHER, AND WING ASSEMBLY
FELT DOT 4 PLACES

Project 7: Note Pad. *The note pad holder is made from bent sheet metal. The paper pads are made from sheets of 8½″ × 11″ paper cut into quarters and drilled near the top.*

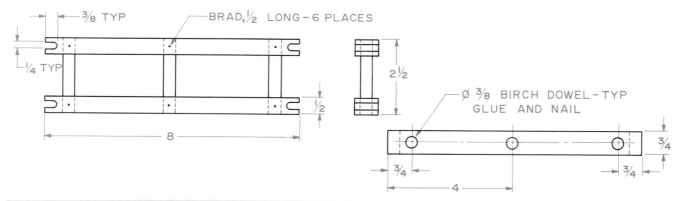

⅜ TYP
BRAD, ½ LONG - 6 PLACES
¼ TYP
½
8
2½
Ø ⅜ BIRCH DOWEL - TYP GLUE AND NAIL
¾
¾
¾
4

Project 8: Cord Reel. *This handy reel can be used to store an extension cord for use in the home. Scrap wood can be used. Finish with oil or varnish.*

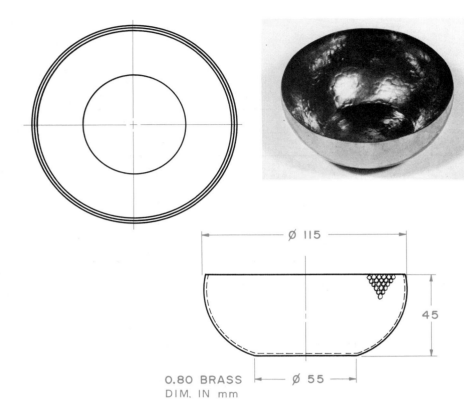

Project 9: Nut Bowl. *Make this bowl from 20-gauge brass. A number of different edge and surface treatments can be used.*

Ø 115

45

0.80 BRASS
DIM. IN mm

Ø 55

Replacement Values*

Millimetre	Inch
45	1¾
55	2⅛
115	4½

*Not exact conversions.

Appendix

Table 1
Approximate Conversions
(Customary to Metric; Metric to Customary)

From	To	Multiply By
LENGTH		
inches	millimetres	25
feet	millimetres	300
yards	metres	0.9
miles	kilometres	1.6
millimetres	inches	0.04
millimetres	feet	0.0033
metres	yards	1.1
kilometres	miles	0.6
AREA		
square inches	square millimetres	650
square inches	square centimetres	6.5
square feet	square metres	0.1
square yards	square metres	0.8
square miles	square kilometres	2.6
acres	square hectometres (hectares)	0.4
square millimetres	square inches	0.001 6
square centimetres	square inches	0.16
square metres	square feet	11
square metres	square yards	1.2
square kilometres	square miles	0.4
square hectometres (hectares)	acres	2.5
VOLUME		
cubic inches	cubic millimetres	16 000
cubic inches	cubic centimetres	16
cubic feet	cubic metres	0.03
cubic yards	cubic metres	0.8
fluid ounces	millilitres	30
pints	litres	0.47
quarts	litres	0.95
gallons	litres	3.8
cubic millimetres	cubic inches	0.000 06
cubic centimetres	cubic inches	0.06
cubic metres	cubic feet	35
cubic metres	cubic yards	1.3
millilitres	fluid ounces	0.03
litres	pints	2.1
litres	quarts	1.06
litres	gallons	0.26
MASS (Weight)		
ounces	grams	28
pounds	kilograms	0.45
tons	metric tons	0.9
grams	ounces	0.04
kilograms	pounds	2.2
metric tons	tons	1.1
TEMPERATURE		
degrees Fahrenheit	degrees Celsius	$(^\circ C \times 1.8) + 32$
degrees Celsius	degrees Fahrenheit	$(^\circ F - 32) \times 0.6$

Table 2
A- and C-Series Paper and Envelopes

A-Series Paper-Trimmed Sizes			C-Series Envelopes		
Size	Millimetres	Inches*	Size	Millimetres	Inches*
A0	841 × 1189	$33\frac{1}{8}$ × $46\frac{3}{4}$	C0	917 × 1297	$36\frac{1}{8}$ × 51
A1	594 × 841	$23\frac{3}{8}$ × $33\frac{1}{8}$	C1	648 × 917	$25\frac{1}{2}$ × $36\frac{1}{8}$
A2	420 × 594	$16\frac{1}{2}$ × $23\frac{3}{8}$	C2	458 × 648	18 × $25\frac{1}{2}$
A3	297 × 420	$11\frac{3}{4}$ × $16\frac{1}{2}$	C3	324 × 458	$12\frac{3}{4}$ × 18
A4	210 × 297	$8\frac{1}{4}$ × $11\frac{3}{4}$	C4	229 × 324	9 × $12\frac{3}{4}$
***					
A5	148 × 210	$5\frac{7}{8}$ × $8\frac{1}{4}$	C5	162 × 229	$6\frac{3}{8}$ × 9
A6	105 × 148	$4\frac{1}{8}$ × $5\frac{7}{8}$	C6	114 × 162	$4\frac{1}{2}$ × $6\frac{3}{8}$
			**DL	110 × 220	$4\frac{5}{16}$ × $8\frac{5}{8}$
A7	74 × 105	$2\frac{7}{8}$ × $4\frac{1}{8}$	C7	81 × 114	$3\frac{1}{4}$ × $4\frac{1}{2}$
A8	52 × 74	2 × $2\frac{7}{8}$	C8	57 × 81	$2\frac{1}{4}$ × $3\frac{1}{4}$
A9	37 × 52	$1\frac{1}{2}$ × 2			
A10	26 × 37	1 × $1\frac{1}{2}$			

* To nearest $\frac{1}{8}$ inch.
** Standard commercial envelope, takes size A4 sheet.
*** Two intermediate sizes, 210 × 280 mm ($8\frac{1}{4}$" × 11") and 210 × 198 mm ($8\frac{1}{4}$" × $7\frac{3}{4}$") are under consideration.

Table 3
B-Series Paper—Trimmed Sizes

Size	Millimetres	Inches*
B0	1000 × 1414	$39\frac{3}{8}$ × $55\frac{5}{8}$
B1	707 × 1000	$27\frac{7}{8}$ × $39\frac{3}{8}$
B2	500 × 707	$19\frac{5}{8}$ × $27\frac{7}{8}$
B3	353 × 500	$13\frac{7}{8}$ × $19\frac{5}{8}$
B4	250 × 353	$9\frac{7}{8}$ × $13\frac{7}{8}$
B5	176 × 250	$6\frac{7}{8}$ × $13\frac{7}{8}$
B6	125 × 176	$4\frac{7}{8}$ × $6\frac{7}{8}$
B7	88 × 125	$3\frac{1}{2}$ × $4\frac{7}{8}$
B8	62 × 88	$2\frac{1}{2}$ × $3\frac{1}{2}$
B9	44 × 62	$1\frac{3}{4}$ × $2\frac{1}{2}$
B10	31 × 44	$1\frac{1}{4}$ × $1\frac{3}{4}$

* To nearest $\frac{1}{8}$ inch.

Table 4
Customary-Metric Drill Sizes and Conversion Chart

Fractional Inch	Decimal Inch	Number or Letter	mm
1/64	0.0156		
	0.0157		0.4
	0.0160	78	
	0.0165		0.42
	0.0173		0.44
	0.0177		0.45
	0.0180	77	
	0.0181		0.46
	0.0189		0.48
	0.0197		0.5
	0.0200	76	
	0.0210	75	
	0.0217		0.55
	0.0225	74	
	0.0236		0.6
	0.0240	73	
	0.0250	72	
	0.0256		0.65
	0.0260	71	
	0.0276		0.7
	0.0280	70	
	0.0292	69	
	0.0295		0.75
	0.0310	68	
1/32	0.0312		
	0.0315		0.8
	0.0320	67	
	0.0330	66	
	0.0335		0.85
	0.0350	65	
	0.0354		0.9
	0.0360	64	
	0.0370	63	
	0.0374		0.95
	0.0380	62	
	0.0390	61	
	0.0394		1.0
	0.0400	60	
	0.0410	59	
	0.0413		1.05
	0.0420	58	
	0.0430	57	
	0.0433		1.1
	0.0453		1.15
	0.0465	56	
3/64	0.0469		
	0.0472		1.2
	0.0492		1.25
	0.0512		1.3
	0.0520	55	
	0.0531		1.35
	0.0550	54	
	0.0551		1.4
	0.0571		1.45
	0.0591		1.5
	0.0595	53	
	0.0610		1.55
1/16	0.0625		
	0.0630		1.6
	0.0635	52	
	0.0650		1.65
	0.0669		1.7
	0.0670	51	
	0.0689		1.75
	0.0700	50	
	0.0709		1.8
	0.0728		1.85
	0.0730	49	
	0.0748		1.9
	0.0760	48	
	0.0768		1.95
5/64	0.0781		
	0.0785	47	
	0.0787		2.0
	0.0807		2.05
	0.0810	46	
	0.0820	45	
	0.0827		2.1
	0.0846		2.15
	0.0860	44	
	0.0866		2.2
	0.0886		2.25
	0.0890	43	
	0.0906		2.3
	0.0925		2.35
	0.0935	42	
3/32	0.0938		
	0.0945		2.4
	0.0960	41	
	0.0965		2.45
	0.0980	40	
	0.0981		2.5
	0.0995	39	
	0.1015	38	
	0.1024		2.6
	0.1040	37	
	0.1063		2.7
	0.1065	36	
	0.1083		2.75
7/64	0.1094		
	0.1100	35	
	0.1102		2.8
	0.1110	34	
	0.1130	33	
	0.1142		2.9
	0.1160	32	
	0.1181		3.0
	0.1200	31	
	0.1220		3.1
1/8	0.1250		
	0.1260		3.2
	0.1280		3.25
	0.1285	30	
	0.1299		3.3
	0.1339		3.4
	0.1360	29	
	0.1378		3.5
	0.1405	28	
9/64	0.1406		
	0.1417		3.6
	0.1440	27	
	0.1457		3.7
	0.1470	26	
	0.1476		3.75
	0.1495	25	
	0.1496		3.8
	0.1520	24	
	0.1535		3.9
	0.1540	23	
5/32	0.1562		
	0.1570	22	
	0.1575		4.0
	0.1590	21	
	0.1610	20	
	0.1614		4.1
	0.1654		4.2
	0.1660	19	
	0.1673		4.25
	0.1693		4.3
	0.1695	18	
11/64	0.1719		
	0.1730	17	
	0.1732		4.4
	0.1770	16	
	0.1772		4.5
	0.1800	15	
	0.1811		4.6
	0.1820	14	
	0.1850	13	
	0.1850		4.7
	0.1870		4.75
3/16	0.1875		
	0.1890		4.8
	0.1890	12	
	0.1910	11	
	0.1929		4.9
	0.1935	10	
	0.1960	9	
	0.1969		5.0
	0.1990	8	
	0.2008		5.1
	0.2010	7	
13/64	0.2031		
	0.2040	6	
	0.2047		5.2
	0.2055	5	
	0.2067		5.25
	0.2087		5.3

(Continued)

Table 4
Customary-Metric Drill Sizes and Conversion Chart (Continued)

Fractional Inch	Decimal Inch	Number or Letter	mm	Fractional Inch	Decimal Inch	Number or Letter	mm	Fractional Inch	Decimal Inch	Number or Letter	mm
	0.2090	4			0.3160	O		17/32	0.5312		
	0.2126		5.4		0.3189		8.1		0.5315		13.5
	0.2130	3			0.3228		8.2	35/64	0.5469		
	0.2165		5.5		0.3230	P			0.5512		14.0
7/32	0.2188				0.3248		8.25	9/16	0.5625		
	0.2205		5.6		0.3268		8.3		0.5709		14.5
	0.2210	2		21/64	0.3281			37/64	0.5781		
	0.2244		5.7		0.3307		8.4		0.5906		15.0
	0.2264		5.75		0.3320	Q		19/32	0.5938		
	0.2280	1			0.3346		8.5	39/64	0.6094		
	0.2283		5.8		0.3386		8.6		0.6102		15.5
	0.2323		5.9		0.3390	R		5/8	0.6250		
	0.2340	A			0.3425		8.7		0.6299		16.0
15/64	0.2344			11/32	0.3438			41/64	0.6406		
	0.2362		6.0		0.3445		8.75		0.6496		16.5
	0.2380	B			0.3465		8.8	21/32	0.6562		
	0.2402		6.1		0.3480	S			0.6693		17.0
	0.2420	C			0.3504		8.9	43/64	0.6719		
	0.2441		6.2		0.3543		9.0	11/16	0.6875		
	0.2460	D			0.3580	T			0.6890		17.5
	0.2461		6.25		0.3583		9.1	45/64	0.7031		
	0.2480		6.3	23/64	0.3594				0.7087		18.0
1/4	0.2500	E			0.3622		9.2	23/32	0.7188		
	0.2520		6.4		0.3642		9.25		0.7283		18.5
	0.2559		6.5		0.3661		9.3	47/64	0.7344		
	0.2570	F			0.3680	U			0.7480		19.0
	0.2598		6.6		0.3701		9.4	3/4	0.7500		
	0.2610	G			0.3740		9.5	49/64	0.7656		
	0.2638		6.7	3/8	0.3750				0.768		19.5
17/64	0.2656				0.3770	V		25/32	0.7812		
	0.2657		6.75		0.3780		9.6		0.7874		20.0
	0.2660	H			0.3819		9.7	51/64	0.7969		
	0.2677		6.8		0.3839		9.75		0.808		20.5
	0.2717		6.9		0.3858		9.8	13/16	0.8125		
	0.2720	I			0.3860	W			0.8268		21.0
	0.2756		7.0		0.3898		9.9	53/64	0.8281		
	0.2770	J		25/64	0.3906			27/32	0.8437		
	0.2795		7.1		0.3937		10.0		0.847		21.5
	0.2810	K			0.3970	X		55/64	0.8594		
9/32	0.2812				0.4040	Y			0.8661		22.0
	0.2835		7.2	13/32	0.4062			7/8	0.8750		
	0.2854		7.25		0.4130	Z			0.886		22.5
	0.2874		7.3		0.4134		10.5	57/64	0.8906		
	0.2900	L		27/64	0.4219				0.9055		23.0
	0.2913		7.4		0.4331		11.0	29/32	0.9062		
	0.2950	M		7/16	0.4375			59/64	0.9219		
	0.2953		7.5		0.4528		11.5		0.926		23.5
19/64	0.2969			29/64	0.4531			15/16	0.9375		
	0.2992		7.6	15/32	0.4688				0.9449		24.0
	0.3020	N			0.4724		12.0	61/64	0.9531		
	0.3031		7.7	31/64	0.4844				0.965		24.5
	0.3051		7.75		0.4921		12.5	31/32	0.9687		
	0.3071		7.8	1/2	0.5000				0.9843		25.0
	0.3110		7.9		0.5118		13.0	63/64	0.9844		
5/16	0.3125			33/64	0.5156			64/64	1.000		
	0.3150		8.0								25.4

Table 5
Drill Chart
(Boring Approximate Metric Holes with Customary Drills)

mm	Customary Drill Size	mm	Customary Drill Size	mm	Customary Drill Size
1.00	60	5.00	9	9.00	T
1.20	$3/64$	5.20	6 or $13/64$	9.20	$23/64$
1.40	54	5.40	3	9.40	U
1.60	$1/16$	5.60	2 or $7/32$	9.60	V
1.80	50	5.80	1	9.80	W
2.00	47	6.00	B or $15/64$	10.00	X or $25/64$
2.20	44	6.20	D	10.20	Y
2.40	$3/32$	6.40	$1/4$ or E	10.50	Z
2.60	38	6.60	G	10.80	$27/64$
2.80	34 or 35	6.80	H	11.00	$7/16$
3.00	31	7.00	J	11.20	*
3.20	$1/8$	7.20	$9/32$	11.50	$29/64$
3.40	29	7.40	L	11.80	$15/32$
3.60	$9/64$	7.60	N	12.00	*
3.80	25	7.80	$5/16$	12.20	$31/64$
4.00	22	8.00	O	12.50	*
4.20	19	8.20	P	12.80	$1/2$
4.40	17	8.40	O or $21/64$	13.00	$33/64$
4.60	14	8.60	R		
4.80	12 or $3/16$	8.80	S		

* no equivalent size

Table 6
Softwood Lumber
(Possible Metric Replacement Sizes)

NOMINAL (inch)	ACTUAL (inch)	REPLACEMENT (mm)
1 × 4	$3/4 × 3 1/2$	19 × 89
1 × 6	$3/4 × 5 1/2$	19 × 140
1 × 8	$3/4 × 7 1/4$	19 × 184
1 × 10	$3/4 × 9 1/4$	19 × 235
1 × 12	$3/4 × 11 1/4$	19 × 285
2 × 4	$1 1/2 × 3 1/2$	38 × 89
2 × 6	$1 1/2 × 5 1/2$	38 × 140
2 × 8	$1 1/2 × 7 1/4$	38 × 184
2 × 10	$1 1/2 × 9 1/4$	38 × 235
2 × 12	$1 1/2 × 11 1/4$	38 × 285

Sheet materials: 4′ × 8′ replaced by 1200 × 2400 mm (47.24″ × 94.48″ or approximately $3/4$″ narrower and $1 1/2$″ shorter).

Metric lumber lengths in metres: 2, 2.4, 3, 3.5, 4, 5, 5.5, and 6.

High-Tech Glossary

address. (1) In computers, a number which identifies a particular part of storage (memory). (2) In telecommunications, a character or group of characters that identifies a data source or destination.

algorithm. A step-by-step procedure for solving a problem.

alphanumeric. Relating to a character set that has letters, numbers, and other characters such as punctuation marks.

alphanumeric display. A unit made up of a typewriter-style keyboard and a display screen (CRT) on which text is seen.

ALU (arithmetic and logic unit). The microprocessor in a computer's central processing unit which carries out arithmetic and logic operations.

annotation. A note or other information placed on a drawing.

application program. A program designed for a user task, such as word processing, accounting, or games.

APT (Automatically Programmed Tools). A computer language used to prepare numerical control tapes for machine tools.

artificial intelligence. The ability of a machine to do things usually related to human intelligence, such as problem-solving, learning, reasoning, etc.

automatic dimensioning. The means by which a CAD system automatically measures distances and places extension lines, dimension lines, and arrowheads.

automation. The process of making a machine self-moving or self-controlling.

BASIC (Beginner's All-Purpose Symbolic Instruction Code). A simplified, algebra-like computer programming language. Most microcomputers use some form of BASIC.

Beta. A type of video cassette recording system, developed by Sony. The Beta system uses slightly smaller cassettes than the VHS system.

binary code. A system that makes use of exactly two distinct characters, usually 0 and 1.

bit (binary digit). The smallest unit of information on a binary system of notation; either of the characters, 0 or 1.

booting. Transferring a disk operating system program from its storage on a disk to a computer's memory. It is necessary to boot a disk in order to begin using its other programs.

bug. A mistake or malfunction in a computer program.

byte. A sequence of adjacent bits used as a unit to specify a letter of the alphabet, an instruction, etc. A byte usually consists of eight bits.

cable television. Sometimes called CATV (for community antenna television), this is a system in which television signals and other information services are transmitted by coaxial cable. With a *reverse channel capability,* viewers can send signals along the cable as well as receive them.

CAD (computer-aided design). Using a computer to assist in the creation or changing of a design.

CAM (computer-aided manufacturing). Using computer technology to plan, manage, and control the operations of a manufacturing plant.

catalog. A list of items arranged for easy reference. For example, a computer disk catalog lists the programs on that disk.

character. A letter, number, or other symbol that is used as part of the organization, control, or representation of data.

chip. *See* **I C.**

circuit board. *See* **printed circuit board.**

CNC (computer numerical control). Using a computer in a numerical control unit to control some or all of the operations of a machine tool.

coaxial cable. A communication cable which consists of a tube of electrically conducting material (usually copper) which surrounds an insulated central conductor (also copper). Usually several such cables are combined into one bundle.

COBOL (Common Business Oriented Language). A programming language designed for business data processing. It uses statements that are similar to everyday English.

code. A system of symbols and rules used to represent information.

command. A control signal, or an instruction in machine language.

communications satellite. A satellite which receives signals from Earth, amplifies and converts them, and retransmits them to ground-based receivers.

compatibility. Used to refer to the ability of two hardware/software devices to work together.

compiler. A program that translates user-written programs into machine language.

computer. An automatic, electronic machine that makes calculations and processes information. It receives information (data), stores it, operates on it according to instructions given in a program, and outputs the results to the user.

computer graphics. Using a computer to create, change, or display pictorial images.

computer language. A language, such as BASIC, used to give instructions to a computer.

computer numerical control. Using a computer to store and send instructions for a numerical control machine tool.

computer program. *See* **program.**

CP/M (control program/microcomputers). A type of operating system widely used by microcomputers.

CPU (central processing unit). The part of the computer which includes the circuits controlling the interpretation and execution of instructions.

CRT (cathode ray tube). A device which presents data in visual form by means of controlled electron beams; similar to a television picture tube.

cursor. A marker on a CRT, used to show the next point at which a character will appear. The cursor can be moved by the computer user.

cybernetics. A science which brings together theories and studies on communication and control systems in living organisms and machines.

daisy wheel printer. A printer which uses an interchangeable typehead called a daisy wheel. The characters on the typehead are on stalks radiating out from a flat, circular center. Daisy wheels are available in many typefaces.

data. (1) A representation of facts, concepts, or instructions in a formalized manner suitable for communication, interpretation, or processing by humans or automatic means. (2) Any representations, such as characters or analog quantities, to which meaning is or might be assigned.

data base. A storage of data on files accessible by computer.

debug. To detect, find, and remove errors from a computer program.

digital. Calculation by discrete (distinct and unconnected) units. A digital computer, for example, operates with numbers (bits) which represent the presence or absence of an electric pulse. A 1 represents a pulse; a 0, no pulse.

digitize. To convert a drawing or picture into numeric form.

digitizer. A device which converts information into numeric form readable by a digital computer.

disk. A circular plate with a magnetic coating. Used to store computer programs and data. Both rigid (hard) and flexible (floppy) disks are available. Disk drives for microcomputers usually use floppy disks.

disk drive. A device which reads from, or writes to, magnetic disks.

display. To show data on a CRT.

DNC (direct numerical control). Using a computer network to send and receive data to and from numerical control machines.

document. A medium and the data on it. Usually refers to print on paper.

dot matrix printer. A printer in which each character is made up of a pattern of dots.

drum plotter. A machine that draws an image on paper or film mounted on a drum.

edit. To modify data by inserting or deleting characters.

electronic mail. The electronic distribution of messages. One feature which distinguishes electronic mail from other forms of communication is the capability for non—real time use; that is, messages can be sent at one time and received or read at a later time.

electrostatic printing/copying. A printing method in which electrostatic charges are produced on paper in the design to be reproduced. Toner is attracted to the charged areas, making them visible. Heat fuses the toner to the paper to make a permanent copy.

EPROM (erasable programmable read-only memory). Read-only memory that can be programmed, erased, and reprogrammed.

facsimile transceiver. A machine which converts documents (such as typewritten copy) into electronic pulses and sends them over telephone lines. It also converts incoming pulses back into images.

facsimile transmission (fax). A system by which documents are transmitted over telephone lines and other telecommunications links, such as satellites. The machines used to do this are a facsimile transceiver and a telephone.

fiber optics. The technique of transmitting light through long, thin, flexible glass fibers.

field. A section of a computer record which is designated for the storage of specified information. A data base containing employee records, for example, would have all the names in one field, the addresses in another field, and so on.

flatbed plotter. A machine that draws an image on paper or film mounted on a flat table.

floppy disk. A flexible plastic disk with a magnetic coating. It is used to store computer data. Floppy disks are usually either 5¼" or 8" in diameter.

FORTRAN (formula translator). A procedure-oriented

computer language used for technical and scientific programing.

function keys. Keys on a computer keyboard which enable the user to issue a series of commands with one stroke. The function keys may be programmed at the factory or by the user.

graphics tablet. A device for inputting line drawings and other graphics to a computer. The tablet usually consists of a wire grid partially embedded in a solid base. When touched by a stylus, the wires in the tablet and stylus are brought together, producing a pulse which enables the computer to register the location of the stylus. That point then appears on the computer screen.

hard copy. Computer output, such as drawings or reports, printed on paper.

hardware. The physical equipment of a computer system, such as mechanical, magnetic, electrical, or electronic devices.

holography. The forming of three-dimensional images by laser light.

IC (integrated circuit). Tiny complex of electronic components and their connections produced on a thin slice of material such as silicon.

impact printer. Any printer in which images are produced by a hard die hitting an inked ribbon onto paper. A typewriter is an example.

ink jet printing. A process in which a cluster of small nozzles squirts droplets of ink onto paper. A computer controls the ink jet printer, directing it to spray in various patterns and densities to form symbols and characters.

input. Information fed into a computer or into its storage devices.

interactive graphics. The ability to perform graphics operations directly on the computer with immediate feedback.

interface. Refers to the connecting link between two systems. For example, in order for a computer and a printer to work together, an *interface card* may be required. This is a printed circuit board which is installed in the computer to enable the computer and printer to communicate with each other.

joystick. A lever used to manually change CRT displays.

K. A unit representing 1024 bytes. Used to refer to a computer's storage capacity. A 64K memory, for example, means the computer can store 65 536 bytes.

language. *See* **computer language.**

laser (light amplification by stimulated emission of radiation). A mechanism that strengthens and directs light. The narrow, high-energy beam that is produced can be used for such varied operations as cutting metal, measuring distances, or carrying messages.

lightpen. A light-sensing device which looks like a pencil and is used to change CRT displays.

machine language. The actual language used by the computer when it performs operations. Machine language is usually a binary code.

magnetic disk. *See* **disk.**

mainframe. (1) A large computer, as distinguished from a minicomputer or microcomputer. Mainframes are large units which are usually kept in a climate-controlled room. (2) A computer's central processing unit.

menu. A display on a CRT that lists options a user may choose.

MICR (magnetic ink character recognition). A technique in which a scanning device recognizes characters by the magnetic particles contained in the ink used to print the characters.

microcomputer. A small, portable computer whose basic element is a microprocessor consisting of a single integrated circuit.

microprocessor. A central processing unit (CPU) that is a single integrated circuit.

minicomputer. A computer of intermediate size and capacity, between a mainframe and a microcomputer.

model. A geometrically accurate representation of a real object stored in a CAD/CAM data base.

modem (modulator/demodulator). A device which makes it possible for a computer to send and receive information over a telephone line.

mouse. A device which a user moves over the surface of a graphics tablet. Its position is recorded by a computer. The mouse is used to record or change the position of text or illustrations on a computer display screen.

NC (numerical control). Prerecorded information providing instructions for the automatic computer control of machine tools.

network. Individually controlled computers connected by telecommunications channels.

OCR (optical character recognition). A process by which characters (such as letters of the alphabet) are examined by a scanner and converted into digital form.

off-line. Equipment or devices in a system which are not under the direct control of a computer's CPU.

on-line. Equipment or devices in a system which are under the direct control of a computer's CPU.

operating system. A type of software which enables a computer to control the sequencing and processing of programs. For example, to use floppy disks with a microcomputer, a disk operating system (DOS) is needed. This program is itself stored on a disk and transferred into the computer's memory.

optical fiber. A thin, flexible strand of pure glass. Optical fibers are used to carry light beams.

output. Information sent out by a computer or its storage devices.

Pascal. A programming language with a highly structured design which facilitates the rapid location and correction of errors.

phototypesetter. A machine which sets type on light-sensitive film or paper. Modern phototypesetters use computers to generate the type characters.

plotter. A device used to make a drawing of a CRT display.

printed circuit board. A board which contains holes or slots for components and integrated circuits and whose back is printed with electronically conductive paths between components.

printer. A computer output mechanism that prints characters one at a time or one line at a time.

program. A series of instructions or statements, in a form acceptable to a computer, directing it to carry out certain operations. The purpose of the program is to achieve a certain result, such as the solution of a problem.

programmer. A person who designs, writes, and tests computer programs.

PROM (programmable read-only memory). Read-only memory that can be programmed into a computer using special equipment.

prompt. A message or symbol from the computer system asking the user for information or telling the user of possible actions or operations. It serves as a guide to the operator in use of the system.

RAM (random access memory). Computer memory from which data can be retrieved regardless of input sequence and which can be altered by the user.

robot. A computer-controlled machine that can be programmed to perform tasks. Most robots do industrial tasks such as welding, cutting, assembling parts, or moving materials.

ROM (read-only memory). Computer memory which cannot be changed.

semiconductor. Any material which conducts electricity only when the voltage across it is above a certain value. Semiconductors are used to make transistors.

servomechanism. An automatic system that uses feedback to control movement or operation, as on a robot.

software. A set of programs used to direct the operation of a computer or other hardware.

stylus. (1) A lightpen. (2) A device used with a graphics tablet to provide coordinate input to the display device.

terminal. A point in a system or communication network at which data can either enter or leave, such as a computer terminal.

toolpath. The center line of a numerically controlled machine tool's path as it performs a specific operation, such as milling or boring.

transistor. An electronic device made from semiconductors and used to amplify or switch electrical signals. On an integrated circuit, numerous transistors and other components are formed on one very small wafer of semiconducting material, usually silicon.

UPC (universal product code). A series of bars printed on the package of a retail product. The code can be read by optical scanners, and the information used to produce a receipt and record the transaction.

VCR (video cassette recorder). A device for recording visual images (such as television programs) onto magnetic tape.

VHS (video home system). A video cassette recording system, developed by JVC.

video camera. A camera which records images on magnetic tape. The tape can be played back on a VCR.

video disk. A disk, usually of plastic, on which visual and sound information is recorded for playback on a television screen.

videotex. Any electronic system which makes computerized information available to users who have appropriately equipped CRTs. Some videotex systems transmit over broadcast television channels; others use cables (usually telephone lines).

word processing. Computer handling of text. Word processing enables the user to create, add, delete, or move text as needed before printing it out on paper. There are word processing *programs,* which are used with computers, as well as *dedicated* word processors, which are preprogrammed at the factory and do only word processing.

Index